ANNUAL REVIEW OF NEUROSCIENCE

ANNUAL REVIEW OF NEUROSCIENCE

VOLUME 5, 1982

W. MAXWELL COWAN, *Editor*
Salk Institute for Biological Studies

ZACH W. HALL, *Associate Editor*
University of California School of Medicine

ERIC R. KANDEL, *Associate Editor*
College of Physicians and Surgeons of Columbia University

ANNUAL REVIEWS INC.   4139 EL CAMINO WAY   PALO ALTO, CALIFORNIA 94306 USA

ANNUAL REVIEWS INC.
Palo Alto, California, USA

International Standard Serial Number: 0147-006X
International Standard Book Number: 0-8243-2405-6

PRINTED AND BOUND IN THE UNITED STATES OF AMERICA

Annual Review of Neuroscience
Volume 5, 1982

CONTENTS

SOME RELATED ARTICLES IN OTHER *ANNUAL REVIEWS*

From the *Annual Review of Biochemistry,* Volume 50 (1981)

Lectins: Their Multiple Endogenous Cellular Functions, S. H. Barondes

Monoclonal Antibodies: A Powerful New Tool in Biology and Medicine, D. E. Yelton and M. D. Scharff

From the *Annual Review of Biophysics and Bioengineering,* Volume 10 (1981)

Nerve Sodium Channel Incorporation in Vesicles, R. Villegas and G. M. Villegas

From the *Annual Review of Medicine,* Volume 32 (1981)

Genetic Component of Alcoholism, D. W. Goodwin

Neuroendocrinological Effects of Alcohol, T. J. Cicero

From the *Annual Review of Pharmacology and Toxicology,* Volume 21 (1981)

Receptor Adaptations to Centrally Acting Drugs, I. Creese and D. R. Sibley

The Pharmacology of Memory: A Neurobiological Perspective, L. R. Squire and H. P. Davis

Hypothalamic Hypophysiotropic Hormones and Neurotransmitter Regulation: Current Views, J. Meites and W. E. Sonntag

From the *Annual Review of Physiology,* Volume 44 (1982)

Structure and Function of the Calcium Pump Protein of Sarcoplasmic Reticulum, N. Ikemoto

Turnover of Acetylcholine Receptors in Skeletal Muscle, D. Pumplin and D. Fambrough

The Development of Sarcoplasmic Reticulum Membranes, A. Martonosi

Ionic Channels in Skeletal Muscle, E. Stefani and D. J. Chiarandini

Bacterial Chemotaxis, A. Boyd and M. Simon

Chemotaxis in Dictyostelium, G. Gerisch

Behavioral Effects of Neuropeptides, G. Koob and F. Bloom

Ann. Rev. Neurosci. 1982. 5:1–31

MORPHOLOGY OF CUTANEOUS RECEPTORS

A. Iggo

University of Edinburgh, Edinburgh, EH9 1QH, United Kingdom

K. H. Andres

Ruhr-Universität Bochum, German Federal Republic

INTRODUCTION

The study of cutaneous receptor morphology can be divided into two main periods: the now classical investigations using light microscopy, and contemporary works using the electron microscope. The dividing line is provided by Pease & Quilliam, who in 1957 published the first electron microscopical study of the Pacinian corpuscle. This was an appropriate receptor since it was also the first to be recognized in the classical era, by Vater in 1741 and by Pacini in 1835 (Pacini 1840).

In this review, the first on cutaneous receptors to appear in the *Annual Review of Neuroscience,* we consider publications in the last two decades (1960–1980), concentrate on mammalian receptors, attempt to present a coherent account of the morphology of cutaneous receptors, and draw on correlative morphological and functional studies where these are appropriate. Current studies are moving beyond the confines of a static mature morphology of the receptors and many of these new aspects are thoroughly explored in a Symposium on Mechanoreception, which covers a wide range of species and receptors (Schwarzkopf 1974).

Technical developments are, inevitably, at the root of much new knowledge which has gone some way to resolving ancient conflicts. The electron microscope, by extending the resolving power available to the morphologists, has been the most important single advance. An associated development was the recognition that fixation artifacts had previously led to

1

0147-006X/82/0301-0001$02.00

mistaken interpretation of structures only just visible in the light microscope. As a result, tissues can be prepared by rapid in situ fixation with glutaraldehyde perfusion, for example, with a minimum of distortion. Nevertheless, such procedures do alter the proteins of the tissues to such an extent that the immunocytochemical labeling of cell components may be greatly reduced.

The pioneering studies of Pease & Quilliam (1957), Cauna (1962), and many others were based on descriptions of single EM sections, and since each section was only 500–1000 Å thick an accurate three-dimensional view of receptors that were several millimeters in length had to await the development of serial sectioning techniques, with which the use of serial ultrathin and semi-thin sections has allowed the laborious reconstruction of whole receptor complexes.

Another major advance, only just being applied to cutaneous receptors, is immunocytochemistry, using either fluorescence- or peroxidase-labeled antibodies to label individual cellular components, such as the catecholamines or the neuropeptides (e.g. Hökfelt et al 1980). Cytochemical studies are also making an impact, for example the presence and location of particular proteins (actin) in sensory hair cells studied by Flock (1971) and coworkers (Flock et al 1981).

Finally, progress has been made in establishing the functional properties of morphologically identified receptors, following on the pioneering investigation of the Pacinian corpuscle by Gray & Matthews (1951). This progression has now led to the recognition that the cutaneous receptors function as selective peripheral encoding devices, able to extract information about various parameters of cutaneous stimuli and to supply the central nervous system with a prepacked analysis of the periphery. This reaffirmation of the old idea of "specific nerve energies" (Müller 1844), albeit in a modern dress, is a significant consequence of recent work and has important consequences for subsequent analyses of sensory and reflex mechanisms in the central nervous system.

GENERAL PRINCIPLES

There are a number of common morphological features that underlie all cutaneous receptors of vertebrates.

1. The receptor is formed from the terminal of an afferent fiber that has its cell body in, or near, a dorsal root ganglion or cranial nerve ganglion.
2. Most spinal afferent fibers enter the spinal cord via the dorsal roots, although a minority enter via the ventral roots (Clifton et al 1976).
3. The receptor terminal shows morphological specialization, including an

accumulation of mitochondria, the presence of clear or granular vesicles or both, and a fine filamentous ground substance (the receptor matrix) under the axonal membrane (Andres & von Düring 1973).

4. Myelin is absent from the receptor terminal.
5. The neural tissue elements often end in special relation to nonneural cells, which may form a complex and characteristic structure.
6. In some receptors, notably the Merkel receptor and the lateral line organs, the nerve terminal has a special relationship to a nonneural cell, which may function as the transducer. In the majority of cutaneous receptors, however, the nerve terminal is the transducer.
7. Particular morphologically distinct receptors occupy typical and regular locations in the skin, e.g. the Meissner corpuscle is present in glabrous (hairless) skin of primates and lies in dermal grooves formed by ridging of the epidermis. The receptors are structurally integrated with their surrounding tissues and may form with them a further characteristic tissue element.

The morphological specialization of cutaneous receptors has allowed them to be placed in groups, according to their structure and location. This was a favorite occupation of nineteenth century morphologists and generated enduring controversies. These have continued into the twentieth century and are being resolved by the new knowledge obtained with the electron microscope. A parallel growth in knowledge of the functional characteristics of skin sensation has a similar history leading from Müller's Doctrine of Specific Nerve Energies (1844), through von Frey's (1895) codification of the sensory functions of certain morphologically distinct receptors, the early single-unit electrophysiological studies of Adrian & Zotterman (1926), combined electrophysiological and morphological studies (Gray & Matthews 1951, Iggo & Muir 1969), to the correlative sensory studies (Merzenich & Harrington 1969, LaMotte & Mountcastle 1975) that have identified the receptors mediating certain kinds of cutaneous sensation.

As a result of the fusion of the two broad streams of morphological and functional studies we can now conclude that the ancient question of "specificity vs nonspecificity" of cutaneous receptors is settled in favor of specificity. In the systematic account that follows we adopt the practice of treating the receptors as functioning as mechanoreceptors, thermoreceptors, or nociceptors. The case for doing so has been argued elsewhere (e.g. Iggo 1974, 1976). The generalization primarily rests on the existence of functionally different sets of cutaneous afferent units and on the strict correlation of morphology and functional characteristics for some of the mechanoreceptors.

CUTANEOUS MECHANORECEPTORS

The mechanoreceptors may have receptors associated with the epidermis (e.g. Merkel receptors, lateral line receptors in fish and amphibia), in the outer layers of the dermis (e.g. Meissner corpuscle, Krause end bulbs), or the deeper layers of the skin (e.g. Pacinian corpuscle). In each location there is a characteristic structural development. The dermal receptors in mammals are encapsulated, quite distinct structures, whereas those in the epidermis have a specialization of the associated epithelial cellular elements, without, however, forming a distinct encapsulated structure. The mechanoreceptors with large (6–16 μm diameter) myelinated afferent fibers form larger, more distinctive receptor structures than do those with smaller myelinated or unmyelinated axons.

Receptive fields Since all the receptors are borne on the ends of afferent fibers it should be possible to define the area of skin supplied by a single axon. This may be difficult to achieve in exclusively morphological studies, since it may require extensive serial sectioning and reconstruction in order to be able to follow the ramifications of an axon that sends branches into several square centimeters of skin. The task is simplified by electrophysiological methods, in which the functional receptive field of a single dorsal root axon is explored. The activity of the axon can be recorded in the peripheral nerve, in the dorsal root or cranial sensory ganglion, or in the dorsal root, with identical results. Even where an axon branches extensively to supply a receptive field of several square centimeters, as is often the case with hair follicle afferents (Brown & Iggo 1967), the branching usually occurs near the termination of the axon, rather than in the peripheral nerve trunk or more centrally.

In general the encapsulated receptors and the epidermal receptors have circumscribed receptive fields, as do the mechanoreceptors with unmyelinated axons. Only the hair follicle afferent units have extensive branching.

Intraepidermal Mechanoreceptors

There are two extensively studied intraepidermal mechanoreceptors, the Merkel cell system and the lateral line organ.

MERKEL CELLS   The Merkel cells and Merkel discs (Merkel 1875) form a characteristic unit in the mammalian epidermis and are present in both glabrous and hairy skin (Figure 1). The myelinated afferent fiber branches extensively in the dermis within 500 μm of the epidermal location of the Merkel cells. The branches lose their myelin sheaths before they penetrate the basement membrane of the epidermis to form the typical flattened nerve plate (Merkel disc) lying along the lower border of the Merkel cell. The nerve terminals and their preterminal parts therefore lack a myelin sheath, although the parent or stem axon is myelinated. Average values for the stem axon diameter, calculated from conduction velocity measurements, are 9.7 ± 0.16 μm (mean ± SE) in the cat and 7.9 ± 0.3 μm (mean ± SE) in the rabbit (Brown & Iggo 1967) for Merkel cell units in hairy skin.

The Merkel cell and nerve terminal were first described electron-microscopically by Cauna (1962). The terminal is an expanded and flattened disc (about 7 μm in diameter and 1 μm thick) and contains numerous mitochondria, which occupy 50% of the terminal. Small, clear vesicles may also be present. The nerve plates may be borne on a single terminal branch, or they may be in tandem, the nerve forming an extension beyond one disc to form a further disc. When this happens the discs are close to each other.

The Merkel disc is invariably associated with a Merkel cell, a unique structure found only in the epidermis in adult mammals (Winkelmann &

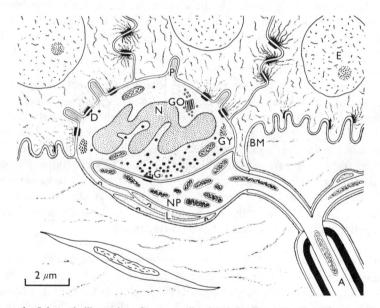

Figure 1 Schematic illustration of a mammalian Merkel cell and disc, lying at the base of the epithelium. The myelinated axon (A) divides at the last node of Ranvier and forms an expanded ending (NP), which is in intimate contact with a Merkel cell. The basement membrane (BM) invests the nerve ending and Schwann cell lamellae (L). The Merkel cell contains a distinctively polylobulated nucleus (N), osmiophilic granular vesicles (G), glycogen (GY), and a Golgi apparatus (GO). It is attached to overlying keratinocytes by desmosomes (D) and sends microvilli (P) into the overlying cells. These processes contain parallel filaments (probably actin-F) (from Iggo & Muir 1969).

Breathnach 1973). Its characteristic features are a polylobulated nucleus, which fills the greater part of the Merkel cell; spherical, osmiophilic, membrane bound granular vesicles (84 ± 13 nm diameter) that are concentrated on the dermal side of the nucleus, between the nucleus and the nerve plate; synaptic-like junctions of the Merkel cell and nerve disc in which the cell membrane is flat and thickened—and the nerve membrane is in apposition to the cell membrane and a gap that contains amorphous ground substance. The granules remain concentrated in their normal cellular location after denervation of the skin, although they may be reduced in number (English 1977). Recent estimates suggest that each Merkel cell in cat skin contains between 125,000 and 200,000 granules (A. Iggo, unpublished). The Merkel cell contains a smooth endoplasmic reticulum in its superficial cytoplasm; the granular vesicles are probably formed in the region of the Golgi apparatus of the cell (Hashimoto 1973) and stored on its dermal side. Other cellular elements include lysosomes and glycogen granules at the poles of the nucleus.

The Merkel cell is attached to the adjacent keratinocytes by desmosomes, but the tonofibrils are finer and fewer in Merkel cytoplasm than in the keratinocytes. These attachments are absent from the typical finger-like projections of the Merkel cell that project into invaginations in the adjacent keratinocytes. These projections, 300 nm in diameter and 2 to 3 μm long (Andres 1966, Iggo & Muir 1969), are usually inserted into invaginations of the adjacent keratinocytes and rarely penetrate between them. Other distinctive features are the absence of desmosomal attachments between these finger-like projections and the enclosing keratinocytes, and the presence of longitudinally oriented filaments within them (Andres 1966). This latter feature assumes particular significance from Flock's studies of lateral-line hair cell cilia in which similar structures were found to be actin filaments (Flock et al 1981). A possible role in the transduction of mechanical stimuli has, quite naturally, been postulated for these filaments, but no experimental validation has yet been forthcoming.

Epidermal location of Merkel cells   These cells are present in glabrous skin and in hairy skin, where they occur in the normal epidermis and in sinus-hair follicles (Merkel 1880, Hoggan 1884, Andres 1966, Patrizi & Munger 1966). The simplest arrangement is in hairy skin epithelium where the Merkel cells lie at the base of the epidermis, in clusters of 50 to 70 cells, borne on the terminals of a single myelinated axon. Each afferent unit forms a compact structure, roughly circular or oval in shape (about 100 × 300 μm), with a thickening of the overlying epidermis and a perimeter marked by the interdigitation of keratinocytes and dermal tissue. The Merkel cells are oriented so that their nuclei are horizontal and the nerve plates are on the dermal side of the cell (Figure 1). This organization, described by Pinkus (1905) and called "haarscheibe" (hair disc) in man, has since been reported in many species and given various names (touch spot, Iggo 1962a; touch corpuscle, Iggo & Muir 1969; tactile pad, Tapper 1965).

A more complex situation in hairy skin is the sinus hair follicle, originally described by Merkel (1875). In these tactile or sinus hairs the Merkel cell lies beneath the glassy membrane and within the basal cell layer (Andres 1966) and, in contrast to the normal epithelial arrangement, the nerve plate is on the opposite side of the cell, corresponding to the epithelial surface. Otherwise the typical cellular relations exist. In a sinus hair a uniform cylinder of Merkel cells surrounds the hair, in the midregion of the follicle below the sebaceous gland. Morphological data do not allow an exact statement of how the Merkel cells are organized with respect to the afferent units, especially since several hundred myelinated axons may innervate a single sinus hair. However, electrophysiological data (Gottschaldt et al 1973) indicate that afferent units are distributed in sets around the perimeter of a follicle and that any one afferent fiber has only a restricted distribution to Merkel cells on only one side of a follicle.

In glabrous skin, with its distinctive ridging, particularly in primates, Merkel cells are contained in rete pegs (Miller et al 1960, Munger & Pubols 1972) at the bottom of the epidermal ridge through which the duct of a sweat gland passes. The Merkel cells, in addition to the typical basal location in the epidermis, are also present in groups in the rete ridge (Halatá 1970, 1972a). In both places they have all the typical Merkel cell and Merkel disc characteristics in human, nonhuman-primate, and nonprimate skin (Winkelmann & Breathnach 1973). Exact morphological data on the number of Merkel cells per afferent unit are not available but electrophysiologically the receptive fields of slowly adapting type I (SA I) units in primate (Iggo 1962a,b) and human (Johansson 1978) finger skin are small, with maximum sensitivity in man restricted to an area 2 to 3 mm in diameter, covering 1 to 3 ridges.

Function of Merkel cells   Correlative electrophysiological and morphological studies in cats (Tapper 1965, Iggo 1966, Iggo & Muir 1969) have established that the Merkel cells in mammals function as slowly adapting mechanoreceptors, and they are now known as slowly adapting type I units (SA I) (Iggo 1966). This identification, after being established in cat and dog hairy skin (Iggo 1962a), has been extended to other species. Similar kinds of physiological properties are found for mechanoreceptor afferents in glabrous skin and it is now generally agreed (e.g. Johansson 1978) that these too are mediated by Merkel cell receptors, although the high density of afferent innervation of glabrous skin has prevented an exact correlation.

Morphogenesis The Merkel "tastfleck" (touch spot) has proved to be a very convenient model for studies of morphogenesis, degeneration, and regeneration of sensory receptors, because it can readily be observed in the epidermis of anaesthetized animals as well as in skin prepared for light and electron microscopy. Merkel cells appear early in the developing human fetus, as early as 16 weeks (Breathnach 1971), and in fetal sheep (Lyne & Hollis 1971). The origin of the Merkel cells continues to be debated: one view asserts that they are epithelial in origin (e.g. Lyne & Hollis 1971, Smith 1970), but on the basis of ultrastructural studies, Breathnach (1971) suggests an origin from the neural crest. English (1974) reports cells that are described as "transitional" between normal epidermal keratinocytes and Merkel cells. English et al (1980) have now reported the existence of Merkel cells in rat fetal skin, before the invading neurite makes contact with the epidermis—a result that suggests an epidermal origin for the Merkel cells.

The Merkel disc (i.e. nerve terminal) degenerates rapidly when the skin is denervated (Brown & Iggo 1963, Smith 1967, Burgess et al 1974), and on permanent denervation the touch dome disappears in the cat (Burgess et al 1974) but not in the rat (Smith 1967). Merkel cells may survive at least 25 days after denervation in the cat, but the receptor organ undergoes extensive regression. If, after nerve crush, regeneration of the peripheral nerve is permitted, axons reappear in the touch spot and reform nerve terminals in association with the Merkel cells (Brown & Iggo 1963, Burgess et al 1974). In a series of quantitatively controlled experiments in the cat, Burgess & Horch (1973), Burgess et al (1974), and English (1977) found that SA I receptors reappeared at old sites after cutting the femoral nerve; they concluded that "some sort of specific guidance mechanism [controlled] the growth of a particular class of cutaneous sensory fibres into denervated skin" (Burgess et al 1974).

MERKEL CELLS IN AMPHIBIA The original description by Merkel (1880) of organized groups of Merkel cells in frog skin has been called in question by ultrastructural studies. The electron microscopical results show that Merkel cells usually occur as isolated cells in the epidermis (Fox & Whitear 1978); in anurans they are immediately above the basal layer, whereas in unrodeles they are usually between the basal layer cells, although only rarely in contact with the basement membrane of the epidermis. The amphibian Merkel cells have most of the characteristics of their mammalian counterparts, such as membrane-bound osmiophilic granules, synapse-like junctions with nerve terminals, and finger-like projections containing fine filaments (Fox & Whitear 1978). In contrast to mammalian Merkel cells, they are not organized in groups. Instead, in the salamander, nerve fibers supply individual Merkel cells (Parducz et al 1977), and the nerve terminal is smaller and makes a much smaller contact area on the dermal surface of the Merkel cell.

Functional characteristics Only the salamander "Merkel" cell receptor has been studied in correlative electrophysiological and morphological studies (Parducz et al 1977). Parducz and co-workers report that in salamanders, in contrast to mammals, the "Merkel" units are associated with rapidly adapting mechanoreceptors. Indeed, Cooper & Diamond (1977) reported that all salamander cutaneous mechanoreceptors are rapidly adapting, from which it would seem to follow that, if the Merkel cells there are involved in mechanoreception, they can only be rapidly adapting. Frog skin, on the other hand, has both rapidly and slowly adapting mechanoreceptors and the morphlogical substrate awaits elucidation.

AVIAN MERKEL CELLS "Merkel" cells are absent from the epidermis in birds, but Saxod (1973, 1978) described small encapsulated structures in the dermis of the hard palate of the beaks of the domestic chicken and Japanese quail; these structures were also reported in chickens by Nafstad (1971). Three components were described: lamellar cells that surrounded Merkel-like cells which were in contact with expanded nerve terminals that were the endings of myelinated afferent fibers. These dermal corpuscles were present in groups of up to 30. The morphological features on the basis of which Saxod (1978) classed the cells as Merkel cells were (*a*) the presence of membrane-bound osmiophilic granules (also present in Grandry corpuscles), (*b*) finger-like axoplasmic processes containing fine filaments (also present in Grandry corpuscles), and (*c*) synaptic-like junctions between the "Merkel" cells and the nerve-endings. In some of his preparations the latter were sandwiched between the Merkel cells. Further cytochemical and electrophysiological work is required to establish the homology that Saxon has proposed on morphological grounds.

LATERAL LINE ORGANS These are present in fish and amphibia, and the receptor elements are the hair cells that are also present in the inner ear of fish and amphibia and in the mammalian organ of Corti and vestibular apparatus. In the lateral line the hair cell is embedded in the epidermis, although it forms a quite distinctive structure (see Flock 1971 for a review). The hair cell is a secondary sensory receptor, bearing on its outer extremity an array of stereocilia and one kinocilium. Its inner end makes contact with both an afferent nerve and an efferent nerve. The particular interest of hair cells for the present review is that ultra-structural and cytochemical studies (Flock 1971, Flock et al 1981), together with associated functional studies, have established some generalizations that are relevant to an understanding of the properties of cutaneous mechanoreceptors. The stereocilia contain longitudinally oriented filaments of actin (Flock et al 1981) which may serve either as a detector of deformation, or they play a more active role in a contractile coupling mechanism. Either way the cilia provide a means for transmitting mechanical energy to the hair cell surface. Micromanipulation of normal sensory hairs shows them to be very stiff, probably due to the inner filamentous core of actin (Flock et al 1981). Filamentous structures in the finger-like processes of Merkel cells are discussed above: there are similar structures in Grandry cells (see below), Pacinian corpuscles (Spencer & Schaumburg 1973), and lanceolate terminals in sinus hair follicles (Andres 1966). This common occurrence of filamentous material in cells involved in mechanoreception has led Flock et al (1981) to speculate that actin and other structural proteins have a general involvement in transduction. Changes in state of the protein could also underlie alterations in the sensitivity of receptors.

Encapsulated Receptors

Encapsulated receptors are found in the dermis (or corium) of the skin and share the following common features (Figure 2).

1. The outer capsule is formed from perineural cells and may be simple as in Ruffini ending with 3 or 4 layers, or complicated as in the Pacinian corpuscle where there may be 70 lamellae.
2. The stratified cells are attached to each other by demosome-like contacts (Chouchkov 1978).
3. Collagen fibers may fill the gaps between successive layers and both sides of a stratified cell may contain pinocytotic vesicles.
4. The outer capsule is separated by a capsular space, containing cells and membranes, from the inner core of the receptor within which the nerve terminal is contained.
5. Highly modified Schwann cells form this core, which completely sheaths the nerve terminal except where there are small cytoplasmic extensions of the nerve terminal that penetrate between the inner cells of the core. The Schwann cell origin of the inner core, although still debated, is now widely accepted (Polácek & Malinovsky 1971, Chouchkov 1978, Andres & von Düring 1973, Saxod 1973). Chouchkov (1978) has reviewed the arguments in detail. He proposes the name "Schwann receptor cell" for these inner core elements, to distinguish them from the Schwann cells that sheath the axons of peripheral nerves. The Schwann receptor cells contain numerous pinocytotic and coated vesicles.
6. The nerve terminal of the encapsulated receptors is never naked, although within the receptor some of its finer projections may not be

completely covered by Schwann cells. In all encapsulated receptors the myelinated parent axon penetrates the capsule and then loses its myelin sheath before forming the nerve terminal.

These common features of the encapsulated receptors are modified in different degrees in the various receptors described. Indeed, the common characteristics and enormous morphological variability have even led to the rejection of the idea of different kinds of receptor—instead they have been regarded as forming a morphological continuum (Weddell 1960). In the following account the various kinds of receptor are reviewed briefly and where possible the functional correlates of the morphologically identified receptors are given.

MEISSNER CORPUSCLES These relatively large (150 μm long, 40–70 μm in diameter) receptors occupy dermal ridges in primate glabrous skin. They are largest in human tissue. Little controversy has surrounded their

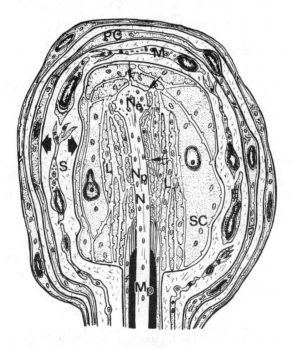

Figure 2 Schematic illustration of an encapsulated cutaneous receptor. The capsule is formed by perineural cells (PC); the capsule space (S) contains fibroblasts (F) and macrophages (M). The large arrows in S show the basal laminae of the capsule layer and the inner core. The inner core contains Schwann receptor cells (SC) and their lamellae (L). At the center lies the nerve terminal (N) with its unmyelinated part (N$_p$), nerve ending (N$_e$), and myelinated part (M$_p$) (from Chouchkov 1978).

existence since Wagner & Meissner's original description in 1852 and they have been fully documented in electron microscopical studies since their first description by Cauna & Ross in 1960 (e.g. Andres & von Düring 1973). The capsule is incomplete, and only encases the basal half of the receptor. It comprises dispersed perineural cell laminae and thick collagen fibers. The capsule space is ill-defined. The receptors occupy the papillary ridges that project in between the epidermal ridges. In a suitable preparation, Andres & von Düring (1973) were able to follow tonofibrils of the epidermal keratinocytes to hemidesmosomes of the basal epidermal cells, thence via collagen fibers to the Meissner corpuscle core or to attachments to the capsule of the receptor. This arrangement could provide an effective mechanical linkage of the epidermis to the receptor.

The inner core, comprised of Schwann cells and nerve terminals, has a distinctive appearance due to the horizontal lamellation of the Schwann cells that sandwich the oval or ellipsoidal nerve terminals. The afferent nerve supply is myelinated and several axons can be seen entering a Meissners corpuscle (2 to 6 in man, Chouchkov 1973).

Functional correlates   The function of Meissner's corpuscles has been debated. There is extensive physiological evidence which, assuming a correct assignment of SA I and SA II responses to Merkel and Ruffini endings (Iggo 1966), argues that the Meissner receptor in monkey (Lindblom, 1965) and in man (Vallbo & Hagbarth 1968, Johansson 1978) is the rapidly adapting (RA) receptor, and that the Pacinian corpuscle is the PC receptor. Munger et al (1979) in correlative experiments in monkeys present convincing evidence that the RA receptors are Meissner corpuscles. These attributions leave the Krause *küglige Endkolben* (see below) out of account, but it may also be a PC-type receptor. The RA and PC afferent units can be separated by their vibratory thresholds, the RA responding to low and PC to high frequencies (LaMotte & Mountcastle 1975), and if the attributions given above are valid, then on the basis of combined psychophysical and electrophysiological studies it becomes possible to assign specific sensory functions to the Meissner's and the Pacinian corpuscles—namely mediation of flutter and vibration, respectively (Merzenich & Harrington 1969, LaMotte & Mountcastle 1975).

KRAUSE ENDINGS   Krause (1860) described two forms of end bulb and clearly distinguished between the cylindrical form (*cylindrische* Endkolben) found in nonprimates and the globular or spherical end bulbs (*küglige Endkolben*) found in man and monkeys.

CYLINDRICAL END BULBS OF KRAUSE   The simpler form (*cylindrical end bulbs* in nonprimates) is present in many cutaneous regions. It lies in the dermis, often close to the epidermis, and has a simple capsule containing one to five layers of perineural receptor cells. There is a small capsular space separating the capsule from the inner core. The latter has a variable number of Schwann cell lamellae, ranging from less than 10 to as many as 30 (Halatá 1972b). Cylindrical Krause endings are present in glabrous skin of nonprimates such as the cat (Jänig 1971), and their morphological variations in cat foot pads have been extensively analyzed by Malinovský (1966). The majority (51%) had simple unbranched terminals, whereas branched terminals were found in 40%, and in the remainder (9%) the corpuscle was more complex, being divided into two or three parts or supplied by more than one axon. The mean length was 86 μm and the mean diameter was 27 μm. These receptors are therefore quite small, which adds to the difficulty of finding them.

The Krause *cylindrische Endkolben* (1876) reappear in the literature with many names, such as "small Vater-Pacini corpuscles" (Boeke 1932) "Mammalian end organ" (Winkelmann 1958), "innominate corpuscles" (Quilliam 1966), "paciniform corpuscles" (Malinovský 1966), and "simple encapsulated end organs" (Halatá 1972b).

The function of Krause endings has excited speculation from the beginning and for 80 years they have appeared in text books as cold receptors, following von Frey's (1895) scheme. Lynn (1969) and Jänig (1971) proposed that they were mechanoreceptors, and Iggo & Ogawa (1977) in correlative electrophysiological and morphological experiments concluded that the RA receptors of the cat's foot-pad were indeed Krause cylindrical end bulbs. The afferent axons are myelinated and 6 ± 1 μm in diameter (based on conduction velocity data). If present in hairy skin, the Krause endings may be the so far elusive field receptors of Burgess et al (1968). Another possibility is that the latter are Golgi-Mazzoni corpuscles.

SPHERICAL END BULBS OF KRAUSE   The *küglige Endkolben* that Krause (1860) originally described in human nasal and oral mucosa and in the human conjunctiva and genitalia are present in glabrous skin of the human extremities as well (Chouchkov 1973). Like the cylindrical end bulbs of non-primates, they have been given many names. The greatest uncertainty surrounds the "genital corpuscles," and although Chouchkov (1978), for example, considered them to be Krause spherical end bulbs, the matter is still in doubt. The typical spherical end bulb is oval or spherical, with a mean diameter of about 100 μm, and is therefore larger than the cylindrical end bulbs of nonprimates. There is a distinct capsule formed of from two to six layers, but a capsular space is not evident. A very characteristic feature of these spherical end bulbs is the tortuous intertwining of the nerve ending within the corpuscle. It more or less completely fills the inner core. The Schwann receptor cells and the nerve terminal do not form a regular structure, but the terminals are always enclosed by Schwann lamellae with which they can make close contact (Chouchkov 1973, 1978). The spherical end bulbs of Krause can be distinguished on the basis of location and structure from Meissner's corpuscles, also present in human glabrous skin.

Genital corpuscles share similar morphological features with Krause spherical end bulbs, but are present in both primates and nonprimates. Polácek & Malinovský (1971) report that in women the external genitalia contain complex encapsulated spherical genital corpuscles, which cannot be distinguished from Krause spherical end bulbs, whereas in the penis of the green monkey (Malinovský 1977) and macaque monkey the receptors have a simpler structure, more like the cylindrical end bulbs of Krause.

Functional characteristics   The functional characteristics of Krause spherical end bulbs has not yet been established. An attempt by Cottrell et al (1978) to correlate the physiology and morphology of receptors in the glans penis of the ram was inconclusive. Rapidly and slowly adapting afferent units were found, and although an assignment of the rapidly adapting units to genital corpuscles was suggested, no definite conclusion could be reached from the information available.

GOLGI-MAZZONI CORPUSCLES   These small lamellated corpuscles are the most difficult to verify. First described in muscle and tendon by Golgi and Mazzoni, they were reported in human subcutaneous tissue by Ruffini in 1893. He considered them to be variants of Pacinian corpuscles and to be at one extreme of size in a continuum from the largest Pacinian corpuscle downward. In the recent literature they have been described in subcutaneous tissue of the forepaw of the tree shrew (*Tuapia*) by Andres (1969). There is a perineural capsule of several layers, an indistinct capsular space, and a lamellated inner core containing at its center a single terminal. This has the typical terminal appearance with prominent cytoplasmic extensions between the poles of the hemilamellae of the inner core. In human skin, according to Chouchkov (1973), there may be a capsule of 10 to 15 layers and the inner core may be asymmetrical. The oval receptor, about 150–250 μm in diameter, is much smaller than the Pacinian corpuscle. The receptors are in the dermis below the epidermal ridges in glabrous skin.

Functional characteristics   The function of primate Golgi-Mazzoni corpuscles has not been established, but their lamellated central core suggests that they are rapidly adapting mechanoreceptors. Golgi-Mazzoni corpuscles have also been described in nonprimate skin (Andres & von Düring 1973). In the cat Sakada & Aida (1971) recorded electrophysiologically from rapidly adapting receptors in the periosteum of cat facial bone. The receptors could follow high frequency vibratory stimuli (200 to 300 /sec). Histological examination in light microscopy revealed simple encapsulated structures, with myelinated fibers, which were identified as Golgi-Mazzoni corpuscles. They were small (mean diameter 32 μm; mean length 83 μm) in contrast to human Golgi-Mazzoni corpuscles (150–250 μm diameter), but very similar both in appearance and size to nonprimate Krause end bulbs (27 μm X 86 μm). Since these latter have been reported by Iggo & Ogawa (1977) to be rapidly adapting receptors in the cat, it is possible that they are the same kind of receptor, especially since removal of the thick stratum corneum resulted in the foot-pad units following frequencies of sinusoidal stimulation up to 300 or 400 Hz (Ogawa & Iggo 1977).

RUFFINI ENDINGS   These were originally described by Ruffini (1893) and after several vicissitudes have now been firmly identified in electron microscopical studies (Goglia & Sklenská 1969, Chambers et al 1972). The whole receptor, which is spindle-shaped, lies in the dermis, both in human glabrous skin where it was first described and in hairy skin, and ranges in length from 0.5 to 2 mm. The distinct outer capsule of 3 to 5 lamellae surrounds a prominent fluid-filled and compartmentalized capsule space. The inner core is supplied by a myelinated axon 6 to 12 μm in diameter in the cat (Chambers et al 1972), which breaks up to form a dense brush-work of finer branches and terminals in the inner core. Collagen fibers enter the poles of the receptor and fuse with collagen fibrils that run continuously through the inner core but do not enter the capsule space. The fine terminals of the nerve ending are covered only by the basal lamina and may be in contact with the collagen fibrils. This arrangement provides effective transmission of mechanical tension in the dermis to the nerve terminals in the core of the receptor. Each Ruffini ending is supplied by one myelinated axon, but there is morphological evidence for branching of an individual stem axon to supply a number of Ruffini endings (Ruffini 1893, Goglia & Sklenská, 1969).

Functional characteristics   The Ruffini endings have been identified as the receptors of the slowly-adapting type II (SA II) cutaneous mechanoreceptor (Iggo 1966, Chambers et al 1972) in the cat. Similar functional properties are found for mechanoreceptors in monkey (Iggo 1962b) and human skin, both hairy and glabrous (Vallbo & Hagbarth 1968, Johansson 1978), and it is assumed that these too are Ruffini endings.

PACINIAN CORPUSCLE   This large (and best known) cutaneous mechanoreceptor, introduced to the scientific literature by Lehman & Vater in 1741, has the typical structure of an encapsulated receptor. The perineural receptor cells form a thick, many layered capsule of 20 to 70 layers; there is a small but definite capsular space, an inner core with many Schwann cell lamellae, and an unmyelinated nerve terminal in the center of the receptor that bears a club-like terminal expansion. The overall size of the receptor varies from 0.5 to 2 mm in length and is about 0.7 mm in diameter. The original electron micrographic description by Pease & Quilliam (1957) has had many successors (e.g. Spencer & Schaumburg 1973, Chouchkov 1978). The nerve terminal is seen in EM preparations to carry axon processes that are long-branched filiform structures, containing 6 nm diameter microfilaments, that penetrate between the hemilamellae of the inner core (Andres & von Düring 1973, Schaumburg et al 1974). These small processes are present in other receptors also (e.g. Golgi-Mazzoni corpuscles; lanceolate terminals of sinus hair follicles) and may contain actin filaments (Flock et al 1981). They can be damaged by noxious chemical agents (Schaumberg et al 1974, Flock et al 1981) which interfere with the normal function of the receptors. A key role for these processes has been suggested in several mechanoreceptors (Andres 1966, Iggo & Muir 1969, Flock 1971, Spencer & Schaumburg 1973).

Morphogenesis The Pacinian corpuscle is a convenient receptor for morphogenetic studies. Chalisova & Ilyinsky (1976) in nerve graft experiments reported that cutaneous, but not muscle, nerves were able to form Pacinian corpuscles when allowed to innervate the cat mesentery. These newly-formed receptors had normal functional characteristics. The normal formation of Pacinian corpuscles has been followed electron-microscopically (Malinovský 1976). At birth the corpuscles are immature and several stages can be found: corpuscles without an inner core, corpuscles with a small number of inner core lamellae that are arranged irregularly, and, finally, corpuscles that have a larger number of lamellae but are nevertheless immature (Malinovský 1976). In all stages the axon is present in the inner core and the axon processes may be present in the second stage receptors. In the neonatal rat, denervation of a hindlimb leads to failure of the normal development of Pacinian corpuscles. The immature corpuscles present at birth disintegrate and disappear within 10 days after nerve section at birth (Zelená 1976).

Functional characteristics The Pacinian corpuscle is a very rapidly adapting mechanoreceptor (Gray & Matthews 1951) capable of following high frequencies of sinusoidal stimulation (see Hunt 1974 for review). The afferent units are present in human skin (Vallbo & Hagbarth 1968, Johansson 1978) and their physiological characteristics indicate that they mediate the human sense of vibration (Merzenich & Harrington 1969, LaMotte & Mountcastle 1975).

The transduction of mechanical stimuli by Pacinian corpuscles is the classical example of cutaneous generator potentials (Gray & Sato 1953, Loewenstein & Rathkamp 1958). The role of the capsule and inner core in acting as a mechanical high-pass filter was established by Loewenstein & Skalak (1966). Detailed mechanisms of the transduction process continue to excite interest; the Pacinian corpuscle is a convenient and manageable example.

Ionic and underlying alterations in membrane structure have been analyzed by Ilyinsky et al (1976) using isolated corpuscles and the response attributed to asymmetrical changes in the conformation of the nerve terminal in the inner core. The extent to which these effects are due primarily to alterations in the configuration of the small axonal processes is not settled (Spencer & Schaumburg 1973) and this is also true for the role of the subcellular components of the nerve terminal.

HERBST CORPUSCLE   This is the avian homologue of the Pacinian corpuscle (Herbst 1848). Herbst corpuscles are present in the skin of the bill of wading and aquatic birds and in other species, and also in other places, including the legs and trunk (Gregory 1973). They have been extensively studied by electron microscopical techniques (see Saxod 1978 for review). The receptors in the bills of wading and aquatic birds lie in the corium, often in a monolayer about 75 μm below the basement membrane of the epidermis (Berkhoudt 1980), as well as in bony lacunae in the tips of upper and lower bills (Quilliam 1966, Gottschaldt & Lausmann 1974). The receptor is ovoid, 150–200 X 90–110μm, and has the typical encapsulated receptor structure; namely, a capsule of 5 to 10 layers, a large capsule space (25–60 μm), and an inner core comprising twin rows of up to 24 Schwann receptor cells and hemilamellae (Quilliam 1966). A single myelinated nerve fiber enters a corpuscle, and the last node of Ranvier, as in the Pacinian corpuscle, is at the beginning of the inner core. The Herbst corpuscle thus shows a variety of differences from the mammalian Pacinian corpuscle, in particular the symmetrical arrangement of the inner core nuclei along the central axon terminal (see Berkhoudt 1980 for review) that give rise to interdigitating hemilamellae (Andres 1966, Saxod 1973).

The anatomical disposition of Herbst corpuscles has been studied in the bill and tongue of the mallard duck (*Anas platyrhynchos L.*) by Berkhoudt (1980), who found them to be particularly numerous toward the side of the bill and toward its tip, and in two regions of the ventral aspect of the upper bill. An especially interesting region is the bill tip organ where Gottschaldt & Lausmann (1974) in the goose and Berkhoudt (1976, 1980) in the duck reported a high concentration of Herbst corpuscles, proximally placed in the dermal papillae of the bill tip.

Morphogenesis   Saxod (1973) has exploited the possibility of homoplastic and xenoplastic association of frontal bud and sensory ganglia in Japanese quail, Pekin ducks, and white leghorn chickens to establish the origin of different cellular components in Herbst corpuscles. Two distinct cellular origins were found: the inner core cells arise from cells that accompany the nerve during its growth, and the capsule cells and majority of capsule space cells arise from the dermal mesenchyme of the beak. These results support the hypothesis (Chouchkov 1978) that the inner core cells of encapsulated receptors are modified Schwann cells.

Functional characteristics   There is general agreement (Dorward & McIntyre 1971, Gregory 1973, Gottschaldt 1974) that Herbst corpuscles are very rapidly adapting receptors, with frequency-following characteristics similar to the Pacinian corpuscles.

GRANDRY CORPUSCLES   Since their first description by Grandry (1869), these receptors have repeatedly been confirmed as existing in the bill and tongue of aquatic birds (Malinovský 1967). They are absent from the Japanese quail and domestic hen (Saxod 1973). The corpuscle (30–80 μm diameter) is spherical and has a single-layered capsule, formed from scattered perineural receptor cells, plus macrophages and fibroblasts (Quilliam 1966, Saxod 1973), as seen electronmicroscopically. Within the capsule lie the Schwann receptor cells that form the lamellae that completely encase the two large hemispherical Grandry (or satellite) cells, that together with the nerve terminal occupy the core of the receptor (Andres 1969, Saxod 1973). The core of the simplest Grandry corpuscle is formed from two Grandry cells, each of which is large and hemispherical (50×15 μm). Their planar faces are separated by the nerve terminal, which is in the form of a flattened disc [40 μm in diameter and 2 μm thick in the duck (Saxod 1973)]. As many as eight Grandry units may be stacked linearly in a common capsule (Andres 1969). Gottschaldt & Lausmann (1974) suggest that the number of such subunits may be species-specific.

There are three features of Grandry sensory cells of special interest. First, there are numerous finger-like projections on the hemispheric surface of the cells, similar to those in Merkel cells, which possibly contain actin filaments. These processes interdigitate with the peripherally placed Schwann satellite cells. Second, the cytoplasm contains osmiophilic dense-cored, membrane-bound vesicles (120–250 nm diameter) clumped in groups scattered throughout the cytoplasm of the sensory cells and not, as in mammalian Merkel cells, concentrated adjacent to the nerve ending. Third, there are junctional zones formed between the sensory cell and the nerve terminal (Saxod 1973). The nerve terminal is formed into an expanded sheet sandwiched between a pair of sensory cells and derives from a myelinated axon.

Grandry sensory cell   Controversy has been endemic in the classification of the large sensory cells as Merkel cells. Although there are several similar morphological features (Saxod 1973, Andres 1969, Munger 1971), the Grandry and Merkel cells are not identical, nor do they occupy similar locations. The mammalian Merkel cell is in the epidermis and is unencapsulated, whereas the Grandry sensory cell is dermal and encapsulated. The similarity in structure may point to analgous functions but not to morphological identity.

DERMAL "MERKEL" CELLS   In the chicken and quail, which lack Grandry cells, Saxod (1973) describes groups of cells in the dermis of the palate immediately below the epidermis. As with Grandry corpuscles there may be an aggregation of these cells and their associated nerve endings. Saxod (1973) has called these structures "les corpuscles de Merkel." Because of their similarity to Grandry corpuscles, Saxod (1973) suggests that they may be homologous; and in view of their dermal location and morphological dissimilarity it is at least necessary to resist equating them with mammalian Merkel cells without further evidence.

Functional characteristics   Gottschaldt (1974) reported only rapidly adapting responses from goose bill afferent fibers. In view of the large numbers of Grandry corpuscles in the bill of the goose (Gottschaldt & Lausmann 1974) and duck (Berkhoudt 1980), and the known ability of Herbst corpuscles to follow high frequencies of stimulation, it is apparent that the Grandry corpuscles do not function as slowly adapting mechanoreceptors. In the bill tip organ, however, there is electrophysiological evidence for an additional slowly adapting unit, whose receptor could not be identified. In nonaquatic birds that have "Merkel corpuscle" receptors in the palate, no evidence for functional characteristics is available.

Hair Follicle Mechanoreceptors

Various descriptions exist for nerve-endings associated with hair follicles. These follicles range in size from simple follicles to the morphologically very complex sinus hairs.

SIMPLE HAIR FOLLICLES   These do not contain erectile tissue and two main subgroups exist (Noback 1951): (*a*) the coarser and longer hairs generally called guard hair and (*b*) the finer underhairs, called down hair. The nerve endings are parallel to the hair follicle long axis and in electron microscopic preparations they are seen as a circumferential array or palisade of unmyelinated nerve terminals derived from myelinated axons (Yamamoto 1966, Andres 1966, Cauna 1969). Each terminal is encased in finger-shaped Schwann cell processes, except at its inner and outer edge.

The interstices between these processes in the palisade are occupied by fine filamentous material (Yamamoto 1966). The number of stem axons supplying one hair follicle is greater in the larger follicles and one stem axon can supply many hair follicles.

Functionally, these mechanoreceptors are rapidly adapting and, according to the kind of hair and diameter of the axon, can be divided into Type D follicle afferents comprising the down hairs and A delta axons (Brown & Iggo 1967, Burgess et al 1968), and Type G follicle afferents, comprising guard hairs and A beta axons (Brown & Iggo 1967). A further subdivision of the Type G units into G_1 and G_2 was proposed by Burgess & Perl (1973), but more recently Tuckett et al (1978) have advanced reasons for doubting the validity of this suggestion.

NONSINUS FACIAL HAIR FOLLICLES   Ruffini-like spray terminations have been reported in nonsinus hair follicles in primate facial skin by Biemesderfer et al (1978) in correlative physiological and morphological studies. Slowly adapting afferent units excited by moving nonsinus facial hairs were isolated as single units. The afferent discharge had the characteristics of SA II receptors of hairy skin (Chambers et al 1972) or sinus type II (St II) units of cat vibrissae (Gottschaldt et al 1973), and in primate hairy skin similar afferent units exist (Merzenich & Harrington 1969). Each hair follicle was supplied by several axons. It is not clear from the morphological analysis whether only Ruffini-like terminals are present in the follicle, although an absence of lanceolate terminals (Andres 1966) and Merkel cells from the SA hair follicles is reported by Biemesderfer et al (1978).

Ruffini originally described his receptors in glabrous skin as distinct spindle-shaped structures in the dermis, a description confirmed in recent articles and extended to hairy skin (Chambers et al 1972). The proposal by Biemesderfer et al (1978) that their structures are "the end organ of Ruffini" can apply only to the ultrastructural characteristics of the nerve terminals, which they describe as encircling the hair follicle just below the sebaceous gland. They propose the term "pilo-Ruffini complex" for this association of nonsinus facial hair and Ruffini-like terminations.

SINUS HAIR FOLLICES   Sinus hair follicles (Merkel 1880) are present in small numbers in the skin of many mammals. In the face they are known as vibrissae or whiskers and elsewhere have been called tactile hairs (Noback 1951) or tylotrich follicles (Straile 1961). Characteristic features are the large diameter and length of the hairs, the presence of a vascular sinus contained in a large bulbous capsule, an associated erectile (smooth or striated muscle) tissue, and a rich nerve supply. A detailed reconstruction based on light and electron microscopes of rat and cat sinus hairs (Figure

3) was made by Andres (1966). Nerve terminals are present in the midregion of the follicle below the sebaceous gland, and the innervation is more complex than in ordinary nonsinus hairs. Andres (1966, 1969) described four kinds of nerve terminal:

1. Merkel cells arranged around the perimeter of the hair in the stratum basale adjacent to the glassy membrane. Patrizi & Munger (1966) reported similar findings.
2. Lanceolate nerve terminals, corresponding to the finger-like terminals in nonsinus hairs and lying in the inner hair follicles on the other side of the glassy membrane from the Merkel cells. These terminals formed thick lancets (lanceolate endings), covered on each side by specialized Schwann cells, and at the edges bearing small finger-shaped axon processes that contained fine filamentous material. Like the cilia of lateral-line hair cells, these filaments may contain actin. The lanceolate terminals occur in three forms, according to their location in the hair follicle (Andres 1966).
3. Small lamellated corpuscles were present in the sinus hair follicle. They had the typical structure of encapsulated dermal receptors: capsule, inner core, and nerve terminal.
4. Fine nerve terminals were present in rat sinus hairs toward the outer end of the inner hair follicle.

The above descriptions apply to facial sinus hairs. In the cat there are also carpal sinus hairs on the forelegs and these differ in that they lack the lamellated corpuscles. Instead, there are Pacinian corpuscles clustered around the outside of the follicle (Nilsson 1969).

Functional characteristics   Several recent reports describe the electrophysiology of sinus hair afferents (Nilsson 1969, Zucker & Welker 1969, Hahn 1971a, Pubols et al 1973, Gottschaldt et al 1973). It has been possible, on the basis of analogous results from single unit analyses, to ascribe functions to the various receptors in cat sinus hairs (Gottschaldt et al 1973). A generally agreed attribution is that the Merkel cells mediate slowly adapting responses, while the high frequency rapidly adapting units are the lamellated corpuscles in vibrissae (and Pacinian corpuscles in carpal sinus hairs).

TYLOTRICH HAIR FOLLICLES   The sinus hairs found on the general body surface, called tylotrich follicles by Straile (1961), have a less complex structure and simpler innervation than the facial and carpal sinus hairs. On the basis of light microscope preparations, Straile (1961) reports palisade endings in the external root sheath as well as circular nerve fibers. In contrast to the facial sinus hairs, there are no Merkel cells at this location.

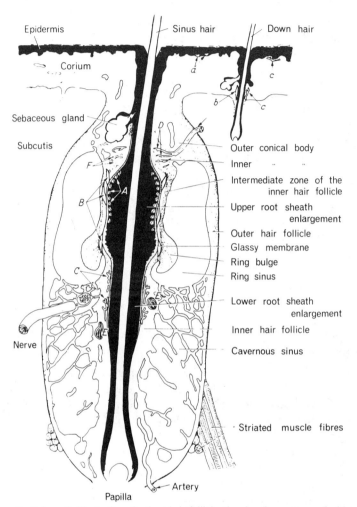

Epidermis

Corium

Sebaceous gland

Subcutis

Nerve

Papilla

Sinus hair   Down hair

Outer conical body
Inner   ..   ..
Intermediate zone of the
        inner hair follicle
Upper root sheath
        enlargement
Outer hair follicle
Glassy membrane
Ring bulge
Ring sinus
Lower root sheath
        enlargement
Inner hair follicle

Cavernous sinus

Striated muscle fibres

Artery

Figure 3   Schematic illustration of a sinus hair follicle, showing the great complexities of this sensory hair follicle and the various receptors that it contains. A, Merkel cells; B, straight lanceolate endings; C, branched lanceolate endings; D, circular lanceolate endings; E, encapsulated receptors; F, unmyelinated nerve fibers (rat only); a, b, c, terminals in normal hair follicle and epidermis corresponding to A, B, and C above (from Andres 1966).

Instead, they are present at the orifice of the follicle, with the characteristic associated structural modification of the skin described above (see Merkel cells).

Functional characteristics   The tylotrich follicles are conspicuous in the rabbit. Brown & Iggo (1967) found that the afferent units were characteristically rapidly adapting, but, in contrast to Type G follicle afferents, they often had a very low critical slope and thus were able to respond to very slow hair deflections, although they were silent when the hair was stationary. These afferent units were placed in a separate category, and called Type T follicle units. The Merkel cells at the orifice of the receptor were typical SA I afferent units (Brown & Iggo 1967).

Free Nerve-Endings

The older literature contains frequent mention of "free" or "naked" sensory endings. Electron microscopists have had great difficulty identifying them. Both Cauna (1976), in his papillary endings in human hairy skin, and Chouchkov (1978) report fine subepidermal nerve terminals partly covered by a basal lamina and Schwann lamellae. These terminations, which usually contain mitochondria and some vesicles, can be traced back to myelinated stem axons and are therefore different from the receptors of the unmyelinated afferent fibers.

It has not yet been possible to make a combined morpho-functional correlative study of these "free" nerve endings.

Unmyelinated Receptors

Unmyelinated receptors continue to be the most difficult cutaneous receptors to study morphologically. Cauna (1973, 1976) has systematically reconstructed unmyelinated fibers as they approach the epidermis. In their dermal location they have the characteristic grouping of several axons invested by a common Schwann cell, each axon with its mesaxon (Gasser 1955). As they approach the epidermis a modified Schwann cell provides the sheath for several unmyelinated axons. The unmyelinated fibers and Schwann cells then form what Cauna calls a penicillate ending. Most of the endings terminate in the subepidermal corium close to the epidermis. Intraepidermal endings are extremely rare. The small size of the terminations (less than 1 μm diameter) causes difficulty in recognizing the endings and Chouchkov (1978) has given a number of criteria to aid in their identification: (a) terminal swellings just below the epidermis, (b) an accumulation of mitochondria and clear vesicles in the axoplasm [although Cauna (1976) found the endings to lack conspicuous subcellular components], (c) pinocytotic vesicles in the Schwann receptor cells round the nerve terminals, (d) the presence of denuded sectors of axolemma covered only with basal lamina (this is the nearest approach to a naked nerve terminal and is very small in extent) and (e) the existence of finger-like axonal processes.

A continuing difficulty is to find a reliable method to identify the terminals of unmyelinated dorsal root afferent fibers and to distinguish them from the unmyelinated terminal extensions of myelinated dorsal root fibers. Since all the cutaneous afferent fibers end in terminals that lack a myelin sheath, it is necessary at present to use serial sectioning (tedious and difficult though it is) and to follow a terminal back to unmyelinated fibers in a common Schwann sheath (Cauna 1976) before reaching a firm conclusion that the afferent fiber is unmyelinated.

Functional characteristics   The cutaneous unmyelinated afferent fibers innervate a variety of receptors: C-mechanoreceptors (Iggo 1960, Bessou et al 1971, Hahn 1971b), C-nociceptors (Iggo 1959, Burgess & Perl 1973, Beck et al 1974), and C-thermoreceptors (Hensel et al 1960, Hensel & Iggo 1971). This variety of functional receptor-types with unmyelinated afferent fibers is not presently matched by a corresponding description of morphological variety. As Zotterman has suggested, it may be necessary to seek the morphological correlates at the molecular level (Iriuchijima & Zotterman 1960).

Reptilian Mechanoreceptors

Reptilian mechanoreceptors have recently been reviewed by von Düring & Miller (1979) and (as with cutaneous receptors in other classes of vertebrate) electron microscopy has helped to clear up morphological problems. Reptilian skin is characterized by epidermal variety but dermal uniformity. Von Düring & Miller (1979) used a system of classification that is broadly in agreement with that used elsewhere in the present article.

THE INTRAEPIDERMAL RECEPTORS   These receptors in reptiles are more elaborate than in amphibia, aves, and mammalia; form large free discoid terminals (von Düring 1974, Landmann & Villiger 1975); are widely distributed in the scales; and are in special relationship to the epithelial tonofibril system. The hinge-region receptors are branched and located in the epidermal fold between scales (von Düring 1974a). The functional characteristics of these receptors awaits elucidation.

MERKEL CELLS   Cells with typical Merkel features are present in the basal layers of the epidermis of several reptilian species (von Düring & Miller 1979), and in one of von Düring & Miller's schematic illustrations a single axon is shown branching to supply two Merkel cells in the epidermis and one in the dermis.

DERMAL RECEPTORS   Reptiles possess relatively simple unencapsulated nerve terminals in the dermis, covered by lamellated Schwann cells only. More complex corpuscles have a thin capsule (von Düring 1973), and in some forms there may, in addition, be a capsule space. The simplest dermal endings are free in the connective tissue. Dermal Merkel cells occur in cell columns in the touch papillae of the upper and lower jaw of the crocodile (von Düring 1974b). The basic structural organization is similar to that reported for avian Merkel cell corpuscles by Saxod (1978). Lamellar cells encase the Merkel cells, and the latter are in close contact with expanded nerve terminals that are interleaved between them.

Functional characteristics of reptilian mechanoreceptors We are unaware of any electrophysiological studies aimed at correlating functional and morphological studies in reptiles. There is evidence for some variety of response to mechanical stimulation, such as two kinds of slowly adapting receptors in alligator skin (Kenton et al 1971) and vibration-sensitive receptors in snake skin (Proske 1969a,b), but insufficient information is available to support any morphofunctional correlations, except by analogy with mammalian and avian mechanoreceptors.

The reptilian cutaneous receptors may be organized into complex tactile sense organs or touch papillae. Various degrees of complexity have been described and indeed even more complex structures have been shown to exist (further details in von Düring & Miller 1979). Various elaborate sense organs also exist in birds [e.g. bill tip organ (Gottschaldt 1974, Berkhoudt 1976)] and in mammals [e.g. Eimer's organ in the mole (Andres & von Düring 1973) and the sinus hair follicle (Andres 1966)].

CUTANEOUS THERMORECEPTORS

The organization of this review implies the existence of mechanoreceptors, thermoreceptors, and nociceptors, although on morphological grounds alone, this would be difficult to justify, because of the difficulty in establishing the function of particular structures. The existence of thermoreceptors is argued on sensory and electrophysiological grounds. In order to recognize the receptors as morphological entities it is necessary to do combined physiological and morphological experiments.

Mammalian Thermoreceptors

Hensel et al (1974) have succeeded in correlating structure and function with the cat's facial cold receptors. First, single unit electrophysiological experiments were used to establish receptive fields. These were all small, and could be delimited by inserting fine marker wires into the skin. Then subsequent histological examination of the marked tissues established that there were distinctive nerve terminals in each identified receptive field. Each spot was supplied by a thin myelinated fiber that branched close to, and ended in, the basal layer of the epidermis. The branching axons were invested by a Schwann cell up to the basement membrane of the epithelium. The nerve terminals were inserted into invaginations of the basal epidermal cells. These cold receptors have thin (1–2 μm in diameter) afferent fibers and it is not known if the more common unmyelinated cold receptors elsewhere in the skin of the cat have similar endings. No comparable reports have been published on mammalian warm receptors.

Reptilian Thermoreceptors

The infrared pit organ in snakes is the best known reptilian thermoreceptor. The organs have been examined electron-microscopically in *Boa, Python,* and *Crotalus* (Terashima et al 1970, von Düring 1974, von Düring & Miller 1979). In *Boa* the receptors are in scales on the surface, whereas in python they are in grooves on the upper and lower lip scales, and in vipers they are in special pit organs. In all examples the afferent nerve fiber, after forming a plexus in the dermis, terminates in expanded terminals rich in mito-chondria. The terminals are intraepidermal in *Boa* and *Python* and intrad-ermal in the pit organ of *Crotalus.* In *C. horridus* (Bullock & Fox 1957) a large number (several thousand) of myelinated axons (3–16 μm in diame-ter) enter the pit organ. The pit organ functions as a warm receptor, and according to Goris & Terashima (1976) it can detect thermal changes in, but not the temperature of, the environment.

CUTANEOUS NOCICEPTORS

If little is known about the morphology of thermoreceptors, even less is known about the nociceptors. On the basis of electrophysiological evidence, they form a separate class (e.g. Burgess & Perl 1973) and in general have small myelinated or unmyelinated axons. (The termination of unmyelinated axons in the skin of mammals is discussed above.) It is not yet possible to make any positive statements about the morphological characteristics of the nociceptors, although Kruger et al (1979) have suggested in a preliminary report that intradermal nerve endings in cat skin are nociceptors.

Literature Cited

Adrian, E. D., Zotterman Y. 1926. The im-pulses produced by sensory nerve-endings. Pt. II. The response of a single end-organ. *J. Physiol.* 61:151–71
Andres, K. H. 1966. Über die Feinstruktur der Rezeptoren an Sinushaaren. *Z. Zellforsch. Mikrosk. Anat.* 75:339–65
Andres, K. H. 1969. Zur Ultrastruktur Ver-schiedener Mechanorezeptoren von hö-heren Wirbeltieren. *Anat. Anz.* 125: 551–65
Andres, K. H., von Düring, M. 1973. Mor-phology of cutaneous receptors. In *Handbook of Sensory Physiology:*

Somatosensory System, ed. A. Iggo. 2:3–28. Heidelberg: Springer-Verlag. 851 pp.
Beck, P. W., Handwerker, H. O., Zimmer-mann, M. 1974. Nervous outflow from the cat's foot during noxious radiant heat stimulation. *Brain Res.* 67:373–86
Berkhoudt, H. 1976. The epidermal structure of the bill tip organ in ducks. *Neth. J. Zool.* 25:561–66
Berkhoudt, H. 1980. The morphology and distribution of cutaneous me-chanoreceptors (Herbst and Grandry corpuscles) in bill and tongue of the

mallard (*Anas platyrhynchos L.*) *Neth. J. Zool.* 30:1–34

Bessou, P., Burgess, P. R., Perl, E. R., Taylor, C. B. 1971. Dynamic properties of mechanoreceptors with unmyelinated (C) fibers. *J. Neurophysiol.* 34:116–31

Biemesderfer, D., Munger, B. L., Binck, J., Dubner, R. 1978. The Pilo-Ruffini complex: A non-sinus hair and associated slowly-adapting mechanoreceptor in primate facial skin. *Brain Res.* 142:197–222

Boeke, J. 1932. Nerve endings, motor and sensory. In *Cytology & Cellular Pathology of Nervous System*, ed. W. S. Penfield, 1:241–315. New York: Hoeber

Breathnach, A. S. 1971. Embryology of human skin: A review of ultrastructural studies. *J. Invest. Dermatol.* 58:381–87

Brown, A. G., Iggo, A. 1963. The structure and function of cutaneous "touch corpuscles" after nerve crush. *J. Physiol.* 165:28–29P

Brown, A. G., Iggo, A. 1967. A quantitative study of cutaneous receptors and afferent fibres in the cat and rabbit. *J. Physiol.* 193:707–33

Bullock, T. H., Fox, W. 1957. The anatomy of the infrared sense organ in the facial pit of pit vipers. *Q. J. Microsc. Sci.* 98:219–34

Burgess, P. R., Horch, K. W. 1973. Specific regeneration of cutaneous fibers in the cat. *J. Neurophysiol.* 36:101–14

Burgess, P. R., Perl, E. R. 1973. Cutaneous mechanoreceptors and nociceptors. See Andres & von Düring, 2:29–78

Burgess, P. R., Petit, D., Warren, R. M. 1968. Receptor types in cat hairy skin supplied by myelinated fibers. *J. Neurophysiol.* 31:833–48

Burgess, P. R., English, K. B., Horch, K. W., Stensaas, L. J. 1974. Patterning in the regeneration of Type I cutaneous receptors. *J. Physiol.* 236:57–82

Cauna, N. 1962. Functional significance of the submicroscopical, histochemical and microscopical organization of the cutaneous receptor organs. *Anat. Anz.* 111:181–97

Cauna, N. 1969. The fine morphology of the sensory receptor organs in the auricle of the rat. *J. Comp. Neurol.* 136:81–98

Cauna, N. 1973. The free penicillate nerve endings of the human hairy skin. *J. Anat.* 115:277–88

Cauna, N. 1976. Morphological basis of sensation in hairy skin. *Prog. Brain Res.* 43:35–45

Cauna, N., Ross, L. L. 1960. The fine structure of Meissner's touch corpuscles of

human fingers. *J. Biophys. Biochem. Cytol.* 8:472–82

Chalisova, N. I., Ilyinsky, O. B. 1976. Process of regeneration and development of the tissue receptors specificity. *Prog. Brain Res.* 43:47–52

Chambers, M. R., Andres, K. H., von Düring, M., Iggo, A. 1972. The structure and function of the slowly adapting Type II mechanoreceptor in hairy skin. *Q. J. Exp. Physiol.* 57:417–45

Chouchkov, C. N. 1973. The fine structure of small encapsulated receptors in human digital glabrous skin. *J. Anat. London* 114:25–33

Chouchkov, C. N. 1978. Cutaneous receptors. *Adv. Anat. Embryol. Cell Biol.* 54:7–62

Clifton, G. L., Coggeshall, R. E., Vance, W. H., Willis, W. D. 1976. Receptive fields of unmyelinated ventral root afferent fibres in the cat. *J. Physiol.* 256:573–600

Cooper, E., Diamond, J. 1977. A quantitative study of the mechanosensory innervation of the salamander skin. *J. Physiol.* 264:695–723

Cottrell, D. F., Iggo, A., Kitchell, R. L. 1978. Electrophysiology of the afferent innervation of the glans penis in the domestic ram. *J. Physiol.* 283:347–67

Dorward, P. K., McIntyre, A. K. 1971. Responses of vibration-sensitive receptors in the interosseous region of the duck's hind limb. *J. Physiol.* 219:77–87

English, K. B. 1974. Cell types in cutaneous Type I mechanoreceptors (Haarscheiben) and their alterations in injury. *Am. J. Anat.* 141:105–24

English, K. B. 1977. The ultrastructure of cutaneous Type I mechanoreceptors (Haarscheiben) in cats following denervation. *J. Comp. Neurol.* 172:137–64

English, K. B., Burgess, P. R., Kavka-Van, D. 1980. Development of rat Merkel cells. *J. Comp. Neurol.* 194:475–96

Flock, Å. 1971. Sensory transduction in hair cells. In *Handbook of Sensory Physiology: Principles of Receptor Physiology*, ed. W. R. Loewenstein, 1:396–441. Berlin: Springer-Verlag. 600 pp.

Flock, Å., Cheung, H. C., Flock, B., Utter, G. 1981. Three sets of actin filaments in sensory cells of the inner ear. Identification and functional orientation determined by gel electrophoresis, immunofluorescence and electron microscopy. *J. Neuroxytol.* 10:133–47

Fox, H., Whitear, M. 1978. Observations on Merkel cells in amphibians. *Biol. Cell.* 32:223–32

Gasser, H. S. 1955. Properties of dorsal root unmedullated fibers on the two sides of the ganglion. *J. Gen. Physiol.* 38:709–28

Goglia, G., Sklenská, A. 1969. Ricerche ultrastrutturali sopra i corpuscoli di Ruffini delle capsule articolari nel coniglio. *Quad. Anat. Prat.* 25:14–27

Goris, R. C., Terashima, S. 1976. The structure and function of the infrared receptors of snakes. *Prog. Brain Res.* 43:159–70

Gottschaldt, K.-M. 1974. Mechanoreceptors in the beaks of birds. In *Mechanoreception*, ed. J. Schwartzkopff, pp. 109–13. Symp. Sonderforschungsbereichs, Bochum, 1973. Opladen: Westdeutscher Verlag. 419 pp.

Gottschaldt, K.-M., Iggo, A., Young, D. W. 1973. Functional characteristics of mechanoreceptors in sinus hair follicles of the cat. *J. Physiol.* 235:287–315

Gottschaldt, K.-M., Lausmann, S. 1974. The peripheral morphological basis of tactile sensibility in the beak of geese. *Cell Tiss. Res.* 153:477–96

Grandry, M. 1869. Recherches sur les corpuscules de Pacini. *J. Anat. Paris.* 6:390–95

Gray, J. A. B., Matthews, P. B. C. 1951. A comparison of the adaptation of the Pacinian corpuscle with the accommodation of its own axon. *J. Physiol.* 144:454–64

Gray, J. A. B., Sato, M. 1953. Properties of the receptor potential in Pacinian corpuscles. *J. Physiol.* 122:610–36

Gregory, J. E. 1973. An electrophysiological investigation of the receptor apparatus of the duck's bill. *J. Physiol.* 229:151–64

Hahn, J. F. 1971a. Stimulus-response relationships in first order sensory fibres from cat vibrissae. *J. Physiol.* 213:215–26

Hahn, J. F. 1971b. Thermal-mechanical stimulus interactions in low-threshold C-fiber mechanoreceptors of cat. *Exp. Neurol.* 33:607–17

Halatá, Z. 1970. Zu den Nervenendigungen (Merkelsche Endigungen) in der haarlosen Nasenhaut der Katze. *Z. Zellforsch. Mikrosk. Anat.* 106:51–60

Halatá, Z. 1972a. Innervation der unbehaarten Nasenhaut des Maulwurfs (*Talpa europaea*). 1. Intraepidermale Nervenendigungen. *Z. Zellforsch. Mikrosk. Anat.* 125:108–20

Halatá, Z. 1972b. Innervation der unbehaarten Nasenhaut des Maulwurfs (*Talpa europaea*) II. Innvervation der dermis (einfache eingekapselte Körperchen). *Z. Zellforsch. Mikrosk. Anat.* 125:121–31

Hashimoto, K. 1973. Fine structure of the Meissner corpuscle of human palmer skin. *J. Invest. Derm.* 60:20–28

Hensel, H., Andres, K. H., von Düring, M. 1974. Structure and function of cold receptors. *Pfluegers Arch.* 352:1–10

Hensel, H., Iggo, A. 1971. Analysis of cutaneous warm and cold fibres in primates. *Pfluegers Arch.* 329:1–8

Hensel, H., Iggo, A., Witt, I. 1960. A quantitative study of sensitive cutaneous thermoreceptors with C afferent fibres. *J. Physiol.* 153:113–26

Herbst, G. E. F. 1848. *Die Pacinischen Körper und ihre Bedeutung.* Göttingen: Badenhoech & Ruprecht

Hoggan, G. 1884. New forms of nerve terminations in mammalian skin. *J. Anat. Physiol.* 18:182–97

Hökfelt, T., Johansson, O., Ljungdahl, Å., Lundberg, J. M., Schultzberg, M. 1980. Peptidergic neurones (Review article). *Nature* 284:515–21

Hunt, C. C. 1974. The Pacinian corpuscle. In *The Peripheral Nervous System*, ed. J. I. Hubbard, pp. 405–20. New York: Plenum. 530 pp.

Iggo, A. 1959. Cutaneous heat and cold receptors with slowly-adapting (C) afferent fibres. *Q. J. Exp. Physiol.* 44:362–70

Iggo, A. 1960. Cutaneous mechanoreceptors with afferent C fibres. *J. Physiol.* 152:337–53

Iggo, A. 1962a. New specific sensory structures in hairy skin. *Acta Neuroveg.* 24:175–80

Iggo, A. 1962b. An electrophysiological analysis of afferent fibres in primate skin. *Acta Neuroveg.* 24:225–40

Iggo, A. 1966. Cutaneous receptors with a high sensitivity to mechanical displacement. In *Ciba Fdn. Symp, Touch, Heat and Pain*, ed. A. V. S. de Reuk, J. Knight, pp. 237–56. Boston: Little, Brown. 389 pp.

Iggo, A. 1974. Cutaneous receptors. See Hunt 1974, pp. 347–404

Iggo, A. 1976. Is the physiology of cutaneous receptors determined by morphology? *Prog. Brain Res.* 43:15–31

Iggo, A., Muir, A. R. 1969. The structure and function of a slowly-adapting touch corpuscle in hairy skin. *J. Physiol.* 200:763–96

Iggo, A., Ogawa, H. 1977. Correlative physiological and morphological studies of rapidly adapting mechanoreceptors in cat's glabrous skin. *J. Physiol.* 266:275–96

Ilyinsky, O. B., Volkova, N. K., Cherepnov, V. L., Krylov, B. V. 1976. Morphofunctional proerties of Pacinian corpuscles. *Prog. Brain Res.* 43:172–86

Iriuchijima, J., Zotterman, Y. 1960. The specificity of afferent cutaneous C fibres in mammals. *Acta Physiol. Scand.* 49:267–78

Jänig, W. 1971. Morphology of rapidly and slowly adapting mechanoreceptors in the hairless skin of the cat's hind foot. *Brain Res.* 28:217–31

Johansson, R. S. 1978. Tactile sensibility in the human hand: receptive field characteristics of mechanoreceptive units in the glabrous skin area. *J. Physiol.* 281:101–23

Kenton, B., Kruger, L., Woo, M. 1971. Two classes of slowly adapting mechanoreceptor fibres in reptile cutaneous nerve. *J. Physiol.* 212:21–44

Krause, W. 1876. *Allemeine und Mikroskopische Anatomie.* Hannover: Hahn' Sche Hofbuchhandlung

Kruger, L., Perl, E. R., Sedivec, M. J. 1979. Electron microscopic identification of cutaneous delta nociceptors of cat. *Neurosci. Lett.* Suppl. 3, p. S262

LaMotte, R. H., Mountcastle, V. B. 1975. Capacities of humans and monkeys to discriminate between vibratory stimuli of different frequency and amplitude: A correlation between neural events and psychophysical measurements. *J. Neurophysiol.* 38:539–59

Landmann, L., Villiger, W. 1975. Glycogen in epidermal nerve terminals of *Lacerta sicula* (Squamata-Reptilia). *Experientia* 31:967–68

Lindblom, U. 1965. Properties of touch receptors in distal glabrous skin of the monkey. *J. Neurophysiol.* 28:965–85

Loewenstein, W. R., Rathkamp, R. 1958. The sites for mechanoelectrical conversion in a Pacinian corpuscle. *J. Gen. Physiol.* 41:1245–65

Loewenstein, W. R., Skalak, R. 1966. Mechanical transmission in a Pacinian corpuscle. An analysis and a theory. *J. Physiol.* 182:346–78

Lyne, A. G., Hollis, D. E. 1971. Merkel cells in sheep epidermis during fetal development. *J. Ultrastruct. Res.* 34:464–72

Lynn, B. 1969. The nature and location of certain phasic mechanoreceptors in the cat's foot. *J. Physiol.* 201:765–73

Malinovský, L. 1966. Variability of sensory nerve endings in foot pads of a domestic cat (*Felis ocreata L., F. domestica*). *Acta Anat.* 64:82–106

Malinovský, L. 1967. Die Nervenendkörperchen in der Haut von Vögeln und ihre Variabilität. *Z. Zellforsch. Mikrosk. Anat.* 77:279–303

Malinovský, L. 1976. Ultrastructural features of Pacinian corpuscles in the early postnatal period. *Prog. Brain Res.* 43:53–58

Malinovský, L. 1977. Ultrastructure of sensory nerve terminals in the penis in green monkey (*Cercopithecus aethiops sabaeus*). *Z. Zellforsch. Mikrosk. Anat.* 91:541–52

Merkel, F. 1875. Tastzellen und Tastkörperchen bei den Haustieren und beim Menschen. *Arch. Microsck. Anat. Entwmech.* 11:636–52

Merkel, F. 1880. *Über die Endigungen der sensiblen Nerven in der Haut der Wirbeltier.* Rostock: H. Schmidt

Merzenich, M. M., Harrington, T. 1969. The sense of flutter-vibration evoked by stimulation of the hairy skin of primates. *Exp. Brain Res.* 9:236–60

Miller, M. R., Ralston, H. J. 3rd, Kasahara, M. 1960. The pattern of cutaneous innervation of the human hand, foot and breast. In *Cutaneous Innervation,* ed. W. Montagna, pp. 1–47. Oxford: Pergamon. 203 pp.

Müller, J. 1844. *Handbuch der Physiologie des Menschen,* p. 667 ff. Coblenz: Hölscher

Munger, B. L. 1971. The comparative ultrastructure of slowly and rapidly adapting mechanoreceptors. In *Oral-Facial Sensory and Motor Mechanisms,* ed. R. Dubner, Y. Kawamura, pp. 83–103. New York: Appleton-Century-Crofts. 384 pp.

Munger, B. L., Pubols, L. M. 1972. The sensorineural organization of the digital skin of the raccoon. *Brain Behav. Evol.* 5:367–93

Munger, B. L., Page, R. B., Pubols, B. H. Jr. 1979. Identification of specific mechanosensory receptors in glabrous skin of dorsal root ganglionectomized primates. *Anat. Rec.* 193:630–31

Nafstad, P. H. J. 1971. On the ultrastructure of neuro-epithelial interactions in the dermal innervation in the snout of the pig. *Z. Zellforsch. Mikrosk. Anat.* 122:528–37

Nilsson, B. Y. 1969. Hair discs and Pacinian corpuscles functionally associated with the carpal tactile hairs in the cat. *Acta Physiol. Scand.* 77:417–28

Noback, C. R. 1951. Morphology and phylogeny of hair. *Ann. NY Acad. Sci.* 53:476–91

Ogawa, H., Iggo, A. 1977. Dependance of the response characteristics of glabrous rapidly-adapting units in the cat on the

stratum corneum. *Brain Res.* 126:
167–71

Pacini, F. 1840. *Nuovi organi Scorperti nel corpo umano.* Pistoja: Ciro

Parducz, A., Leslie, R. A., Cooper, E., Turner, C. J., Diamond, J. 1977. The Merkel cells and the rapidly adapting mechanoreceptors of the salamander skin. *Neuroscience* 2:511–21

Patrizi, G., Munger, B. L. 1966. The ultrastructure and innervation of rat vibrissae. *J. Comp. Neurol.* 126:423–25

Pease, D. C., Quilliam, T. A. 1957. Electron microscopy of the Pacinian corpuscle. *J. Biophys. Biochem. Cytol.* 3:331–43

Pinkus, F. 1905. Über Hautsinnesorgane neben dem menschlichen Haar (Haarscheiben) und ihre verleichend-anatomische Bedeutung. *Arch. Mikr. Anat.* 65:121–79

Polácek, P., Malinovsky, L. 1971. Die ultrastruktur der genitalkörperchen in der clitoria. *Z. Zellforsch. Mikrosk. Anat.* 84:293–310

Proske, U. 1969a. Vibration-sensitive mechanoreceptors in snake skin. *Exp. Neurol.* 23:187–94

Proske, U. 1969b. An electrophysiological analysis of cutaneous mechanoreceptors in a snake. *Comp. Biochem. Physiol.* 29:1039–46

Pubols, B. H. Jr., Donovick, P. J., Pubols, L. M. 1973. Opossum trigeminal afferents associated with vibrissa and rhinarial mechanoreceptors. *Brain Behav. Evol.* 7:360–81

Quilliam, T. A. 1966. Unit design and array patterns in receptor organs. See Iggo 1966, pp. 86–116

Ruffini, A. 1893. Di un nuovo Organo nervoso terminale e sulla presenza die corpuscoli Golgi-Mazzoni nel connettivo sottocutaneo dei polpastrelli delle dita dell, uomo. *Atti d. r. Accad. d. Lincei. Cl. d. Sc. Fis. Matemat. e. Nat.* ser. 4a, 7:398–409

Sakada, S., Aida, H. 1971. Electrophysiological studies of Golgi-Mazzoni corpuscles in the periosteum of the cat facial bones. *Bull. Tokyo Dent. Coll.* 12: 255–72

Saxod, R. 1973. Organisation ultrastructurale des corpuscules sensoriels cutanes des oiseaux. *Sci. Nat.* 1:79–98

Saxod, R. 1978. Ultrastructure of Merkel corpuscles and so-called "Transitional" cells in the white leghorn chicken. *Am. J. Anat.* 151:453–74

Schaumburg, H. H., Wisniewski, H. M., Spencer, P. S. 1974. Ultrastructural studies of the dying-back process. *J. Neuropathol. Exp. Neurol.* 33:260–84

Schwarzkopf, J. 1974. *Symposium: Me chanoreception.* Opladen: Westdeutscher Verlag. 419 pp.

Smith, K. R. 1967. The structure and function of the haarscheibe. *J. Comp. Neurol.* 131:459–74

Smith, K. R. 1970. The ultrastructure of the human Haarscheibe and Merkel Cell. *J. Invest. Derm.* 54:150–59

Spencer, P. S., Schaumburg, H. H. 1973. An ultrastructural study of the inner core of the Pacinian corpuscle. *J. Neurocytol.* 2:217–35

Straile, W. E. 1961. The morphology of tylotrich follicles in the skin of the rabbit. *Am. J. Anat.* 109:1–13

Tapper, D. N. 1965. Stimulus-response relationships in the cutaneous slowly-adapting mechanoreceptor in hairy skin of the cat. *Expl. Neurol.* 13:364–85

Terashima, S., Goris, R. C., Katsuki, Y. 1970. Structure of warm fiber terminals in the pit membrane of vipers. *J. Ultrastr. Res.* 31:494–506

Tuckett, R. P., Horch, K. W., Burgess, P. R. 1978. Response of cutaneous hair and field mechanoreceptors in cat to threshold stimuli. *J. Neurophysiol.* 41: 138–49

Vallbo, Å. B., Hagbarth, K.-E. 1968. Activity from skin mechanoreceptors recorded percutaneously in awake human subjects. *Exp. Neurol.* 21:270–89

von Düring, M. 1973. The ultrastructure of lamellated mechanoreceptors in the skin of reptiles. *Z. Anat. Entwickl. Gesch.* 145:81–94

von Düring, M. 1974a. The radiant receptor and other tissue receptors in the scales of the upper jaw of *Boa constrictor. Z. Anat. Entwickl. Gesch.* 145:299–319

von Düring, M. 1974b. The ultrastructure of cutaneous receptors in the skin of *Caiman crocodilus.* See Gottschaldt 1974, pp. 123–34

von Düring, M., Miller, M. R. 1979. Sensory nerve endings of the skin and deeper structures of reptiles. In *Biology of Reptilia,* ed. C. Gans, 9:407–41. New York: Academic. 462 pp.

von Frey, M. 1895. Beiträge zur sinnesphysiologie der haut, III. *Ber. Verh. K. Saechs. Ges. Wiss. Leipzig. (Math.-Phys. Kl.).* 47:166–84

Wagner, R., Meissner, G. 1852. Über Vorhandsein bischer unbekannten eizenhumlichen Körperchen (Corpuscula tactus). *Gött. Nachr.* 2:17–30

Weddell, G. 1960. Studies related to the mechanism of common sensibility. In *Cutaneous Innervation,* ed. W. Montagna, pp. 112–160. Oxford: Pergamon. 203 pp.

Winkelmann, R. K. 1958. The sensory endings in the skin of the cat. *J. Comp. Neurol.* 109:221–32

Winkelmann, R. K., Breathnach, A. S. 1973. The Merkel cell. *J. Invest. Derm.* 60:2–15

Yamamoto, T. 1966. The fine structure of the palisade-type sensory endings in rela-tion to hair follicles. *J. Electron Microsc.* 15:158–66

Zelená, J. 1976. The role of sensory innervation in the development of mechanoreceptors. *Prog. Brain Res.* 43:57–64

Zucker, E., Welker, W. I. 1969. Coding of somatic sensory input by vibrissae neurons in the rat's trigeminal ganglion. *Brain Res.* 12:138–56

Ann. Rev. Neurosci. 1982. 5:33–56

INHERITED METABOLIC
STORAGE DISORDERS*

Roscoe O. Brady

Developmental and Metabolic Neurology Branch, National Institute of
Neurological and Communicative Disorders and Stroke, National Institutes of
Health, Bethesda, Maryland 20205

INTRODUCTION

A critical review of recent developments in our understanding of the heritable storage disorders that involve the nervous system is highly appropriate at this time, since a number of new topics that deserve inclusion in this category have not been previously covered by a comprehensive overview. For many years when one considered storage disorders involving the brain, one dealt primarily with the sphingolipidoses, for the following reasons:

1. The extraordinary prevalence of these disorders compared with most of the other storage diseases.
2. The early descriptions of patients suffering from sphingolipidoses (perhaps a consequence of their comparative frequency).
3. The successes achieved by analytical chemists in identifying the stored lipids.
4. The antecedent discovereies of the metabolic defects underlying these disorders.

The classic example of this is seen in Gaucher's disease:

1. There are between 4000 and 5000 patients with the adult (Type I) form of this disorder in the United States—by far the greatest number of patients with a heritable storage disorder of any type.
2. The first patient was reported 100 years ago.
3. The accumulating lipid in this disorder was identified in 1933 as glucocerebroside.
4. The basic enzymatic defect in the lipidoses was demonstrated first in this condition in 1965.

A related development was the progressive assembly of information concerning a second group of metabolic storage disorders, called the mucopolysaccharidoses. The elaboration of the stored materials and enzyme defects occurred later than in the lipidoses; however, the principle of a deficiency of a catabolic hydrolytic enzyme was also established for these disorders.

Based on these developments, it seemed likely that in time one would eventually encounter patients in whom the catabolism of proteins was compromised—this was particularly likely with regard to glycoproteins, because it had been shown that the enzymes that were lacking in the lipidoses and mucolysaccharidoses were generally involved in the cleavage of a component from the nonreducing terminus of oligosaccharide chains. (Niemann-Pick disease and Farber's disease are, of course, exceptions because the accumulating lipids in these conditions are not glycolipids.) This review therefore includes a consideration of the nosology and pathogenesis of glycoproteinoses.

To complete a survey of this expanding area, two additional categories of storage disorders will be considered. The first is the group of four disorders known as the mucolipidoses. There is some overlap concerning the nosology of these disorders; I offer some suggestions to reduce the confusion. The second group of disorders comprise the neuronal ceroidlipofuscinoses. There has been a great deal of interest recently in this area, and the pertinent new findings are summarized.

CLASSIFICATION

The salient aspects of the clinical presentations, types of compounds stored, and the enzymatic lesions in the neuronal storage disorders are summarized in Tables 1–5 in order to bring these interwoven conditions into perspective. Situations where there is overlap and uncertainty concerning the classification of a disorder are indicated. I have followed the suggestion of Beaudet (1981) that several disorders previously classified as oligosaccharidoses—viz. fucosidosis, mannosidosis, nonganglioside sialidosis, and aspartylglycosaminuria—be considered in the category of glycoproteinoses (Table 3). Note that fucosidosis might also be properly included as a sphingolipidosis.

The nosology of the mucolipidoses is in flux at present. Thus, a patient in the Mucolipidosis I group (Table 4-A) should probably be better considered as having a sialidase-deficient glycoproteinosis (Table 3-C). However, precise classification is difficult at the moment because of the extensive organomegaly that these patients display. Most patients with glycoprotein sialidosis have little enlargement of visceral organs except for those with the congenital hydropic form. This dilemma will probably have to be settled by an appropriate commission on nomenclature.

There is also some uncertainty concerning the classification of Mucolipidosis IV patients (Table 4-D). Two reports (Bach et al 1979, Hahn et al 1980) indicate that this disorder might well be a ganglioside sialidosis (Table 1-J). If this view can be confirmed, reclassification of this disease as a glycosphingolipidosis would be appropriate.

Although the metabolic lesion in none of the diseases included in the neuronal ceroid-lipofuscinoses (Table 5) is known, this should not preclude their inclusion in this consideration of neuronal storage disorders. In fact, it emphasizes the need for informed deductions and the development of reasonable hypotheses concerning the etiology of these disorders.

PATHOGENESIS OF STORAGE DISORDERS

Lipid Storage Disorders (Table 1)

GANGLIOSIDOSES

Sialidoses (Table 1-J; Table 4-D)   Gangliosides are sialic acid-containing glycolipids that are particularly concentrated in the gray mattter of the brain although they are also present in many other tissues. Because the rate of ganglioside turnover in the central nervous system is maximal in the neonatal period, just prior to myelination, considerable thought has been devoted to the role that gangliosides might play in neuronal recognition (Roth 1978), myelination (Brady & Quarles 1973), and synaptogenesis (Brady 1976). Turnover implies both synthesis and, particularly, catabolism. Because impairment of lipid catabolizing enzymes had been demonstrated in several heritable neurodegenerative disorders, and because the catabolism of polysialogangliosides is initiated through the action of sialic acid-cleaving enzymes, it seemed reasonable to expect that eventually patients with ganglioside storage disorders would be identified in whom the metabolic defect was a deficiency of sialidase (neuraminidase) activity (Brady 1966). Recent observations seem to substantiate this speculation. Bach et al (1979, 1980) have reported that the catabolism of gangliosides G_{M3} (N-acetylneuraminyl-galactosylglucosylceramide) and G_{D3} (N-acetyl-neuraminyl-N-acetylneuraminyl-galactosylglucosylceramide), which must be initiated by the hydrolytic splitting of the sialic acid residues, is impaired in cultured skin fibroblasts and cultured aminiocytes from patients with Mucolipidosis IV. Hahn et al (1980) observed an even more drastic reduction in sialidase activity in cultured fibroblasts by the use of gangliosides G_{D1a} and G_{D1b} (cf Table 1-J) that were labeled in the sialic acid

Table 1 Lipid storage disorders (sphingolipidoses)

Disease	Major signs and symptoms	Major accumulating lipid	Enzyme defect	Normal metabolic product
A. Farber's disease	Hoarseness, dermatitis, skeletal deformation, mental retardation	Ceramide (N-Fatty–acylsphingosine)	Ceramidase	Sphingosine + fatty acid
B. Gaucher's disease	Hepatosplenomegaly, erosion of long bones and pelvis, mental retardation only in infantile form (Type II)	Glucocerebroside (glucosylceramide)	Glucocerebrosidase	Glucose + ceramide
C. Niemann-Pick	Hepatosplenomegaly, mental retardation, emaciation	Sphingomyelin (ceramide-phosphocholine)	Sphingomyelinase	Phosphocholine + ceramide
D. Krabbe's disease	Mental retardation, diminished myelin, globoid bodies in white matter of brain	Galactocerebroside (galactosylceramide)	Galactocerebrosidase	Galactose + ceramide
E. Metachromatic leukodystrophy	Mental retardation, psychological disturbances in adult form, nerves stain yellow-brown with cresyl violet dye	Sulfatide (galactocerebroside-sulfate)	Sulfatidase	Sulfuric acid + galactocerebroside
F. Fabry's disease	Reddish-purple skin rash, kidney failure, burning pains in hands and feet	Ceramide trihexoside (galactosylgalactosylglucosylceramide)	Ceramide trihexosidase	Galactose + lactosylceramide

Disease	Clinical features	Enzyme	Substrate	Products
G. Tay-Sachs disease	Mental retardation, red spot in retina, blindness, muscular weakness	Hexosaminidase A	Ganglioside G_{M2} (N-acetylgalactosaminyl-(N-acetylneuraminyl)-galactosylglucosyl-ceramide)	N-acetylgalactosamine + ganglioside G_{M3}
H. Sandhoff's disease	Same as Tay-Sachs disease but more rapid progression	Hexosaminidase A and Hexosaminidase B	Ganglioside G_{M2} + asialo-G_{M2} (N-acetylgalactosaminylgalactosylglucosylceramide) + globoside (N-acetylgalactosaminylgalactosylgalactosylglucosylceramide)	N-acetylgalactosamine + ganglioside G_{M3}; N-acetylgalactosamine + ceramidetrihexoside
I. Generalized (G_{M1})-gangliosidosis	Mental retardation	β-Galactosidase	Ganglioside G_{M1} [galactosyl-N-acetylgalactosaminyl-(N-acetylneuraminyl)-galactosylglucosylceramide]	Galactose + ganglioside G_{M2}
J. Sialidosis (Mucolipidosis IV)[a]	Psychomotor retardation, corneal opacities	Ganglioside sialidase	Ganglioside G_{D1a} [N-acetylneuraminylgalactosyl-N-acetylgalactosaminyl-(N-acetylneuraminyl)-galactosylglucosylceramide]; Ganglioside G_{D1b} [Galactosyl-N-acetylgalactosaminyl-(N-acetylneuraminyl-N-acetylneuraminyl)-galactosylglucosylceramide]	Sialic acid + ganglioside G_{M1}
K. Fucosidosis[b]	Cerebral degeneration, muscle spasticity, thickened skin	α-Fucosidase	H-sioantigen (Fucosylgalactosyl-N-acetylglucosaminyl galactosylglucosylceramide)	Fucose + neolactotetraosyl-ceramide

[a]See Table 4-D. Note: Sialidase = N-acetylneuraminidase.
[b]See Table 3-A.

Table 2 Mucopolysaccharide storage disorders (mucopolysaccharidoses)

Type	Name of disorder	Major signs and symptoms				Major accumulating materials	Enzyme defect	Site of defect on Fig. 1
		Dys-morphism	Mental retarda-tion	Corneal cloud-ing	Organ-omegaly			
IH	Hurler	+	+	+	+	Dermatan sulfate, heparan sulfate	α-Iduronidase	B
IS	Scheie	+	0	+	0	Dermatan sulfate, heparan sulfate	α-Iduronidase	B
IH/S	Hurler/Scheie compound	+	±	+	+	Dermatan sulfate, heparan sulfate	α-Iduronidase	B
II	Hunter	+	+	0	+	Dermatan sulfate, heparan sulfate	Idurondsulfate sulfatase	A
IIIA	Sanfilippo syndrome	±	+	0	±	Primarily heparan sulfate	Heparan N-sulfatase	F
IIIB	Sanfilippo syndrome	+	+	0	±	Primarily heparan sulfate	α-N-Acetylglucosaminidase	I
IIIC	Sanfilippo syndrome	+	+	0	±	Primarily heparan sulfate	Acetyl-CoA: α-glucosaminide N-acetyltransferase	
IIID	Sanfilippo syndrome	+	++	0	+	Primarily heparan sulfate	N-Acetylglucosamine 6-SO$_4$ Sulfatase	H
IVA	Morquio syndrome	±	0	±	0	Keratan sulfate, chondroitin sulfate	N-Acetylgalactosamine 6-SO$_4$ Sulfatase	M
IVB	Morquio syndrome	±	0	+	0	Keratan sulfate, chondroitin sulfate	β-Galactosidase	J
VI	Maroteaux-Lamy	+	0	+	0	Dermatan sulfate	N-Acetylgalactosamine 4-SO$_4$ sulfatase (arylsulfatase B)	C
VII	β-Glucuronidase deficiency	+	+	0	0	Heparan sulfate, dermatan sulfate, chondroitin sulfate	β-Glucuronidase	E
VIII	Mucosulfati-dosis	+	+	0	+	Dermatan sulfate, keratan sulfate, heparan sulfate	Multiple sulfatase deficiency	C, F, L, M

Table 3 Glycoproteinoses

Disorder	Major signs and symptoms					Accumulating metabolities	Enzyme defect
	Facies	Dysmorphism (dysostosis multiplex)	Mental retardation	Organomegaly	Ocular involvement		
A. Fucosidosis	Mild coarsening	++	+++ Seizures	++	±	H-isoantigen[a], fucosyl-N-acetylglucosaminide, fucosyl-dekasaccharide and other fucose-containing oligosaccharides	α-Fucosidase
B. Mannosidosis	Coarse	+++	Severe	+++	Cataracts, corneal opacities	Mannosylmannosyl-N-acetylglucosamine, mannose-containing oligosaccharides	α-Mannosidase
C. Sialidosis							
Type I	Normal	0	Progressive, myoclonic seizures, gait disturbance	0	Cherry red spot, blindness	N-acetylneuraminylgalactosyl-N-acetyl-glucosaminylmannosylmannosyl-N-acetyl-glucosamine and other sialic-acid-containing oligosaccharides	α-Neuraminidase
Type II Juvenile onset	Mildly coarse	++	++, myoclonus	0	Cherry red spot, decreased activity	N-acetylneuraminylgalactosyl-N-acetyl-glucosaminylmannosylmannosyl-N-acetyl-glucosamine and other sialic-acid-containing oligosaccharides	α-Neuraminidase
Infantile onset	Coarse	+++	+++	±	Cherry red spot	N-acetylneuraminylgalactosyl-N-acetyl-glucosaminylmannosylmannosyl-N-acetyl-glucosamine and other sialic-acid-containing oligosaccharides	α-Neuraminidase
Congenital (hydropic form)	Coarse	+++	+++	++	Unknown	Unknown	α-Neuraminidase
D. Aspartylglycosaminuria	Coarse, sagging	+	++	±	Lens opacities	N-acetylglucosaminylasparagine, galactosyl-N-acetylglucosaminylasparagine, mannosylmannosyl-N-acetylglucosaminyl-N-acetylglucosaminylasparagine, other glycoasparagines	Aspartylglycosaminidase

[a] See Table 1–K.

Table 4 Mucolipidoses

Name of disorder	Major signs and symptoms					Pathological chemistry	Enzyme defect
	Facies	Dysostosis multiplex	Mental retardation	Organo-megaly	Ocular manifestations		
A. Mucolipidosis I[a] (pseudo-hurler)	Coarse	++	++	++	Cherry-red spot	Excess sialated oligosaccharides in urine	Non-ganglioside neuraminidase
B. Mucolipidosis II (I-cell disease with early onset)	Coarse	+++	+++	+++	Corneal clouding	Increased lysosomal enzymes in serum; reduced levels in fibroblasts; inclusion bodies in fibroblasts	Decreased phosphorylation of glycoprotein hydrolases
C. Mucolipidosis III (milder form of I-cell disease)	Can be normal	++	+	+	Clouding by split-lamp examination	Elevated serum lysosomal enzymes; low fibroblast enzymes; inclusion bodies	Probably similar to mucolipidosis II.
D. Mucolipidosis IV[b]	Normal	0	++	0	Corneal clouding	Increased gangliosides in fibroblasts	Ganglioside neuraminidase

[a] This designation probably will be reclassified as a glycoprotein sialidosis (Table 3–C).
[b] This designation probably will be reclassified as a ganglioside sialidosis (Table 1–J).

Table 5 Neuronal ceroid lipofuscinoses (NCL)

Disorder	Onset	Clinical manifestations	Neuronal accumulation	Metabolic defect
A. Santavuori-Haltia	Infantile	Delay in motor and intellect, visual loss, hypotonia, ataxia, myoclonus, microcephaly	Greenish yellow autofluorescent material (all forms of NCL)	Unknown
B. Batten's disease				
Jansky-Bielschowsky	Late infantile	Seizures, atazia, mental retardation spasticty, blindness	Curvilinear bodies; increased dolichols in gray matter	Dolichol metabolism? (vitamin A metabolism ?)
Spielmeyer-Vogt (Spielmeyer-Sjögren)	Juvenile	Visual deterioration, seizures, progressive dementia, motor apraxia	Fingerprint (rectilinear) profiles	Unknown; no increase in dolichols in gray matter
C. Kuf's disease	Adult	Seizures, myoclonus, cerebellar ataxia, progressive dementia	Marked neuronal dropout; ceroid storage in histiocytes	Unknown

moieties. While these observations are intellectually satisfying, Bach et al (1979) emphasize the preliminary nature of their observations and a detailed report by Hahn and co-workers has not yet appeared. There are two particularly disquieting aspects of the findings reported by these and other investigators who have examined cells and tissues from Mucolipidosis IV patients:

1. There was only a partial deficiency of ganglioside neuraminidase in the cells analyzed by Bach's group. However, these authors believed that they were able to identify cells with intermediate deficiencies from heterozygotes, although there was some overlap between the neuraminidase activity in these cells and that in the controls.
2. Mucopolysaccharides such as dermatan sulfate and herparan sulfate (cf Figure 1) accumulated in the cells from patients with Mucolipidosis IV and the extent of this accumulation exceeded that of the gangliosides (Hahn et al 1980).

Although I have included gangliosidosis due to neuraminidase deficiency in the lipid storage disorders, it may be prudent to withold final categorization of these patients until these points are resolved. This caveat seems particularly appropriate since it is well established that gangliosides G_{M3} and G_{M2} accumulate in the brain of patients with mucopolysaccharide storage diseases where the enzymatic lesions have been clearly identified as defects in mucopolysaccharide degradation (Constantopoulos & Dekaban 1978).

Generalized Gangliosidosis (Table 1-I)   Although it is well known that the major metabolic derangement in patients with generalized (G_{M1})-gangliosidosis is a deficiency of ganglioside-β-galactosidase activity (O'Brien 1978), several less well publicized observations merit consideration. 1. There is a ten-fold accumulation of asialo-G_{M1} (galactosyl-N-acetylgalactosaminylgalactosylglucosylceramide), as well as the predominant storage of G_{M1}, in the brain of these patients. This observation is consistent with the existence of an alternative pathway for monosialoganglioside catabolism through neuraminidase activity (Tallman et al 1972). 2. Unusually high levels of galactose-terminated oligosaccharides appear in the brain, liver, and urine of these patients (Wolfe et al 1977, Tsay and Dawson 1976, Brunngraber 1978). The pathogenetic importance of the accumulation of oligosaccharides vs ganglioside in this disorder is unresolved. The origins of these substances are distinct: oligosaccharides are derived from glycoproteins and gangliosides arise from glycolipids. Thus, the quantity and rate of turnover of these precursors differentially contribute to the accumulation of metabolites. A priori, one would expect that gangliosides would remain within neurons as part of incompletely digested membranes, and that the water-soluble oligosaccharides might diffuse out. One reason for the accu-

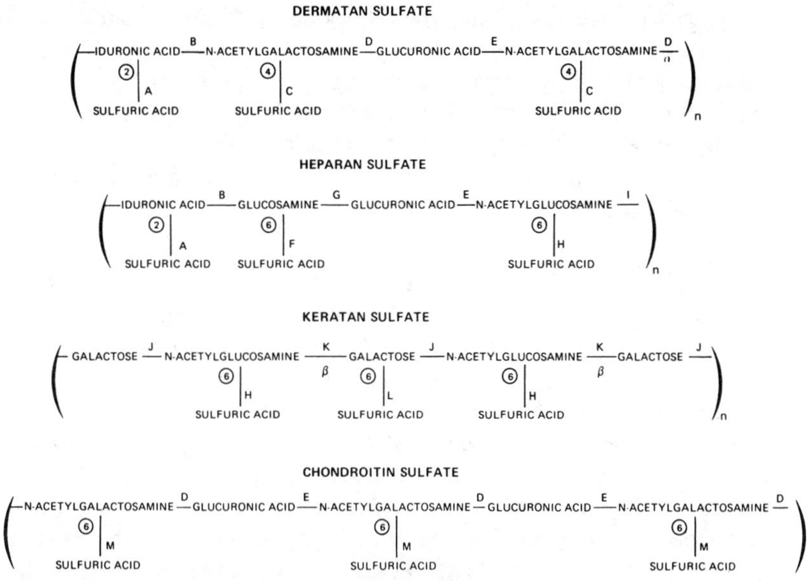

DERMATAN SULFATE

$$\left(\begin{array}{c} \text{IDURONIC ACID} \xrightarrow{B} \text{N-ACETYLGALACTOSAMINE} \xrightarrow{D} \text{GLUCURONIC ACID} \xrightarrow{E} \text{N-ACETYLGALACTOSAMINE} \xrightarrow[a]{D} \\ \quad \end{array}\right)_n$$

②|A SULFURIC ACID   ④|C SULFURIC ACID   ④|C SULFURIC ACID

HEPARAN SULFATE

$$\left(\begin{array}{c} \text{IDURONIC ACID} \xrightarrow{B} \text{GLUCOSAMINE} \xrightarrow{G} \text{GLUCURONIC ACID} \xrightarrow{E} \text{N-ACETYLGLUCOSAMINE} \xrightarrow{I} \end{array}\right)_n$$

②|A SULFURIC ACID   ⑥|F SULFURIC ACID   ⑥|H SULFURIC ACID

KERATAN SULFATE

$$\left(\begin{array}{c} \text{GALACTOSE} \xrightarrow{J} \text{N-ACETYLGLUCOSAMINE} \xrightarrow{K} \text{GALACTOSE} \xrightarrow{J} \text{N-ACETYLGLUCOSAMINE} \xrightarrow{K} \text{GALACTOSE} \xrightarrow{J} \end{array}\right)_n$$

⑥|H SULFURIC ACID   β ⑥|L SULFURIC ACID   ⑥|H SULFURIC ACID   β

CHONDROITIN SULFATE

$$\left(\begin{array}{c} \text{N-ACETYLGALACTOSAMINE} \xrightarrow{D} \text{GLUCURONIC ACID} \xrightarrow{E} \text{N-ACETYLGALACTOSAMINE} \xrightarrow{D} \text{GLUCURONIC ACID} \xrightarrow{E} \text{N-ACETYLGALACTOSAMINE} \xrightarrow{D} \end{array}\right)_n$$

⑥|M SULFURIC ACID   ⑥|M SULFURIC ACID   ⑥|M SULFURIC ACID

Figure 1 Principal repeating oligosaccharide units that accumulate in patients with mucopolysaccharide storage disorders. The letters indicate sites of enzymatic defects in the diseases listed in Table 2.

mulation of galactose oligosaccharides may be the presence of surface (and intracellular) receptors for galactose-containing ligands of the type that has been found on synaptic membranes (Kusiak et al 1979, 1980). These membrane components cycle from the interior of cells to the surface and back (Steer & Ashwell 1980), which may account in part for the exodus of oligosaccharides from the intracellular stores.

The metabolic lesion in several patients diagnosed as having G_{M1}-gangliosidosis seems actually to have been a defect in sialidase (neuraminidase) activity (Lowden & O'Brien 1979). It has been proposed that the observed decrease in β-galactosidase activity results from the inhibition of this enzyme by accumulating sialylated oligosaccharides. This deduction has been strengthened by genetic complementation experiments in which high levels of β-galactosidase activity were demonstrated after neuraminidase was partially restored by fusion of cells from patients previously thought to have an abnormal β-galactosidase gene.

The interesting report of Farrell & Ochs (1981) of infantile and juvenile forms of G_{M1} gangliosidosis in a single family strongly supports the concept of allelic mutations in the β-galactosidase gene. In fact, the authors deduced that the patients were actually compound heterozygotes. Farrell & MacMartin (1981) investigated the isoelectric focusing of β-galactosidase isoenzymes from the two phenotypes. They observed three residual bands of acid β-galactosidase in tissues from the patient with the juvenile form of the disorder, whereas there was only a single band in the patients with the infantile form. They inferred that differences in isoenzyme patterns obtained by isoelectric focusing could be correlated with allelic mutations. A similar deduction was made by Ginns et al (1980) from an investigation of the pathologic enzymology of Gaucher's disease.

Tay-Sachs Disease and Sandhoff's Disease (Table 1-G; 1-H)   The metabolic defect in Tay-Sachs disease is a deficiency of a hexosaminidase that is required for the hydrolytic cleavage of the terminal molecule of N-acetyl-galactosamine from G_{M2} (Kolodny et al 1969, Sandhoff et al 1971, Tallman et al 1972). In spite of these established observations, the pathogenesis of Tay-Sachs disease remains exceptionally complex. It has been demonstrated that the catabolism of the accumulating G_{M2} may be initiated by two enzymes—either through the activity of a neuraminidase that cleaves N-acetylneuraminic acid, forming asialo-G_{M2} (G_{A2}), or via hexosaminidase, forming G_{M3}. In fact, G_{M2} sialidase appears to be more active than hexosaminidase in many human tissues (Kolodny et al 1969, Tallman & Brady 1972, Tallman et al 1974). This is significant because only one of the two major normal hexosaminidase isozymes (Hex B) occurs in the tissues of most patients with Tay-Sachs disease (Okada & O'Brien 1969). The activity of this residual Hex B may be increased many-fold over normal in the brain of Tay Sachs patients. Hex B readily catalyzes the cleaveage of G_{A2} produced by neuraminidase to ceramide lactoside, whose further catabolism is unimpaired. Several issues remain to be resolved. Even with the increased Hex B activity, G_{A2} accumulates approximately 20-fold over normal in the brains of patients with this form of Tay-Sachs disease (G_{M2} accumulation may be over 100-fold). Patients with the Sandhoff form of Tay-Sachs disease, with diminished Hex A and Hex B, accumulate G_{M2} much more rapidly, and there is also a greater quantity of cerebral G_{A2} than is seen in the conventional form of the disorder. These observations provide considerable support for the view that ganglioside catabolism in the brain normally occurs to an appreciable extent by the neuraminidase pathway. Because ganglioside turnover is particularly rapid in the neonatal period, the neuraminidase route may be inadequate, in the absence of Hex A, to prevent G_{M2} accumulation, even though Hex B is greatly increased. Although this hypothesis is satisfying, one disquieting aspect is the finding that neuraminidase activity is not increased in Tay-Sachs disease. There are several possibilities that explain this finding:

1. In contrast to the induction of glycolipid hydrolases in organs such as the liver and spleen when they are presented with an excess of substrate

(Kampine et al 1967a), this enzyme may not be regulated by the quantity of G_{M2} that the brain is required to catabolize.

2. Neuronal cells may not be able to respond by enzyme induction when confronted with excess catabolite.

3. Enzyme induction in the liver and spleen may be the result of an increase in the number of cells, particularly of the reticuloendothelial system. Since neurons do not proliferate one would not expect an increase in neuraminidase activity by such a mechanism.

A potentially important development in ganglioside storage disorders is the demonstration by Conzelman & Sandhoff (1978) that patients with the AB-form of Tay-Sachs disease (in which Hex A and Hex B activities with artificial substrates are normal although there is still a pathologic accumulation of G_{M2}) may have a diminished amount of a heat-stable "activator" required for the catabolism of glycolipids. The existence of such compounds has been anticipated for a number of years, and their discovery and characterization has been summarized elsewhere (Brady 1978). The experiments of Conzelmann & Sandhoff strongly suggest that the absence of a factor that is unaffected by heating to 60° in the crude form may be the metabolic defect in patients with the AB form of Tay-Sachs disease. A recent report of similar findings in another patient appears to confirm this pathogenetic mechanism (Hechtman 1980).

Some intriguing additional observations have recently been made by Li et al (1981), who investigated the catabolism of G_{M2} in tissues from these patients as well as from two additional AB-variant cases (Goldman et al 1980). Hexosaminidase isozymes A and B and activator material appeared to be present in the tissues from both patients. In one patient, G_{M2} was catabolized in vitro upon the addition of an activator from normal human tissue to a homogenate of the patient's tissue. This observation implies that a mutation had occurred in the gene coding for the activator, itself a protein. In the second patient, no hydrolysis of G_{M2} occurred when normal activator was added, which suggests that a mutation had occurred in Hex A at the site of interaction of the enzyme with the activator. The analysis of these metabolic interactions has obvious importance for the selection of the proper therapeutic agent.

Mucopolysaccharidoses (Table 2)

Recent evidence indicates that the accumulation of heparin sulfate in the central nervous system may be pathogenetically unique and a primary factor in the brain damage that occurs in many of the patients with mucopolysaccharidoses. Furthermore, there appears to be an important correlation between the concomitant accumulation of gangliosides and central nervous system damage in these patients. For example, in patients with

Mucopolysaccharidoses Types IH, II, and III, there are significant alterations of the ganglioside patterns in the brain and in isolated neuronal preparations, as well as an accumulation of mucopolysaccharide (Constantopoulos & Dekaban 1978, Constantopoulos et al 1978). In particular, significant increases occurred in the quantity of minor monosialogangliosides G_{M1}, G_{M2}, and G_{M3} in the brain, that sometimes match the levels seen in lipid storage disorders. Levels of G_{D3} and the neutral glycolipid ceramide lactoside also increased in isolated neurons (Constantopoulos et al 1980a). The significance of these observations is heightened by the fact that there is similar accumulation of mucopolysaccharides and gangliosides in a pharmacological model of mucopolysaccharidosis produced by the trypanocide suramin (Constantopoulos et al 1980b). Suramin is a noncompetitive inhibitor of iduronosulfate sulfatase, which suggests that this animal model resembles the Hunter syndrome (Mucopolysaccharidosis Type II) in humans. Furthermore, suramin was found to be a potent inhibitor of Hex A and, to a lesser extent, Hex B (Constantopoulos et al 1981). Suramin is a heavily sulfated compound and therefore has a formal resemblance to mucopolysaccharides. These polyanions appear to inhibit lysosomal enzymes involved in the degradation of gangliosides. Support for this contention is derived from the work of Kint (1973), who demonstrated a secondary deficiency of β-galactosidase activity in the liver of patients with mucopolysaccharidosis. He believed that the inhibition was due to the formation of a complex between mucopolysaccharides and the enzyme, which resulted in its inactivation (Kint 1973, Rushton & Dawson 1977). Furthermore, Heijlman (1974) reported that glycosaminoglycans (precursors of mucopolysaccharides) inhibited brain sialidase activity. However, Kint (1973) also observed activation of sialidase by mucopolysaccharides in vitro and the precise mechanism of ganglioside accumulation in mucopolysaccharide storage disorders thus remains to be resolved. A possible alternative explanation is that the polyanionic mucopolysaccharides form complexes between membrane structures—hence the appearance of multilamellated, intraneuronal inclusions (Rees et al 1979)—and gangliosides in these membranous structures may be inaccessible to catabolic enzymes. Nevertheless, it seems that there is a correlation between the severity of CNS damage and the quantity of accumulated gangliosides.

Glycoproteinoses (Table 3)

Most of the patients classified as having glycoproteinoses do not excrete intact glycoproteins or glycopeptides in their urine. If the stored oligosaccharides and those excreted in the urine are derived from glycoproteins, it is probable that mammalian cells contain an enzyme that cleaves fairly large oligosaccharide chains from glycoproteins. The most likely site of this reaction is the chitobiosyl bond between the two molecules of *N*-acetyl-glucosamine (Figure 2). Although enzymes of this type are well known in bacteria and in the hen oviduct (Tarentino & Maley 1976), little is known about their characteristics in mammalian tissues except for the brief report by Nishigaka et al (1974). Further information concerning such an enzyme will probably be forthcoming (Steele et al 1981). Furthermore, one may postulate that a deficiency of such a catalyst might occur in patients with a heritable metabolic disorder with some of the clinical features of both glycoproteinoses and mucopolysaccharidoses. I believe that such patients will eventually be described and clinicians should be aware of this possibility when confronted with mentally retarded patients of unknown etiology.

Mucolipidoses (Table 4)

MUCOLIPIDOSES II AND III (I-CELL DISEASE) (TABLE 4-B AND 4-C)
Significant progress has been made in our understanding of the pathophysiology of Mucolipidoses II and III. The disorder is characterized by dramatic increases in the activity of a number of lysosomal enzymes in the serum and in the media of cultured cells. It was predicted a number of years ago that the basic defect in these disorders is a lack of an uptake signal on enzymes from these patients and this hypothesis has been substantiated by a number of investigations (Neufeld 1981). In particular, the oligosaccharide portion of glycoprotein enzymes released from cultured fibroblasts from patients with this disorder is significantly less phosphorylated than normal (Bach et al 1979, Hasilik & Neufeld 1980). Since mannose-6-phosphate seems likely to be the signal for the uptake and packaging of enzymes by lysosomes (Kaplan et al 1977a,b, Sly et al 1981), the underphosphorylated enzymes escape compartmentation and are released into the culture medium in vitro and appear in the serum and urine in vivo. More specific information concerning the nature of the defect in phosphorylation

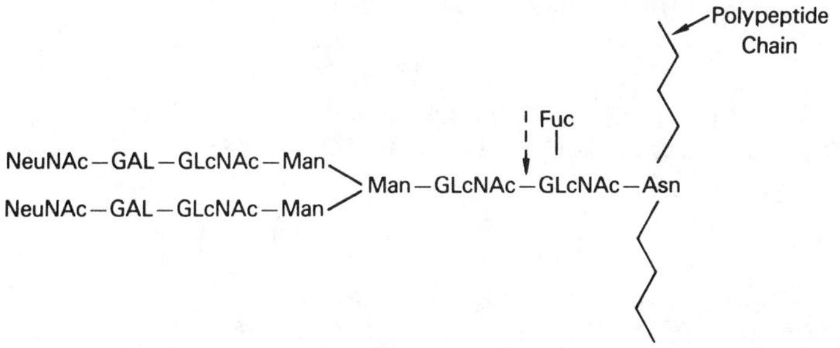

Figure 2 Proposed site of endoglycosidase that catalyzes the cleavage of oligosaccharide chains from glycoproteins. NeuNAc = N-acetylneuraminic acid; GAL = galactose; GlcNAc = N-acetyl-glucosamine; Man = mannose; Fuc = fucose; Asn = asparagine

has appeared, based on the observation of Tabas & Kornfeld (1980) that the phosphorylation of mannose residues of lysosomal enzyme oligosaccharides occurs through the transfer of N-acetylglucosamine phosphate and that the phospate exists as a diester until N-acetylglucosamine is enzymatically released (Figure 3). Hasilik and co-workers (1981) and Reitman et al (1981) recently reported that the uridinediphosphate-N-acetylglucosaminyl phosphate transferase is deficient in fibroblasts derived from patients with I-cell disease and that the lack of this enzyme is the primary metabolic defect in this disorder.

Neuronal Ceroid-Lipofuscinosis (Table 5)

There has been a great deal of activity in the disorder known as Batten's disease (Table 5-B) since the report by Wolfe et al (1977) that retinoyl complexes constituted the autofluorescent component of the neuronal storage material in the brain of these patients. The relationship of retinoic acid derivatives to the pathogenesis of this disorder has been disputed, and recent attention has focused on the increases in the quantity of dolichol derivatives in curvilinear bodies isolated from the brain of patients with the late infantile form (Table 5-B1) (Wolfe et al 1981a). The pathogenetic relationship of the increase in dolichols is uncertain because patients with the juvenile form do not show increased amounts of these polyisoprenoid compounds, whereas there are strikingly elvated levels of these substances in the brains of aged individuals (Wolfe et al 1981b). Furthermore, lipofuscin-ceroid-like pigments accumulate under a large number of conditions (Wolfe et al 1981a). The significance of these accumulations in late infantile Batten's disease requires clarification, and nothing is conclusively known about a specific metabolic abnormality in these disorders. Deficiency of leukocyte and thyroid peroxidase activity has been reported (Armstrong et al 1974a,b, 1975); however, these findings have been fervently contested (cf Wolfe et al 1981a). Further insight is required concerning the metabolic lesion(s) and pathogenesis in this group of storage disorders.

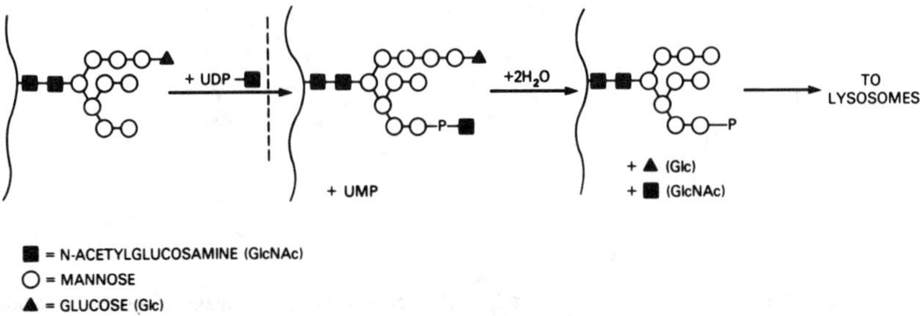

■ = N-ACETYLGLUCOSAMINE (GlcNAc)
○ = MANNOSE
▲ = GLUCOSE (Glc)

Figure 3 Pathway for the phosphorylation of glycoproteins and postulated site of defective N-acetylglucosaminylphosphate transferase (*dashed vertical line*) in I-Cell disease (Mucolipidoses II and III). Modified with permission, from Figure 5 of Tabas & Kornfeld (1980).

NEW DIAGNOSTIC METHODS AND REAGENTS

Lipid Storage Disorders

NIEMANN-PICK DISEASE   For many years after the discovery of the metabolic defect in Niemann-Pick disease, the use of radioactively labeled sphingomyelin was required for the enzymatic diagnosis of patients with this disorder (Kampine et al 1967b, Sloan et al 1969), for the detection of carriers (Brady et al 1971), and for the monitoring of pregnancies at risk for this condition (Epstein et al 1971). A major development in this area was the synthesis of a reliable chromogenic analogue of sphingomyelin in which the ceramide portion of the molecule was replaced with 2-acylamido-p-nitrophenol (Gal et al 1975, Gal & Fash 1976). This use of this reagent has several advantages: it eliminates the requirement for radioactive counting equipment; it is commercially available; it is inexpensive; and it is water soluble. This substrate has also been shown to be eminently satisfactory for the detection of the carriers of Niemann-Pick disease Types A and B (Gal et al 1980) as well as for prenatal diagnosis (Brady 1977, Patrick et al 1977).

KRABBE'S DISEASE   A similar chromogenic reagent has been developed for the diagnosis of Krabbe's disease; the ceramide again is replaced with a 2-acylamido-p-nitrophenyl group (Gal et al 1977). Although the usefulness of this reagent has been substantiated by other laboratories (Besley 1978, Besley & Bain 1978), measurement of galactocerebroside-β-galactosidase with this reagent requires exquisite attention to details of substrate prepartion and the conditions of incubation (Brady 1978). More recently, chromogenic and fluorogenic derivatives of galactocerebroside itself have been reported to be useful for the diagnosis of Krabbe's disease (Besley & Gatt 1981). The fatty acid was replaced with either the chromogen 2,4,6-trinitrophenylaminolauric acid, or 11-(9-anthroyloxy) undecanoic acid, a fluorogenic compound. After incubation with an appropriate source of enzyme, the chromogenic or fluorogenic ceramide produced by the cleavage of the galactose was separated from unreacted substrate by solvent partition and quantitated spectrophotometrically or fluorometrically. It is contended that these substrates provide discrimination for the diagnosis of Krabbe's disease and identification of heterozygotes better than the chromogenic substrate developed by Gal and his associates.

SIALIDOSES   The diagnosis of ganglioside sialidosis is probably most reliably made by using authentic gangliosides labeled in the sialic acid portion of the molecule as substrate (Bach et al 1979, Hahn et al 1980). Procedures for the biosynthetic labeling of gangliosides in the sialic acid portions of the molecule have been available for some time (Kolodny et al 1970). Other substrates such as N-acetylneuraminosyl-N-acetylgalactosaminitol, neuraminyl-lactose, methoxyphenyl-N-acetylneuraminic acid, and urinary oligosaccharides have also been employed (cf Beaudet 1981). Because ganglioside neuraminidase activity is exceptionally low in most tissues compared to other lysosomal hydrolases (Tallman & Brady 1973), a substrate that provides reliability and high sensitivity is required. 4-Methylumbelliferyl-N-acetylneuraminic acid may be the reagent of choice for identifying patients with a neuraminidase deficiency. This substrate probably will not permit discrimination between patients with neuraminidase-deficient glycoproteinosis and gangliosidosis; however, screening and genetic counseling should be facilitated through the use of this material. Procedures have been published for the preparation of this compound (Warner & O'Brien 1979, Myers et al 1980), and it has been manufactured commercially by Koch-Light in England. Patent litigation has prevented its distribution in the United States, but it is currently available in Europe and Canada. It is hoped that this impediment will be swiftly overcome since an increasing number of patients with sialidoses are being identified (Strecker et al 1977, Spranger et al 1977, O'Brien 1977, Thomas et al 1978, Johnson et al 1980).

THERAPY

Enzyme Replacement

A number of strategies have been proposed for the treatment of heritable metabolic disorders, including (a) organ transplantation, (b) direct enzyme replacement, and (c) genetic engineering (Brady 1966, 1973). The most encouraging observations have been made in a prospective trial of enzyme replacement in young patients with the noncerebral form of Gaucher's disease (Brady et al 1980, 1981a,b). Organ transplants have been performed in patients with the infantile (Desnick et al 1973) and juvenile forms of Gaucher's disease (Groth et al 1971, 1980) and with Niemann-Pick disease (Daloze et al 1977). These attempts and enzyme replacement trials in disorders in which the central nervous system is damaged, such as metachromatic leukodystrophy (Austin 1967, Greene et al 1969) and Tay-Sachs disease (Johnson et al 1973, von Specht et al 1979), have been routinely unsuccessful. The enzymes simply did not enter or penetrate into the substance of brain in therapeutically significant quantities after intravenous infusion (Johnson et al 1973) or intrathecal and intracisternal injection (Austin 1967). One might expect that similar difficulties would preclude successful attempts to supply enzyme by subcutaneous implantation of fibroblasts in patients with mucopolysaccharidoses (Dean et al 1980, Gibbs, el 1980).

At the present time, enzyme replacement is likely to be ineffectual in the treatment of disorders in which the central nervous system is damaged, unless procedures are developed for the delivery of effective quantities of enzymes to the brain. The central problem is the blood-brain barrier that prevents molecules as large as enzymes from entering the brain. We therefore began a series of investigations to determine access to the brain if we temporarily modified the barrier by the intracarotid infusion of hyperosmolar solutions of mannitol and arabinose. We observed that this procedure would effectively allow the passage of physiologically significant quantities of mannosidase in experimental animals (Barranger et al 1979). Furthermore, it was observed that injected enzymes were taken up by neuronal cells and that they eventually became associated with lysosomal particles—which is precisely the site of the subcellular localization of the accumulating metabolites (Tallman et al 1971). These studies, which were initiated in rats, have been extended to larger animals such as dogs and primates where it has been shown that repeated temporary alteration of the blood-brain barrier in this fashion does not cause pathologic changes in the brain (Smith et al 1980). In fact, successful barrier modification has been carried out in humans with metastatic lesions in the CNS (Neuwelt et al 1980).

Additional aspects of enzyme replacement in the nervous system deserve comment. Specific, high-affinity receptors have been demonstrated on the surfaces of neuronal cells for hexosaminidase A, the enzyme lacking in Tay-Sachs patients (Kusiak et al 1979). This finding is extraordinarily important: it means that if the enzyme can be delivered to the brain in sufficient quantity, it is highly likely that it will be specifically taken up by neurons—the cells in which the pathologic metabolite is accumulating. It is well known that many components of the brain are turned over very rapidly in the neonatal period, and at only a small fraction of this rate as the brain matures. Therefore, one might not have to supply large amounts of enzyme throughout life, but only during the critical period of rapid metabolic activity. However, the recent reports of patients with late onset Tay-Sachs disease presenting with cerebellar ataxia and spinal cord involvement (Johnson et al 1977, Oonk et al 1979) cause one to temper the optimisim that might have followed the preceding deduction. This development seems a bit ominous since alteration of the blood-brain barrier by intracarotid infusion might not enable one to deliver exogenous enzyme to the areas that may be damaged in these patients.

Genetic Engineering

Speculations concerning the treatment of hereditary disorders by genetic modification are receiving an extraordinary amount of publicity at the present time, and the topic will arise frequently in the years to come. It seems to me that the effectiveness of such a strategy will be limited for a long time to the modification and replacement of proliferating cells such as those derived from the bone marrow. Since neurons are not renewed throughout life, successful genetic modification for the treatment of a heritable metabolic disorder would require the delivery of an appropriate genetic message to all or at least a highly significant number of cells in the nervous system. There are no known strategies by which this goal can be achieved, although the possibility of a viral vector has been mentioned frequently. Any consideration of this type must be concerned with the use of a totally innocuous agent, otherwise a disastrous result is inevitable.

CONCLUSIONS

A number of new clinical entities have been described recently that are characterized by the accumulation of deleterious quantities of metabolites in the nervous system. It seemed useful, therefore, to provide a rational classification of these disorders since their categorization is in flux and has not hitherto been subjected to a comprehensive analysis. I fully expect that some of the assignments made above will have to be revised and certainly other disorders will be added in time. However, I believe the framework provided will facilitate the nosology of heritable storage disorders.

Literature Cited

Armstrong, D., Dimmitt, S., Van Wormer, D. E. 1974a. Studies in Batten disease. I. Peroxidase deficiency in granulocytes. *Arch. Neurol.* 30:144–52

Armstrong, D., Dimmitt, S., Boehme, D. H., Leonberg, S. C., Vogel, W. 1974b. Leukocyte peroxidase deficiency in a family with a dominant form of Kuf's disease. *Science* 186:155–57

Armstrong, D., Van Wormer, D. E., Neville, H., Dimmitt, S., Clingan, F. 1975. Thyroid peroxidase deficiency in Batten-Spielmeyer-Vogt disease. *Arch. Pathol.* 19:430–35

Austin, J. H. 1967. Some recent findings in leukodystrophies and in gargoylism. In *Inborn Disorders of Sphingolipid Metabolism*, ed. S. M. Aronson, B. W. Volk, pp. 359–87. New York: Academic

Bach, G., Zeigler, M., Schaap, T., Kohn, G. 1979. Mucolipidosis type IV: Ganglioside sialidase deficiency. *Biochem. Biophys. Res. Commun.* 90:1341–47

Bach, G., Zeigler, M., Kohn, G. 1980. Biochemical investigations of cultured amniotic fluid cells in mucolipidosis type IV. *Clin. Chim. Acta* 106:121–28

Barranger, J. A., Rapoport, S. I., Fredricks, W. R., Pentchev, P. G., MacDermot, K. D., Steusing, J. K., Brady, R. O. 1979. Modification of the blood-brain barrier: Increased concentration and fate of enzymes entering the brain. *Proc. Natl. Acad. Sci. USA* 76:481–85

Beaudet, A. L. 1981. Disorders of glycoprotein degradation: Mannosidosis, fucosidosis, sialidosis and aspartylglycosaminuria. In *The Metabolic Basis of Inherited Disease*, ed. J. B. Stanbury, J. B. Wyngaarden, D. S. Frederickson, M. S. Brown, J. L. Goldsten. New York: McGraw-Hill. In press. 5th ed.

Besley, G. T. N. 1978. The use of natural and artificial substrates in the prenatal diagnosis of Krabbe's disease. *J. Inher. Metab. Dis.* 1:115–18

Besley, G. T. N., Bain, A. D. 1978. Use of a chromogenic substrate for the diagnosis of Krabbe's disease, with special reference to its application in prenatal diagnosis. *Clin. Chim. Acta* 88:229–36

Besley, G. T. N., Gatt, S. 1981. Spectrophotometric and fluorimetric assays of galactocerebrosidase activity, their use in the diagnosis of Krabbe's disease. *Clin. Chim. Acta* 110:19–26

Brady, R. O. 1966. The sphingolipidoses. *N. Engl. J. Med.* 275:312–18

Brady, R. O. 1973. The abnormal biochemistry of inherited disorders of lipid metabolism. *Fed. Proc.* 32:1660–67

Brady, R. O. 1976. Inherited metabolic diseases of the nervous system. *Science* 193:733–39

Brady, R. O. 1977. Heritable catabolic and anabolic disorders of lipid metabolism. *Metabolism* 26:329–45

Brady, R. O. 1978. Spingolipidoses. *Ann. Rev. Biochem.* 47:687–713

Brady, R. O., Quarles, R. H. 1973. The enzymology of myelination. *Mol. Cell. Biochem.* 2:23–29

Brady, R. O., Johnson, W. G., Uhlendorf, B. W. 1971. Identification of heterozygous carriers of lipid storage diseases. *Am. J. Med.* 51:423–35

Brady, R. O., Barranger, J. A., Gal, A. E., Pentchev, P. G., Furbish, F. S. 1980. Status of enzyme replacement therapy for Gaucher's disease. In *Enzyme Therapy in Genetic Diseases,* ed. R. J. Desnick, 2:361–68. New York: Liss. 544 pp.

Brady, R. O., Barranger, J. A., Gal, A. E., Pentchev, P. G., Furbish, F. S., Kusiak, J. W. 1981a. Treatment of lipidoses by enzyme infusion. In *Lysosomes and Lysosomal Storage Diseases,* ed. J. W. Callahan, J. A. Lowden, pp. 373–79. New York: Raven. 434 pp.

Brady, R. O., Barranger, J. A., Pentchev, P. G., Furbish, F. S., Gal, A. E. 1981b. Prospects for enzyme replacement therapy in heritable metabolic disorders. In *Inborn Errors of Metabolism in Humans,* ed. H. Aebi, N. N. Herschkowitz. London: MTP Press. In press

Brunngraber, E. G. 1978. Lysosomal enzyme deficiency diseases—Glycoprotein catabolism in brain tissue. In *Glycoproteins and Glycolipids in Disease Processes,* ed. E. L. Walborg, Jr., pp. 135–49. Washington DC: Am. Chem. Soc. 480 pp.

Constantopoulos, G., Dekaban, A. S. 1978. Neurochemistry of the mucopolysaccharidoses: Brain lipids and lysosomal enzymes in patients with four types of mucopolysaccharidosis and in normal controls. *J. Neurochem.* 30:965–73

Constantopoulos, G., Eiben, R. M., Schafer, I. A. 1978. Neurochemistry of the mucopolysaccharidoses: Brain glycosaminoglycans, lipids and lysosomal enzymes in mucopolysaccharidosis type IIIB (α-N-acetylglucosaminidase deficiency). *J. Neurochem.* 31:1215–22

Constantopoulos, G., Iqbal, K., Dekaban, A. S. 1980a. Mucopolysaccharidosis types IH, IS, II, and IIIA: Glycosaminoglycans and lipids of isolated brain cells and other fractions from autopsied tissues. *J. Neurochem.* 34:1399–1411

Constantopoulos, G., Rees, S., Cragg, B. G., Barranger, J. A., Brady, R. O. 1980b. Experimental animal model for mucopolysaccharidosis: Suramin-induced glycosaminoglycan and sphingolipid accumulation in the rat. *Proc. Natl. Acad. Sci. USA* 77:3700–4

Constantopoulos, G., Rees, S., Cragg, B. G., Barranger, J. A., Brady, R. O. 1981. *In vitro* inhibition of N-acetyl-β-hexosaminidase A by suramin. *Trans. Am. Soc. Neurochem.* 12:169 (Abstr.)

Conzelmann, E., Sandhoff, K. 1978. AB variant of infantile G_{M2} gangliosidosis: Deficiency of a factor necessary for stimulation of hexosaminidase A-catalyzed degradation of ganglioside G_{M2} and glycolipid G_{A2}. *Proc. Natl. Acad. Sci. USA* 75:3979–83

Daloze, P., Delvin, E. E., Glorieux, F. H., Corman, J. L., Bettez, P., Toussi, T. 1977. Replacement therapy for inherited enzyme deficiency: Liver orthotopic transplantation in Niemann-Pick disease Type A. *Am. J. Med. Genet.* 1:229–39

Dean, M. F., Muir, H., Benson, P., Button, L. 1980. Enzyme replacement therapy in the mucopolysaccharidoses by fibroblast transplantation. See Brady et al 1980, pp. 445–56

Desnick, S. J., Desnick, R. J., Brady, R. O., Pentchev, P. G., Simmons, R. L., Najarian, J. S., Swaiman, K., Sharp, H. L., Krivit, W. 1973. Renal transplantation in type 2 Gaucher's disease. In *Enzyme Therapy in Genetic Diseases,* ed. R. J. Desnick, R. W. Bernlohr, W. Krivit, pp. 109–19. Baltimore: Williams & Wilkins. 236 pp.

Epstein, C. J., Brady, R. O., Schneider, E. L., Bradley, R. M., Shapiro, D. 1971. *In utero* diagnosis of Niemann-Pick disease. *Am. J. Hum. Genet.* 23:533–35

Farrell, D. F., MacMartin, M. P. 1981. G_{M1} gangliosidosis: Enzymatic variation in a single family. *Ann. Neurol.* 9:232–36

Farrell, D. F., Ochs, U. 1981. G_{M1} gangliosidosis: Phenotypic variation in a single family. *Ann. Neurol.* 9:225–31

Gal, A. E., Fash, F. J. 1976. Synthesis of 2-N-(hexadecanoyl)-amino-4-nitrophenyl phosphorylcholine-hydroxide, a chromogenic substrate for assaying sphingomyelinase activity. *Chem. Phys. Lipids* 16:71–79

Gal, A. E., Brady, R. O., Hibbert, S. R., Pentchev, P. G. 1975. A practical chromogenic procedure for the detection of homozygotes and heterozygous carriers of Niemann-Pick disease. *N. Engl. J. Med.* 293:632–36

Gal, A. E., Brady, R. O., Pentchev, P. G., Furbish, F. S., Suzuki, K., Tanaka, H., Schneider, E. L. 1977. A practical chromogenic procedure for the diagnosis of Krabbe's disease. *Clin. Chim. Acta* 77:53–59

Gal, A. E., Brady, R. O., Barranger, J. A., Pentchev, P. G. 1980. The diagnosis of Type A and Type B Niemann-Pick disease and detection of carriers using leukocytes and a chromogenic analogue of sphingomyelin. *Clin. Chim. Acta* 104:129–32

Gibbs, D. A., Spellacy, E., Roberts, A. A., Watts, R. W. E. 1980. The treatment of lysosomal storage disorders by fibroblast transplantation: Some preliminary observations. See Brady et al 1980, pp. 457–74

Ginns, E. I., Brady, R. O., Stowens, D. W., Furbish, F. S., Barranger, J. A. 1980. A new group of glucocerebrosidase isozymes found in human white blood cells. *Biochem. Biophys. Res. Commun.* 97:1103–7

Goldman, J. E., Yamanaka, T., Rapin, I., Adachi, M., Suzuki, K., Suzuki, K. 1980. The AB-variant of G_{M2} gangliosidosis. *Acta Neuropathol.* 52:189–202

Greene, H. L., Hug, G., Schubert, W. K. 1969. Metachromatic leukodystrophy. Treatment with arylsulfatase A. *Arch. Neurol.* 20:147–53

Groth, C. G., Hagenfeldt, L., Dreborg, S., Lofstrom, B., Ockerman, P. A., Samuelsson, K., Svennerholm, L., Werner, B., Westberg, G., 1971. Splenic transplantation in a case of Gaucher's disease. *Lancet* 1:1260–64

Groth, C. G., Collste, H., Drebrog, S., Hakansson, G., Lundgren, G., Svennerholm, L. 1980. Attempt at enzyme replacement in Gaucher disease by renal transplantation. *Acta Paediatr. Scand.* 68:475–79

Hahn, L. C., Ben-Joseph, Y., Nadler, H. L. 1980. Glycoprotein and ganglioside α-neuraminidases in sialidosis and mucolipidoses. *Am. J. Hum. Genet.* 32:41A

Hasilik, A. Neufeld, E. F. 1980. Biosynthesis of lysosomal enzymes in fibroblasts. Phosphorlyation of mannose residues. *J. Biol. Chem.* 255:4946–50

Hasilik, A., Waheed, A., von Figura, K. 1981. Enzymatic phosphorylation of lysosomal enzymes in the presence of UDP-*N*-acetylglucosamine. Absence of the activity in I-Cell fibroblasts. *Biochem. Biophys. Res. Commun.* 98:761–67

Hechtman, P. 1980. Deficiency of the Hex A activator protein in Tay-Sachs AB variant. *Am. J. Hum. Genet.* 32:42A

Heijlman, L. 1974. The inhibitory action of glycosaminoglycons on brain sialidase. *Biochem. Soc. Trans.* 2:638–39

Johnson, W. G., Desnick, R. J., Long, D. M., Sharp, H. L., Krivit, W., Brady, B., Brady, R. O. 1973. Intravenous injection of purified hexosaminidase A into a patient with Tay-Sachs disease. See Desnick et al 1973, pp. 120–24

Johnson, W. G., Chutorian, A., Miranda, A. 1977. A new juvenile hexosaminidase deficiency disease presenting as cerebellar ataxia. *Neurology* 27:1012–18

Johnson, W. G., Thomas, G. H., Miranda, A. F., Driscoll, J. M., Wigger, J. H., Yeh, M. N., Schwartz, R. C., Cohen, C. S., Berdon, W. E., Koenigsberger, M. R. 1980. Congenital sialidosis: Biochemical studies; clinical spectrum in four sibs; two successful prenatal diagnoses. *Am. J. Hum. Genet.* 32:43A (Abstr.)

Kampine, J. P., Brady, R. O., Kanfer, Jh. N., Feld, M., Shapiro, D. 1967a. The diagnosis of Gaucher's disease and Niemann-Pick disease using small samples of venous blood. *Science* 155:86–88

Kampine, J. P., Kanfer, J. N., Gal, A. E., Bradley, R. M., Brady, R. O. 1967b. Response of sphingolipid hydrolases in spleen and liver to incrased erythrocytorrhexis. *Biochim. Biophys. Acta* 137:135–39

Kaplan, A., Achord, D. T., Sly, W. S. 1977a. Phosphohexosyl components of a lysosomal enzyme are recognized by pinocytosis receptors on human fibroblasts. *Proc. Natl. Acad. Sci. USA* 74:2026–30

Kaplan, A., Fischer, D., Achord, D. T., Sly, W. S. 1977b. Phosphohexosyl recognition is a general characteristic of pinocytosis of lysosomal glycosidases by human fibroblasts. *J. Clin. Invest.* 60:1088–93

Kint, J. A. 1973. Antagonistic action of chondroitin sulfate and cetylpyridinium chloride on human liver β-galactosidase. *FEBS Lett.* 36:53–56

Kolodny, E. H., Brady, R. O., Volk, B. W. 1969. Demonstration of an alteration of ganglioside metabolism in Tay-Sachs disease. *Biochem. Biophys. Res. Commun.* 37:526–31

Kolodny, E. H., Brady, R. O., Quirk, J. M., Kanfer, J. N. 1970. Preparation of radioactive Tay-Sachs ganglioside labeled in the sialic acid moiety. *J. Lipid Res.* 11:144–49

Kusiak, J. W., Toney, J. H., Quirk, J. M., Brady, R. O. 1979. Specific binding of ^{125}I-β-hexosaminidase A to rat brain synaptosomes. *Proc. Natl. Acad. Sci. USA* 76:982–85

Kusiak, J. W., Quirk, J. M., Brady, R. O. 1980. Specific binding of β-hexosaminidase A to rat brain synaptic plasma membranes. See Brady et al 1980, pp. 93–102

Li, S.-C., Hirabayashi, Y., Li, Y.-T. 1981. A protein activator for the enzymic hydrolysis of G_{M2}-ganglioside. *Trans. Am. Soc. Neurochem.* 12:210 (Abstr.)

Lowden, J. A., O'Brien, J. S. 1979. Sialidosis: A review of human neuraminidase deficiency. *Am. J. Hum. Genet.* 31:1–18

Myers, R. W., Lee, R. T., Lee, Y. C., Thomas, G. H., Reynolds, L. W., Uchida, Y. 1980. The synthesis of 4-methylumbelliferyl α-ketoside of N-acetylneuraminic acid and its use in fluorometric assay for neuraminidase. *Anal. Biochem.* 101:166–74

Neufeld, E. F. 1981. Recognition and processing of lysosomal enzymes in cultured fibroblasts. See Brady et al 1981a, pp. 115–29

Neuwelt, E. A., Frenkel, E. P., Diehl, J., Vu, L. H., Rapoport, S., Hill, S. 1980. Reversible osmotic blood brain barrier disruption in man: Implications for chemotherapy of malignant brain tumors. *Neurosurgery* 7:44–52

Nishigaka, M., Muramatsu, T., Kobata, A. 1974. Endoglycosidases acting on carbohydrate moieties of glycoproteins: Demonstration in mammalian tissues. *Biochem. Biophys. Res. Commun.* 59:638–45

O'Brien, J. S. 1977. Neuraminidase deficiency in the cherry-red spot-myoclonus syndrome. *Biochem. Biophys. Res. Commun.* 79:1136–41

O'Brien, J. S. 1978. The gangliosidoses. In *The Metabolic Basis of Inherited Disease,* ed. J. B. Stanbury, J. B. Wyngaarden, D. S. Frederickson, pp. 841–65. New York: McGraw-Hill. 1862 pp. 4th ed.

Okada, S., O'Brien, J. S. 1969. Tay-Sachs disease: Generalized absence of a β-D-N-acetylhexosaminidase component. *Science* 165:698–700

Oonk, J. G. W., van der Helm, H. J., Martin, J. J. 1979. Spinocerebellar degeneration: Hexosaminidase A and B deficiency in

two adult sisters. *Neurology* 29:380–84

Patrick, A. D., Young, E., Kleijer, W. J., Niermeijer, M. F. 1977. Prenatal diagnosis of Niemann-Pick disease Type A using chromogenic substrate. *Lancet* 2:144

Rees, S., Cragg, B. G., Constantopoulos, G., Brady, R. O. 1979. Neuronal inclusions and attempts to identify them. In *Muscle, Nerve and Brain Degeneration,* ed. A. D. Kidman, J. K. Tomkins, pp. 121–221. Amsterdam-Oxford: Excerpta Medica

Reitman, M. L., Varki, A., Kornfeld, S. 1981. Fibroblasts from patients with I-cell disease and pseudo-Hurler polydystrophy are deficient in uridine 5′-diphosphate-N-acetylglucosamine: glycoprotein N-acetylglucosaminyltransferase activity. *J. Clin. Invest.* 67:1574–79

Roth, S. 1978. Molecular analysis of neural recognition. A potential role for the Tay-Sachs ganglioside (G_{M2}) in the development of retinotectal projection. *Int. Congr. Ser. Excerpta Med.* 432:196–207

Rushton, A. R., Dawson, G. 1977. Effect of glycosaminoglycans on the in vitro activity of human skin firbroblast glycosphingolipid β-galactosidases and neuraminidases. *Clin. Chim. Acta* 80:133–39

Sandhoff, K., Harzer, K., Wassle, W., Jatzkewitz, H. 1971. Enzyme alterations and lipid storage in three variants of Tay-Sachs disease. *J. Neurochem.* 18:2469–89

Sloan, H. R., Uhlendorf, B. W., Kanfer, J. N., Brady, R. O., Frederickson, D. S. 1969. Deficiency of sphingomyelin-cleaving enzyme activity in tissue cultures derived from patients with Niemann-Pick disease. *Biochem. Biophys. Res. Commun.* 34:582–88

Sly, W. S., Natowicz, M., Gonzalez-Noriega, A., Grubb, J. H., Fisher, H. D. 1981. The role of mannose-6-phosphate recognition marker and its receptor in the uptake and intracellular transport of lysosomal enzymes. See Brady et al 1981a, pp. 131–46

Smith, M. T., Girton, M., Rapoport, S. I., Brady, R. O., Barranger, J. A. 1980. Pathology of reversible blood-brain barrier opening. *J. Neuropathol. Exp. Neurol.* 39:389 (Abstr.)

Spranger, J., Gehler, J., Cantz, M. 1977. Mucolipidosis I—a sialidosis. *Am. J. Med. Genet.* 1:21–29

Steele, R., Hirabayashi, Y., Li, Y.-T., Li, S.-C. 1981. Isolation and characterization of endo-β-N-acetylglucosaminidase from human liver and brain. *Fed. Proc.* 40:1565 (Abstr.)

Steer, C. J., Ashwell, G. 1980. Studies on a mammalian hepatic binding protein specific for asialoglycoproteins. Evidence for receptor recycling in isolated rat hepatocytes. *J. Biol. Chem.* 255:3008–13

Strecker, G., Peers, M.-C., Michalski, J. C., Hondi-Assah, T., Fournet, B., Spik, G., Montreuil, J., Farriaux, J. P., Maroteaux, P., Durand, P. 1977. Structure of nine sialyloligosaccharides accumulated in urine of eleven patients with three different types of sialidosis: Mucolipidosis II and two new types of mucolipidosis. *Eur. J. Biochem.* 75:39–403

Tabas, I., Kornfeld, S. 1980. Biosynthetic intermediates of β-glucuronidase contain high mannose oligosaccharides with blocked phosphate residues. *J. Biol. Chem.* 255:6633–39

Tallman, J. F., Brady, R. O. 1972. The catabolism of Tay-Sachs ganglioside in rat brain lysosomes. *J. Biol. Chem.* 247:7570–75

Tallman, J. F., Brady, R. O. 1973. The purification and properties of a mammalian neuraminidase (sialidase). *Biochim. Biophys. Acta* 193:1–10

Tallman, J. F., Brady, R. O., Suzuki, K. 1971. Enzymic activities associated with membranous cytoplasmic bodies and isolated brain lysosomes. *J. Neurochem.* 18:1775–77

Tallman, J. F., Johnson, W. G., Brady, R. O. 1972. The metabolism of Tay-Sachs ganglioside: Catabolic studies with lysosomal enzymes from normal and Tay-Sachs brain. *J. Clin. Invest.* 51:2339–45

Tallman, J. F., Brady, R. O., Quirk, J. M., Villalba, M., Gal, A. E. 1974. Isolation and relationship of human hexosaminidases. *J. Biol. Chem.* 249:3489–99

Tarentino, A. L., Maley, F. 1976. Purification and properties of endo-β-N-acetylglucosaminidase from hen oviduct. *J. Biol. Chem.* 251:6537–43

Thomas, G. H., Tipton, R. E., Chien, L. T., Reynolds, L. W., Miller, C. S. 1978. Sialidase (α-N-acetylneuraminidase deficiency: The enzyme defect in an adult with macular cherry-red spots and myoclonus without dementia. *Clin. Genet.* 13:369–79

Tsay, G. C., Dawson, G. 1976. Oligosaccharide storage in brains from patients with focusidosis, G_{M1}-gangliosidosis, and G_{M2}-gangliosidosis (Sandhoff's disease). *J. Neurochem.* 27:733–40

von Specht, B. U., Geiger, B., Arnon, R., Passwell, J., Keren, G., Goldman, B., Padeh, G. 1979. Enzyme replacement in Tay-Sachs disease. *Neurology* 29:848–54

Warner, T. G., O'Brien, J. S. 1979. Synthesis of 2'-(4-methylumbelliferyl)-α-D-N-acetylneuraminic acid and detection of skin fibroblast neuraminidase in normal humans and in sialidosis. *Biochemistry* 18:2783–87

Wolfe, L. S., Ng Ying Kin, N. M. K., Baker, R. R., Carpenter, S., Andermann, F. 1977. Identification of retinoyl complexes as the autofluorescent component of the neuronal storage material in Batten disease. *Science* 195:1360–62

Wolfe, L. S., Ng Ying Kin, N. M. K., Baker, R. R. 1981a. Batten disease and related disorders: New findings on the chemistry of the storage material. See Brady et al 1981a, pp. 315–30

Wolfe, L. S., Ng Ying Kin, N. M. K. 1981b. Dolichols in storage cytosomes and urine sediment of Batten disease cases. *Trans. Am. Soc. Neurochem.* 12:194 (Abstr.)

Ann. Rev. Neurosci. 1982. 5:57–106
Copyright © 1982 by Annual Reviews Inc. All rights reserved

THE MOLECULAR FORMS
OF CHOLINESTERASE AND
ACETYLCHOLINESTERASE
IN VERTEBRATES

Jean Massoulié and Suzanne Bon

Laboratoire de Neurobiologie, Ecole Normale Supérieure, Paris, 75005, France

INTRODUCTION

The mechanism of cholinergic neurotransmission requires the rapid inactivation of acetylcholine (Dale 1914). Loewi & Navratil showed in 1926 that acetylcholine can be destroyed by an enzyme that exists in aqueous extracts of frog tissues. An esterase that specifically hydrolyzes choline esters was characterized in horse serum by Stedman et al (1932), who called it cholinesterase. It was later found that blood cells also contain a high level of acetylcholine-hydrolyzing activity (Stedman & Stedman 1935). Alles & Hawes (1940) subsequently found that in human blood the serum and cell enzymes are qualitatively different; Mendel et al (1943a) showed that, although the serum enzyme hydrolyzes butyrylcholine or propionylcholine faster than acetylcholine [as already noted by Stedman et al (1932)], the cell-bound enzyme acts preferentially on acetylcholine, at low substrate concentration. The particulate enzyme also presents a characteristic excess substrate inhibition, so that its activity varies, in a bell-shaped manner, as a function of substrate concentration (Mendel & Rudney 1943).

These two activities exist in all classes of vertebrates. The serum enzyme (EC 3.1.1.8) varies somewhat in its specificity, notably in the relative rates of hydrolysis of propionylcholine and butyrylcholine (Augustinsson 1959a,b). The serum enzyme was called originally "nonspecific" cholinesterase, or "pseudocholinesterase" (Mendel et al 1943, Mendell & Rudney 1943). In contrast, the erythrocyte enzyme (EC 3.1.1.7) was considered the

57

0147-006X/82/0301-0057$02.00

specific, or true cholinesterase. Nachmansohn & Rothenberg (1945) showed that nerve and muscle tissue contain the same type of enzyme, i.e. specific cholinesterase. Augustinsson & Nachmansohn (1949) introduced the term acetylcholinesterase (AChE) for this enzyme and proposed to restrict the term cholinesterase to the "nonspecific" enzyme. This cholinesterase is also called butyrylcholinesterase or propionylcholinesterase, depending on the species. We maintain the terminology introduced by Augustinsson & Nachmansohn in the present paper, although "cholinesterase" is also frequently used as a general name for two types of enzyme.

The catalytic properties of acetylcholinesterase (EC 3.1.1.7) and cholinesterase (EC 3.1.1.8), as well as their occurrence and histochemical localization in the tissues of many animal species, have been extensively investigated. They have been the subject of a number of books and reviews (e.g. Koelle 1963, Silver 1974, Rosenberry 1975) and their molecular heterogeneity has been discussed (Rosenberry 1975, Massoulié et al 1980). Because cholinesterase (EC 3.1.1.8) was first characterized in blood serum, while acetylcholinesterase (EC 3.1.1.7) was bound to erythrocytes and was difficult to solubilize quantitatively from muscle and nerve tissue, the two enzymes were initially considered to be soluble and membrane-bound proteins, respectively. However, both enzymes have long been known to be polymorphic, as demonstrated for example by electrophoretic methods, and both occur in soluble and membrane-bound states. They present, in fact, homologous sets of molecular forms, which possess characteristic quaternary structures and interaction properties.

The molecular state of these enzymes and the nature of their interactions obviously determine their positioning in cellular structures, and hence their action in cholinergic neurotransmission. It is therefore likely that some forms are physiologically active while others correspond to their biosynthetic precursors or degradation products. In the present paper we discuss the properties of the molecular forms of acetylcholinesterase and cholinesterase, their distribution in the tissues of various vertebrate species, and their cellular localization. We define molecular forms as stable entities that differ in molecular size and shape. These differences are reflected in the hydrodynamic properties of the forms, thus allowing them to be separated and analyzed by gel filtration chromatography and zone sedimentation. This type of analysis, which we first introduced for acetylcholinesterase from fish electric organ, is being used more and more systematically in studies of acetylcholinesterase and cholinesterase. It is interesting at this point to discuss the numerous findings based on molecular forms that have been reported; however, because not all studies have used such analysis we also refer to histochemical or biochemical studies made on total enzyme activity.

ACETYLCHOLINESTERASE MOLECULAR FORMS IN FISH ELECTRIC ORGANS

In their early investigations, Marnay & Nachmansohn (1937, 1938) found that in muscles a high concentration of acetylcholinesterase activity is concentrated in the region of the neuromuscular junctions, or endplates. This led to the discovery that the electric organs of the marine electric ray, *Torpedo,* which correspond to phylogenetically modified muscles and therefore possess a high density of synapses, contain an exceptionally high level of acetylcholinesterase (Marnay 1937, Nachmansohn & Lederer 1939). The most detailed biochemical studies have therefore been done on the enzyme obtained from electric organs of *Torpedo* and the freshwater teleost, *Electrophorus* (electric eel). In this first section, we describe the molecular forms of acetylcholinesterase from *Torpedo* and *Electrophorus* electric organs. (It should be noted that these organs contain only negligible amounts of cholinesterase, as compared to acetylcholinesterase.)

Electrophorus and Torpedo Acetylcholinesterase: Collagen-tailed and Globular Forms

Acetylcholinesterase may be solubilized from *Electrophorus* electric organs by homogenization in the presence of a high salt concentration (e.g. 1 *M* NaCl). Such extracts contain three major enzyme forms, as shown by sedimentation analysis in sucrose gradient (Massoulié & Rieger 1969) or by gel filtration chromatography (Bon et al 1973). These forms are highly asymmetric molecules in which globular catalytic subunits are associated with a rod-shaped element, or "tail," approximately 50 nm long, which is clearly visible in electron micrographs (Dudai et al 1973, Rieger et al 1973, Cartaud et al 1975). They contain one, two, or three tetrameric groups of subunits (Bon et al 1976) corresponding to 4, 8, and 12 active sites, respectively (Vigny et al 1978a). We call these forms A_4, A_8, and A_{12}, respectively (cf. Figure 1). The mass of each catalytic subunit is about 80,000 daltons (Dudai & Silman 1974, Bon & Massoulié 1976b, Rosenberry & Richardson 1977).

The "tail" has a mass of approximately 100,000 daltons (Bon et al 1976). It has a collagen-like structure, as demonstrated by the presence of hydroxyproline and hydroxylysine (Anglister & Silman 1978, Rosenberry & Richardson 1977) and by its sensitivity to collagenase (Johnson et al 1977, Webb 1978, Anglister & Silman 1978, Bon & Massoulié 1978). Collagenase digests the tail in a temperature-dependent manner; at 20°C it removes about 20% of its mass, producing molecules that possess a decreased Stokes radius, and thus sediment faster than the native forms. At a higher temperature (37°C)

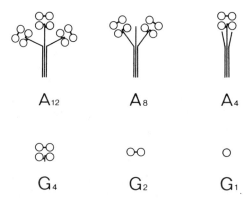

Figure 1 Schematic diagram showing the quaternary structure of the asymmetric (collagen-tailed) and globular forms of acetylcholinesterase. The proposed nomenclature distinguishes between the two types of forms (A and G) and identifies each form by its number of catalytic subunits, as established for *Electrophorus*. The intersubunit disulfide bridges are symbolized by *dashes* and the helical collagen-like tails by the *triple lines*. We have extended the nomenclature to acetylcholinesterase and cholinesterase forms of other species on the basis of the clear correspondence that exists between the sets of molecular forms, considering their molecular parameters and their interaction properties. This scheme does not take into account the differences that may exist for each quaternary structure in the distribution of intra- and intersubunit disulfide bonds, in glycosidic residues, in intrasubunit cleavages, in the presence of hydrophobic peptides, or of noncatalytic components, as shown in Figure 2. As illustrated, the structure of the tetrameric G_4 form corresponds to the product of autolysis, which retains a noncollagenic peptide from the tail. In the trypsin-dissociated tetramers, this peptide is digested so that the two subunits are no longer covalently linked (Rosenberry et al 1980).

collagenase produces molecules that have lost most of the tail, so that the enzyme dissociates into individual tetramers (Bon & Massoulié 1978).

Conversely, a rod-like fragment of the tail can be recovered after pepsin digestion in acidic conditions (Anglister & Silman 1979, Mays & Rosenberry 1981). The circular dichroic spectrum of this pepsin-resistant fragment is characteristic of a collagen-like triple helical structure (Rosenberry et al 1980). The tail peptides appear to constitute a doublet, whose mass is on the order of 30,000 daltons (Anglister & Silman 1978). Rosenberry and co-workers (1980) recently isolated the intact tails, after reduction of the A forms and denaturation of the catalytic subunits.

In the collagen-tailed molecules, each peptide of the three-stranded tail is covalently linked by disulfide bonds to a dimer which is itself associated through quaternary interactions with another disulfide-linked dimer (Anglister & Silman 1978, McCann & Rosenberry 1977, Rosenberry & Richardson 1977). It should be noted that these complex molecules remain stable and active, even after reduction of all interchain disulfide bonds (Bon & Massoulié 1976b). This demonstrates the existence of quaternary interactions, particularly between the tail and the catalytic tetramers.

The existence of the A_4, A_8, and A_{12} molecules implies that there are only weak interactions among the tetramers themselves: this is consistent with the ease of their release upon digestion of the collagen-like tail. The collagen-tailed forms may be dissociated by various proteases as well as by collagenase. Because of their asymmetric structure, they are also mechanically disrupted by sonication (Bon et al 1973). These degradation procedures yield active tail-less forms. Trypsin, for instance, sequentially converts the major A_{12} form into A_8 and A_4 and releases tetramers which constitute the end product of the digestion (Massoulié & Rieger 1969, Massoulié et al 1970a,b), confirming that they represent particularly stable assemblies of catalytic subunits. However, active dimers and monomers may also be obtained (Massoulié et al 1971, Bon & Massoulié 1976a), thus *Electrophorus* acetylcholinesterase can exist as six distinct molecular forms.

We have proposed that the asymmetric collagen-tailed and the globular tailless molecules be designated A and G forms, respectively, and that a subscript be used to indicate the number of catalytic subunits, as schematically illustrated in Figure 1 (Bon et al 1979). This nomenclature describes only the overall quaternary structure of the molecules; it must be stressed that more subtle variations occur, either in the noncatalytic tail component or in the catalytic subunits themselves. The first case may be illustrated by the difference between tetramers obtained by proteolytic digestion and by autolysis of crude electric organ extracts. In the trypsinized tetramers, the tail peptides are split between the two cysteine residues that are linked to the catalytic subunits, whereas the autolyzed tetramers retain a proximal 8000 daltons noncollagenic peptide from the tail (Rosenberry et al 1980).

Modifications of the subunits themselves may also occur as a result of degradation procedures. For example, autolysis and trypsin digestion cleave them progressively into peptides of various sizes (Morrod et al 1975, Bon & Massoulié 1976b, Rosenberry et al 1974, Dudai & Silman 1974). In an initial step, these proteolytic attacks generate a peptide of approximately 60,000 daltons, which carries the active site serine residue and can therefore be labeled by radioactive phosphorylating agents that inhibit the enzyme, such as diisopropylfluorophosphonate (DFP). This peptide remains linked through disulfide bridges to a complementary 30,000 daltons fragment (Dudai & Silman 1974, Rosenberry et al 1974, Bon & Massoulié 1976b). Other differences may also arise in the saccharidic moieties, since acetylcholinesterase is a glycoprotein, containing about 15% carbohydrates, mainly as mannose, galactose, hexosamines, and sialic acid (Bon et al 1976).

It is important to realize that although the catalytic subunits may differ because of proteolytic cleavages and because of their insertion in different quaternary associations, in the monomer and the collagen-tailed triple tet-

ramer (note that all subunits are not equivalent in this structure because of
the tail linkages), their catalytic properties appear essentially identical
(Vigny et al 1978a, Barnett & Rosenberry 1979). Interestingly, the isolated
monomer itself differs only very slightly from the polymeric forms in its
K_m value, and it displays the characteristic excess substrate inhibition (Bon
& Massoulié, 1976a).

The study of *Electrophorus* acetylcholinesterase has thus led us to recognize that six catalytically equivalent forms of the enzyme can exist, differing
in their quaternary structure. The three globular, or G, forms appear as
degradation products of the collagen-tailed, asymmetric A forms, the most
abundant of which is the triple tetrameric enzyme, A_{12}.

Asymmetric molecular forms of acetylcholinesterase that resemble the
Electrophorus forms have been identified in *Torpedo* electric organs (Massoulié & Rieger 1969, Rieger et al 1976a), and have been shown to contain
a collagen-like tail (Lwebuga-Mukasa et al 1976). The asymmetric molecules appear very similar in electron micrographs to *Electrophorus* A forms,
and there is an obvious correspondence between the molecular forms of the
two species. However, Taylor and co-workers recently found that the quaternary structure of these molecules is more complex than that of *Electrophorus* collagen-tailed acetylcholinesterase (Lee et al 1981). Electrophoretic
analyses of the nonreduced, denatured molecules reveal the presence of a
number of heavy components, together with disulfide-linked catalytic dimers. These components contain not only catalytic subunits and tail peptides,
but also a noncatalytic peptide of 100,000 daltons. This heavy structural
subunit appears to replace randomly about a third of the tail-linked active
subunits. The significance of this structural peptide is at present unknown.
It has been proposed that it derives from a procollagenic telopeptide of the
tail. If this is the case, interactions between such telopeptides and acetylcholinesterase subunits might explain how the complex structure of the
collagen-tailed molecules has evolved. The *Torpedo* molecules would thus
represent a primitive form of these structures. In addition to the occurrence
of this heavy, noncatalytic subunit—and the heterogeneity it introduces
among collagen-tailed molecules—*Torpedo* and *Electrophorus* acetyl-
cholinesterases also differ in the arrangement of disulfide bonds in the lytic
G_4 enzyme. The *Torpedo* catalytic subunits are associated mainly as dimers,
but also as trimers and tetramers. The additional bonds that are responsible
for such linkages may exist in the native molecules, or arise because of
secondary rearrangements during proteolysis.

Molecular forms that appear structurally homologous to the *Electrophorus* A_{12}, A_8, and A_4 forms exist in other vertebrates, notably in skeletal
muscles, as discussed below. In the next section we see that these collagen-
tailed forms possess characteristic ionic interaction properties.

Low Salt Aggregation, Ionic Interactions, and Possible Hydrophobic Character of the Collagen-tailed Forms

It has long been known that acetylcholinesterase from *Torpedo* and *Electrophorus* electric organs reversibly aggregates in low-salt solutions (Grafius & Millar 1965, 1967, Changeux 1966). We have studied this phenomenon extensively for *Electrophorus* acetylcholinesterase and have shown that it is characteristic of the collagen-tailed forms (Bon et al 1978, Cartaud et al 1978). Aggregation seems to require the integrity of the tail, since even a limited digestion by collagenase modifies its characteristics. The cleavage of a sufficient fraction of the tail eventually produces nonaggregating molecules (Bon & Massoulié 1978).

After extensive purification, collagen-tailed acetylcholinesterase no longer aggregates, unless it is supplemented with an "aggregating agent" (Bon et al 1978). This aggregating agent is a thermostable, nonproteinaceous macromolecule, which exists in large excess in electric organ extracts. The agent promotes in vitro formation of polydisperse aggregates, sedimenting around 80 S, which appear in the electron microscope as groups of about ten molecules with the tails associated as a central bundle and with the catalytic globular subunits at both extremities (Cartaud et al 1978).

Treatment of the enzyme with maleic anhydride (Dudai et al 1973) or acetic anhydride (Cartaud et al 1975, Bon et al 1978) abolished its aggregation, probably by alkylating ϵ amino groups. This suggested that the aggregating agent might carry negative charges, and indeed we found that it could be replaced by various polyanions, including chondroitin sulfate. We also found that the chondroitin sulfate content of *Electrophorus* electric organ extracts could account quantitatively for their aggregating capacity.

Collagen-tailed acetylcholinesterase forms of other species also aggregate, but instead of forming discrete molecular assemblies, they are in general recovered in a precipitate that forms upon dialysis of crude saline extracts from nerve or muscle tissue against a low salt buffer (Bon et al 1979). This precipitation does not occur in purified enzyme preparations; it therefore requires the presence of an aggregating agent, or of several factors, the nature of which is not known. These factors allow the coprecipitation of *Electrophorus* A forms, indicating that similar interactions are involved for enzymes from different species.

The aggregates are dissociated, or redissolved, by increasing the salt concentration. Divalent cations are more efficient than monovalent cations at equivalent ionic strength in crude extracts, i.e. in the presence of saturating aggregating agent. The half-dissociation occurs at about 50–100 mM MgCl$_2$ for *Electrophorus,* as well as for amphibian, avian, and mammalian enzymes (Bon et al 1979, Massoulié et al 1980; S. Bon and J. Massoulié, unpublished results) and at a notably higher concentration for *Torpedo* (Bon & Massoulié 1980).

Since aggregation occurs in physiological ionic conditions, similar interactions probably exist in situ and we suggest that they could anchor the collagen-tailed enzymes within the extracellular basal lamina, as discussed below. The interactions of the collagen-tailed forms with other cellular components appear to be, in fact, more complex than expected for ionic binding of polyanions to positive charges of their collagen-like tail. In *Torpedo* extracts, we observed a dual effect of divalent cations on the solubilization of these molecules: a small proportion of A forms was soluble in the absence of divalent cations (10 mM tris-HCl pH 7 buffer), but these forms quantitatively precipitated in the presence of 50 mM MgCl$_2$ (Bon & Massoulié 1980). At low concentration, divalent cations therefore appear to favor the association of the collagen-tailed molecules with other components. Dudai & Silman (1973) also reported experiments that suggest that the *Electrophorus* A forms bind, in a Ca^{2+} dependent manner, to some particulate proteins.

We must consider the possibility that acetylcholinesterase A forms may be attached not only to the basal lamina, but also to the plasma membranes, as suggested by Silman & Dudai (1975), who observed that the non-ionic detergent Triton X 100 enhances their solubilization from *Electrophorus* electric tissue. Watkins et al (1977) indeed reported that these collagen-tailed forms interact in vitro with phospholipid liposomes. Kaufman & Silman (1980) recently showed that lecithin liposomes bind the collagen-tailed A$_{12}$ and A$_8$ forms, but not the tailless G$_4$ forms at high ionic strength (1 M NaCl), whereas all forms are bound at low ionic strength (0.1 M NaCl). Binding to liposomes prepared with an uncharged lecithin analogue occurred only at low ionic strength. These results suggest the existence of interactions, possibly hydrophobic in nature, between the tail and the lipids, in addition to ionic interactions between the positively charged lipid polar heads and negative groups of the enzyme (e.g. anionic groups involved in substrate and ligand binding). The difference between the binding of G$_4$ and that of A$_8$ and A$_{12}$ may be due, however, at least in part, to the number of sites and the size of the molecules: Webb & Clark (1978) observed a similar differential binding to an affinity column and ascribed it to multiple site effects. A comparison between the A$_4$ and G$_4$ forms would allow an unambiguous estimate of the influence of the collagen-like tail, but this has not been done because the A$_4$ form is generally a minor acetylcholinesterase component.

In any case, the collagen-tailed forms do not appear to possess any obvious hydrophobic character, since their hydrodynamic properties (sedimentation coefficient, Stokes radius) are not modified in the presence of

Triton X 100. Moreover, in the case of *Electrophorus*, these forms, as well as the lytic tetrameric G_4 form, do not bind significant amounts of radioactive Triton X 100 (Millar et al 1978).

We discuss below the insertion of the collagen-tailed forms in different tissues, particularly in the basal lamina of skeletal muscles. We must keep in mind in considering the anchorage of these molecules that in no case do we know the relative contributions of ionic interactions with acidic polysaccharides, of ionic interactions with other proteins, or of possible hydrophibic interactions.

Hydrophobic Acetylcholinesterase in Torpedo Electric Organs

We have found that *Torpedo* electric organs contain not only collagen-tailed A forms, but also a large pool of acetylcholinesterase that is insensitive to salt concentration and that interacts with nondenaturing detergents (Bon & Massoulié 1980) [as also reported by Witzemann (1980) and by Viratelle & Bernhard (1980)]. We have distinguished two fractions of hydrophobic acetylcholinesterase: a low salt soluble (LSS) fraction and a detergent soluble (DS) fraction. Each of these fractions corresponds to an activity approximately equal to that of the A forms.

In absence of detergent, the LSS fraction apparently includes tetramers and possibly higher polymers, but its major component, sedimenting around 8 S, corresponds to noncovalent associations of catalytic dimers (linked by disulfide bridges) with heterogeneous noncatalytic elements, the mass of which may be as high as 100,000 daltons. Both LSS and DS fractions are converted into a monodisperse species in the presence of detergents such as Triton X 100 or bile salts. The apparent sedimentation coefficient of this enzyme, about 6 S in Triton X 100, varies significantly with the nature of the detergent and with the density of the medium, thus indicating that it is a detergent-dimer complex.

Trypsin digests a fraction of the noncatalytic components, but it neither appears to modify the catalytic subunits nor, in particular, to affect their hydrophobic character. In contrast, pronase or proteinase K produce isolated dimers, which no longer interact with detergents. It is therefore likely that a pronase-sensitive hydrophobic polypeptide sequence is responsible for the specific properties of the LSS and DS acetylcholinesterase dimers. This peptide must be absent in collagen-tailed forms, which do not appear to interact with nondenaturing detergents.

Digestion of the A forms by pronase yields nonhydrophobic catalytic dimers that are indistinguishable from those obtained from the hydrophobic fractions. This suggests that the two types of enzyme correspond to products of the same gene that differ in their post-translational processing. Viratelle & Bernhard (1980) found that tetramers derived from the A forms and hydrophobic dimers possess the same turnover number per active site. This argues strongly in favor of the genetic identity of the catalytic subunits in the two pools of enzyme. In this view, the A forms might derive from the hydrophobic enzyme fractions by removal of a hydrophobic peptide and attachment to the collagen-like tail. It should be noted, however, that during embryonic development of *Torpedo* electric organ the accumulation of the collagen-tailed A_{12} forms begins earlier than that of the hydrophobic dimeric enzyme (S. Bon, in preparation).

The polymorphism of *Torpedo* acetylcholinesterase appears more complex than that of *Electrophorus* in several respects. First, the catalytic dimers, in the LSS fraction, may be associated with noncatalytic components which probabaly interact with their hydrophobic domains. Noncatalytic subunits, which may or may not be related to these components, have also been shown by P. Taylor to replace partially the catalytic subunits in the structure of the collagen tailed forms (P. Taylor, personal communication). In addition, we must distinguish between hydrophobic tetramers, which may be identified in the LSS fractions, and nonhydrophobic tetramers, which may be obtained by degradation of the A forms. The two types of molecules possess nearly identical molecular parameters, but they are clearly differentiated by their interactions with detergents, since the LSS tetramers dissociate into detergent-dimer complexes. Similarly, dimers occur both as detergent-bound and lytic, nonhydrophic molecules. These different types of molecules are schematically represented in Figure 2.

This complex polymorphism is not encompassed by the nomenclature proposed in Figure 1, which only indicates the general quaternary structure of the different molecules. In spite of such reservations, we believe this nomenclature is useful because it expressed homologies that clearly exist between the molecular forms of both acetylcholinesterase and cholinesterase throughout the range of vertebrates.

MOLECULAR FORMS OF CHOLINESTERASE AND ACETYLCHOLINESTERASE IN OTHER VERTEBRATES

Aggregating and Soluble Forms: Homology with Molecular Forms from Electric Organs

Almost all of the cholinesterase and acetylcholinesterase activity from the muscles or nervous tissues of vertebrates can be recovered in a soluble extract, after homogenization in a detergent-containing saline medium (e.g. 1% Triton X 100, 1 M NaCl, 50 mM MgCl$_2$, 10 mM tris HCl pH 7) (Hall, 1973). We have found that the solubilized enzymes can be separated into a low salt aggregating fraction and a low salt soluble fraction, which corre-

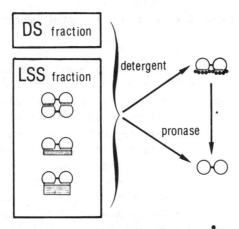

Figure 2   Hypothetical structure of low-salt soluble, hydrophobic *Torpedo* acetylcholinesterase. The low salt soluble fraction of *Torpedo* acetylcholinesterase consists of disulfide-bridged
dimers that are associated with noncatalytic components of various sizes, or with themselves,
e.g. as tetramers, so that their hydrophobic domain is probably occluded. In the presence of
nondenaturing detergents, the dimers form homogeneous detergent-complexes. Pronase converts them into detergent-insensitive dimers. The *dashes* represent hydrophobic domains of
the proteins, the *shaded rectangles* symbolize noncatalytic subunits, and the *black dots,*
detergent molecules.

spond to distinct molecular forms. We have studied in detail the properties
of these different forms in the case of mammalian acetylcholinesterase (Bon
et al 1979, Vigny et al 1979). The molecular parameters of the low salt
soluble forms suggest that they represent monomers, dimers, and tetramers
of catalytic subunits, which are generally in the range of 80,000 daltons. We
have therefore proposed that these forms be called G_1, G_2, and G_4. As
shown below, these forms occur both as soluble and hydrophobic molecules.

The low salt insoluble forms appear homologous to the *Electrophorus*
collagen-tailed A_4, A_8, and A_{12} forms. This conclusion is based on the
following points:

1. There is a one-to-one correspondence so that for each *Electrophorus*
   form there is, for instance, a bovine form with similar hydrodynamic
   properties—notably, a large Stokes radius. These parameters are not
   affected by the presence of Triton X 100.
2. As with the *Electrophorus* enzyme, treatment of the bovine forms with
   collagenase produces modified molecules that sediment faster and possess a reduced Stokes radius.
3. The low salt insoluble bovine forms aggregate in practically identical
   ionic conditions as the *Electrophorus* A forms, and their aggregation is
   abolished in the same way by acetylation or by treatment with collagenase.

4. Trypsin degrades these forms in a sequential manner, producing G_4 forms.

Although the number of catalytic subunits and active sites has only been determined in the case of *Electrophorus* acetylcholinesterase, these converging lines of evidence constitute a very strong argument in favor of a structural homology between mammalian and *Electrophorus* acetylcholinesterase forms, for both collagen-tailed and globular forms. This conclusion may be extended to cholinesterase: we have shown that in rat tissues, for example, cholinesterase exists in multiple forms that closely parallel the acetylcholinesterase multiple forms and possess similar solubility properties. They are, however, clearly distinct molecules, as indicated by their slightly different sedimentation coefficients, by their thermal stability, and by the lack of immunocrossreactivity (Vigny et al 1978b). The quaternary structure of the major globular form of cholinesterase that occurs in human serum has been found to be very similar to that of the *Electrophorus* G_4 acetylcholinesterase: in this enzyme, subunits of 90,000 daltons are linked in pairs by an intersubunit disulfide bond (Lockridge et al 1979). Like the acetylcholinesterase form, this structure is resistant to disulfide reduction, which only lowers its thermal stability. Cholinesterase is a glycoprotein, and has long been known to present a number of electrophoretic variants, which probably differ by their sialic acid content.

We have characterized aggregating and nonaggregating acetylcholinesterase forms that clearly correspond to the *Electrophorus* A_{12}, A_8, and A_4 and G_4, G_2, G_1 forms, not only in mammals, including man (Carson et al 1979), but also in tissues of amphibians (the frog *Rana temporaria*) and reptiles (the snake *Natrix natrix* and the turtle *Pseudemys picta*) (unpublished results).

The homology of chicken enzymes with *Electrophorus* acetylcholinesterase forms appears less obvious than in the case of other species, because (*a*) only the two largest aggregating forms have been identified in muscle extracts and (*b*) their hydrodynamic properties are markedly different: e.g. the heaviest acetylcholinesterase form sediments at 19–20 S (Vigny et al 1976b, Rotundo & Fambrough 1979), while the bovine A_{12} form sediments at 17 S (cf Table 1). We have, however, established unambiguously the correspondence of the chicken enzymes with those of other species by studying their tryptic degradation (Allemand et al 1981). Trypsin degrades the 20 S molecules in a stepwise manner, like the A_{12} forms of other species; it produces two smaller aggregating forms that clearly correspond to A_8 and A_4 and an apparently tetrameric G_4 form. Apart from this sequential degradation of the asymmetric forms, we have observed that a very low trypsin concentration also partially digests all asymmetric and globular forms in a progressive manner, converting them into a set of molecules that resemble much more closely the mammalian forms in their sedimentation coefficients. Thus, both chicken cholinesterase and acetylcholinesterase possess a trypsin-sensitive fragment, this is particularly large in the case of acetylcholinesterase, where trypsin treatment reduces the mass per catalytic subunit from about 120,000 to 85,000 daltons. These trypsin-sensitive components may correspond either to fragments of the catalytic subunits or to distinct polypeptides. In any case, it is remarkable that the removal of such a large fraction of the protein neither modifies its catalytic properties significantly nor changes the low salt aggregation properties of the collagen-tailed forms or the hydrophobic character of the globular forms. The possible meaning of the trypsin-sensitive peptides thus remains entirely unknown.

The existence of such dispensable peptides again illustrates the complexity of acetylcholinesterase and cholinesterase polymorphism. In the following section we further show that the globular forms may be quite heterogeneous in their solubility properties and hydrophobic character.

Hydrophobic and Nonhydrophobic Globular Forms

The globular forms of acetylcholinesterase and cholinesterase exist both as hydrophobic proteins, which interact with detergents, and as detergent-insensitive, nonhydrophobic proteins. This has been illustrated in a murine cell line (T_{28}, a neuroblastoma X sympathetic neuron hybrid) (Lazar & Vigny 1980). These cells produce mostly G_1 and G_4 forms, together with a very low level of G_2. These forms are present in the cells themselves and are also secreted into the culture medium. The secreted forms are of course soluble proteins and do not interact with detergents. The cellular enzyme, however, can be partially solubilized in a detergent-free buffer, yielding molecular forms that sediment like the secreted enzymes in the absence of detergent, but that are modified by Triton X 100. The apparent sedimentation coefficient of G_1 is thus shifted from 4.5 S to 3.2 S, and that of G_4 from 10.2 S to 9.7 S. This cellular soluble fraction contains about 40% of the G_1 activity but only less than 5% of the G_4 activity. The remaining enzyme may be solubilized by Triton X 100. The corresponding molecular forms are indistinguishable from the soluble cellular form when analyzed in the presence of the detergent, but aggregate in detergent-free buffers. This phenomenon may be due either to an intrinsic property of these molecules, or, more probably, to other components of the detergent-solubilized extracts.

We have made similar observations for acetylcholinesterase molecular forms of bovine caudate nucleus (Grassi et al 1981). In this case, however, the soluble hydrophobic G_4 form (11.5 S) differs somewhat from the nonhydrophobic form (10.5 S), thus suggesting that it is associated with non-

Table 1 Forms of acetylcholinesterase and cholinesterase in different species

	Collagen-tailed forms			Tail-less forms		
	A_{12}	A_8	A_4	G_4	G_2	G_1^c
Mass[a]	1,150,000	796,000	410,000	331,000	165,000	70,000
			Sedimentation coefficient[b]			
Electrophorus acetylcholinesterase (Bon et al 1976)	18.4 S	14.2 S	9.1 S	11.8 S	7.7 S	5.3 S
Torpedo acetylcholinesterase (Bon & Masoulié 1980)	17.4 S	13.4 S	9.1 S	11.1 S	7.8 S	5 S
Bovine acetylcholinesterase (Bon et al 1979)	17.1 S	13.0 S	8.7 S	10.9 S	7.2 S	4.8 S
Chicken acetylcholinesterase (Allemand et al 1981)						
Native	19.2 S	—	—	11.8 S	7.9 S	5.2 S
Trypsin-treated	17.3 S	13.3 S	8.5 S	10.9 S	7.2 S	—
Chicken cholinesterase (Allemand et al 1981)						
Native	18.6 S	—	—	11.1 S	7.1 S	4.9 S
Trypsin-treated	17.3 S	13.3 S	8.5 S	11 S	—	4.7 S

[a] The mass of the major molecular forms from *Electrophorus* were determined by sedimentation diffusion equilibrium and that of the minor G_2 and G_1 forms were obtained by comparison, using the products of sedimentation coefficients and Stokes radius (Bon et al 1976).

[b] All sedimentation coefficients have been determined by sucrose gradient centrifugation, in the absence of detergent in saline buffer (1 M NaCl; 50 mM MgCl$_2$; 10 mM tris HCl pH7).

[c] The data concerning the G forms refer to lytic molecules in the case of *Electrophorus* and *Torpedo* (obtained by tryptic dissociation of collagen-tailed forms and by pronase digestion of hydrophobic form, respectively) and to naturally occurring molecules in the case of the other species.

catalytic components that mask its hydrophobic region. As observed in the case of *Torpedo,* the hydrophobic forms may be converted into detergent-insensitive molecules by pronase, which again probably digests a hydrophobic peptide.

Human erythrocytes possess a membrane-bound, dimeric acetylcholinesterase form, which has been shown by Brodbeck and co-workers to be highly hydrophobic. It may be solubilized by detergents, but aggregates into various polymers if detergent is removed (Ott et al 1975, Ott & Brodbeck, 1978). The enzyme can be reinserted, after detergent-solubilization, into phospholipid vesicles (Hall & Brodbeck 1978, Römer-Lüthi et al 1980). In addition, the maintenance of its catalytic activity requires a hydrophobic environment (Wiedmer et al 1979, Frenkel et al 1980).

It is worth noting that the hydrophobic erythrocyte acetycholinesterase has an amino acid composition similar to that of the *Electrophorus* enzyme (Grossman & Lieflander 1979) and possesses an equivalent carbohydrate content (Niday et al 1977).

The molecular parameters of the hydrophobic acetylcholinesterase or cholinesterase forms in higher vertebrates are quite similar to those of the corresponding nonhydrophobic forms, in spite of the possible presence of hydrophobic peptides and of noncatalytic components. Therefore, we believe that as a first approximation it is legitimate to consider them generally as monomers, dimers, and tetramers and to use the same terminology for all types of molecules.

Molecular Forms in Solution and State of the Enzyme in situ: Solubilization Artifacts?

Before discussing the possible biological significance of acetylcholinesterase and cholinesterase multiple molecular forms, we must ask whether the different molecules identified in solution necessarily correspond to distinct pools of enzyme in situ.

Acetylcholinesterase and cholinesterase molecular forms may be modified or dissociated by proteolytic enzymes. Such effects have been observed during storage of tissue extracts, as a result of the action of endogenous proteases. The complex collagen-tailed forms in particular tend to be converted into smaller globular forms, so that their apparent proportion may appear lower than it would be in the absence of autolysis. Silman and co-workers (1978) demonstrated that this spontaneous degradation can be prevented when muscle extracts are prepared in the presence of antiproteolytic agents. The degradation of the high molecular weight forms may also be minimized if the extracts are analyzed immediately after solubilization or kept frozen at low temperature, e.g. −80°C. In nerve extracts, Couraud et al (1980b) did not observe such a degradation of the asymmetric forms, so the use of protease inhibitors does not seem to be necessary in this case.

In contrast, Hollunger & Niklasson (1973) reported that in calf brain extracts during storage monomeric acetylcholinesterase (G_1) was converted into "aggregates": a similar observation was made by Adamson et al in mouse brain extracts (Adamson et al 1975, Adamson 1977). Adamson et al concluded that the multiple molecular forms of acetylcholinesterase arose from association-dissociation processes. Chang & Blume (1976) analyzed the enzyme from murine neuroblastoma cells in culture, in various conditions, and reached the similar conclusion that monomers associate into dimers and tetramers in different fashions, depending upon their environment.

We recently analyzed the acetylcholinesterase molecular forms in bovine caudate nucleus: we obtained consistent results by sedimentation analysis and gel filtration chromatography, indicating that the extracts contain mostly G_4 and G_1 forms (Grassi et al 1981). We observed that in extracts prepared in Tris-HCl buffer (10 mM tris-HCl pH 7), G_4 remained unmodified during storage at 4°C for up to ten days, but G_1 (4.5 S) was progressively converted into a 7.5 S form. This process does not seem to correspond to a simple dimerization, but probably involves other proteins. It did not occur in the presence of Triton X 100, was blocked by the alkylating agent N-ethylmaleimide, and could be prevented or reversed by reducing agents. It is therefore possible to find conditions in which the composition of molecular forms in an extract remains stable. It is, moreover, possible to isolate and purify individual molecular forms: we can therefore consider them stable molecular entities.

It remains to be established, however, whether the different forms represent distinct species in situ or whether they are formed or modified during the solubilization process itself. Hollunger & Niklasson (1973) already noted that it is possible to increase the yield of solubilized monomeric acetylcholinesterase by repeated extractions of calf caudate nucleus in a detergent-free buffer. Similarly, Gisiger et al (1978) found that a direct quantitative solubilization of acetylcholinesterase from rat superior cervical ganglia in the presence of Triton X 100 yielded less monomeric G_1 form than was obtained if the ganglia were first extracted in a detergent-free buffer. In bovine caudate nucleus, we found that the proportion of G_1 could be increased from less than 10% in a direct Triton X 100 solubilization to over 40% when the tissue was first extracted three times in detergent-free buffer. Although the total yield of acetylcholinesterase activity was 20% lower in the second procedure than in the first one, this difference represents almost a four-fold increase in the absolute amount of G_1, which is apparently obtained at the expense of G_4.

These experiments show that interconversion of molecular forms may indeed occur during solubilization procedures. We believe, however, that the different forms represent distinct cellular components and that under appropriate conditions of extraction and analysis it is possible to obtain a profile that accurately reflects their relative amounts in the cell. The most direct demonstration was given by Lazar & Vigny (1980), who analyzed the acetylcholinesterase molecular forms in murine nerve cell cultures (T_{28}), after solubilization in the presence of 1% Triton X 100. They showed that a short exposure of the cells to the cationic phosphorylating inhibitor echothiopate (phospholine) inactivated the G_4 form while G_1 was essentially unaffected, although in solution both forms reacted with the inhibitor at the same rate. This experiment clearly shows that G_1 and G_4 correspond to distinct enzyme pools in the cells: G_1 is probably intracellular and G_4 is associated with the plasma membrane and is externally accessible. Taylor et al (1981) reached similar conclusions by studying the protection of the ectoenzyme with the nonpermeating reversible inhibitor, BW 284C51, against irreversible inactivation by DFP, for both T_{28} murine cells and chick sympathetic ganglion cells in culture. In chicken cells, G_4 is the external form as in murine cells, but G_2 is the major intracellular enzyme instead of G_1.

We conclude from these studies that under the experimental conditions used, the composition of molecular forms in solution reflected the state of the enzymes in situ. This is consistent with numerous studies in which molecular forms have been analyzed in different tissues and under different physiological conditions.

DISTRIBUTION AND LOCALIZATION OF MOLECULAR FORMS IN HIGHER VERTEBRATES

That acetylcholinesterase is not inserted in the same manner at different sites is well established. For example, histochemical staining shows that acetylcholinesterase may be solubilized from muscle endplates by collagenase (Hall & Kelly 1971, Betz & Sakmann 1973, Sketelj & Brzin 1979) but not by incubation with the detergent Triton X 100 (Koelle et al 1970), whereas the opposite is true for the enzyme associated with neuronal membranes in ganglia (Koelle et al 1970, Klinar & Brzin 1980). The collagenase sensitivity and detergent solubility of the enzyme at these two sites obviously recall the respective properties of the asymmetric collagen-tailed forms and the hydrophobic, globular forms. No such identification should be proposed, however, without great caution. In the next section we discuss the distribution of the forms in various tissues of birds and mammals.

Distribution of Globular and Collagen-tailed Forms in Vertebrate Tissues

Electric organs and skeletal muscles of *Electrophorus* appear exceptional because they contain exclusively, or almost exclusively, collagen-tailed acetylcholinesterase forms (mostly as the A_{12} form). In contrast, *Torpedo* electric organs contain a large proportion of noncollagenous enzyme (Bon & Massoulié 1980, Viratelle & Bernhard 1980, Witzemann 1980), together with asymmetric forms. Because the electric organs are derived from muscles, it is perhaps not surprising to find that skeletal muscles from all vertebrates contain "heavy" forms (Hall 1973, Vigny et al 1976a) that have been recognized as collagen-tailed molecules (Bon et al 1979). Although these forms are particularly interesting in view of their localization and regulation, as discussed below, it must be emphasized that they generally constitute only a small fraction of the enzyme. Most of the activity in the muscle corresponds to globular forms.

Globular forms indeed represent the major fraction of acetylcholinesterase and cholinesterase in most vertebrate tissues. In blood, for example, one finds only globular forms. In bovine erythrocytes, acetylcholinesterase is a membrane-bound, disulfide-linked dimeric form (Grossman & Lieflander 1979). Human erythrocytes also contain a hydrophobic dimeric form of acetylcholinesterase (G_2) (Niday et al 1977, Römer-Luthi et al 1979, 1980), and human serum contains a tetrameric form (G_4) of cholinesterase. Acetylcholinesterase also exists as a serum-soluble form in some species. In rat, for example, the serum contains G_4, G_2, and G_1 cholinesterase forms (M. Vigny, unpublished result), together with G_4 acetylcholinesterase (Gisiger & Vigny 1977, Fernandez et al 1979b), and the erythrocytes contain a detergent-soluble monomeric acetylcholinesterase form (G_1) (Rieger et al 1976b, Fernandez et al 1979b). In chicken, there is little activity in the erythrocytes, and the three globular forms of both cholinesterase and acetylcholinesterase occur in soluble form in the serum (Lyles et al 1980).

The cerebrospinal fluid contains soluble acetylcholinesterase (Chubb et al 1974, 1976), which corresponds mainly to G_4 in man (F. Muller and J. Massoulié, unpublished result), and it contains essentially no cholinesterase.

Globular forms seem to be the only type of the enzyme in tissues such as placenta (Vigny et al 1976a) and liver. The liver contains principally cholinesterase (G_1 and G_4 in chicken; mostly G_1 in rat), but also some acetylcholinesterase (G_1 and G_4 in chicken; G_1, G_2, and G_4 in rat) (S. Bon, M. Vigny, and J. Massoulié, unpublished results).

Collagen-tailed forms exist not only in skeletal muscles: in rat heart the A_{12} form has been observed in the atria, but not in the ventricles (Skau & Brimijoin 1980). We have found that in rat atria the acetylcholinesterase activity corresponds mostly to G_1 and G_4, with about 5% of A_{12}; and the cholinesterase activity corresponds mostly to G_1. In chicken heart, A_{12} forms of both enzymes can be detected in the atria (S. Bon and J. Massoulié, unpublished results).

Collagen-tailed forms also occur in peripheral ganglia: e.g. in mammalian superior cervical ganglia (Gisiger et al 1978); in chicken ciliary ganglia (Vigny et al 1976b, Scarsella et al 1978); and, at low levels, in motor nerves of birds and mammals (Di Giamberardino & Couraud 1978, Fernandez et al 1979a,b 1980), as well as in the vagus nerve of rat (Skau & Brimijoin 1980); but not in sensory nerves (Fernandez et al 1979a). In the brain, the proportion of collagen-tailed acetylcholinesterase is relatively high in fishes [e.g. *Electrophorus,* (Tsuji et al 1972), *Carassius auratus* (Guillon & Massoulié, 1976)], and in amphibians and reptiles (S. Bon and J. Massoulié, unpublished results). In the central nervous system of *Torpedo,* the electric lobes that command the electric discharge possess a high level of A_{12} form. In other brain regions, however, globular forms (tetramers, dimers, and a minor proportion of monomers) are predominant (S. Bon and J. Massoulié, unpublished results). The collagen-tailed forms were initially considered to be absent from the brains of higher vertebrates; however, Villafruela et al (1980) recently demonstrated that such forms can be solubilized by collagenase from chicken optic lobes, where they represent about 0.5% of the acetylcholinesterase activity.

A small proportion of A_{12} form has also been found in the central nervous system of mouse and rat, principally in the cerebellum, but not in the brain itself (Rieger et al 1980a). We have detected, however, trace amounts (on the order of 0.002%) of A_{12} and A_8 in the bovine caudate nucleus (Grassi, Vigny & Massoulié 1981).

Definite patterns of molecular forms therefore exist in different tissues for acetylcholinesterase and, when it is present, for cholinesterase. It must be pointed out that the proportions of these forms in a given tissue may present marked interspecific, and even intraspecific, variations. We have, for example, observed widely variable patterns of molecular forms in human muscles from different individuals (Carson et al 1979).

Insertion of Acetylcholinesterase in Skeletal Muscles: Collagen-tailed Forms and the Basal Lamina

Neuromuscular endplates are characterized by a very high concentration of acetylcholinesterase, as first indicated by biochemical (Marnay & Nachmansohn 1937) and histochemical methods (Koelle & Friedenwald 1949, Couteaux and Taxi 1952). The neuromuscular junction of skeletal muscles is usually characterized by deep infoldingsof the muscle cell's membrane which are perpendicular to the fiber's axis. This membrane is followed by an extracellular basal lamina, which penetrates the synaptic folds (Couteaux 1960, Heuser 1980). It appears to consist mostly of acidic glycoproteins and possesses distinct staining characteristics at the endplate (Zacks et al 1973). Whereas the acetylcholine receptor sites are localized in the postsynaptic membrane at the top of the junctional folds (Fertuck & Salpeter 1976, Matthews-Bellinger & Salpeter 1978), the acetylcholinesterase activity is distributed along the postsynaptic membrane and includes its folds (Eränkö and Teräväinen 1967, Salpeter et al 1972). There have been many hypotheses about the physiological significance of these folds; in particular it has been suggested that they allow acetylcholinesterase to be located near the receptor sites, without being immediately adjacent to them. It is interesting in this respect that the number of acetylcholinesterase and cholinesterase active sites has been found to be approximately equal to the number of receptor sites, at the neuromuscular end-plates of various species (Barnard et al 1971, 1973). The level of cholinesterase at the endplates is generally low (Brzin & Pucihar 1976), and both cholinesterase and acetylcholinesterase activities appear to be localized in the same way in the pre- and postsynaptic membranes and in the synaptic basal lamina (Brzin et al 1980).

Collagenase solubilizes acetylcholinesterase activity from the endplates of rat (Hall & Kelly 1971) and frog muscles (Betz & Sakmann 1973). This observation suggested that the enzyme may be anchored within the extracellular basal lamina (Silman & Dudai 1975) and indeed a direct demonstration was given by McMahan and collaborators, who observed the persistence of acetylcholinesterase activity in the endplate basal lamina of frog muscle, after degeneration of both pre- and postsynaptic cellular elements (Marshall et al 1977, McMahan et al 1978).

It is very tempting to assume that the basal lamina linked activity corresponds to the asymmetric forms of acetylcholinesterase, both in view of the collagenous nature of their "tail," and because their aggregation properties would be consistent with an attachment to acidic components of this structure (Bon et al 1978). In addition, these forms are specifically localized at the endplates, at least in adult rat muscles (Hall 1973, Vigny et al 1976a). We have indeed shown that treatment of frog muscles with collagenase solubilizes lytic teteramers, but also heavy molecules that unambiguously derive from collagen-tailed forms by a partial digestion of the tail: these molecules sediment faster than the native forms and do not aggregate (S. Bon and J. Massoulié, unpublished results).

Ionic interactions of the collagen-tailed forms with polyanionic components, which we have shown to be involved in the low ionic strength aggregation of these molecules, may not be their only type of interaction with the basal lamina. For example, interaction of the collagen tails with components such as fibronectin must also be considered (Emmerling et al 1981). In addition, covalent bridges similar to the well-known collagen cross-links may occur between the collagen-like tails and basal lamina components, although this has not yet been demonstrated. If such bonds exist, one might expect that with age a larger fraction of enzyme would become immobilized, so that only collagenase or proteases could solubilize it.

In adult rat muscles, the A forms are exclusively localized at the endplate (Hall 1973, Vigny et al 1976a, Bon et al 1979), but this is not the case in neonatal rats, where they also exist in endplate-free segments (Sketelj & Brzin 1980, Koenig et al 1980). In frog muscles, we have found that the three A forms represent about a third of the actylcholinesterase activity that is concentrated at the endplates and that they also occur at a low level in the tendinous region, but not in the aneural part of the fibers, which contains only G_2 and G_4 forms (S. Bon and J. Massoulié, unpublished results). In human muscle (Carson et al 1979), and probably in chicken muscle (Koenig & Vigny 1978b), the collagen-tailed forms also exist in endplate-free sections. For example, these forms have been found to accumulate, together with the globular G_2 forms, in the perijunctional zone of dystrophic chicken muscles (Jedrzejczyk et al 1981). The collagen-tailed forms cannot therefore be considered as biochemical markers of neuromuscular contacts, as originally suggested.

Although endplate sections of fast twitch muscles from normal chicken contain a major proportion of collagen-tailed forms (Jedrzejczyk et al 1981), this is generally not the case in other species. In rat, for example, the globular forms apparently amount to as much as 75% of the enzyme accumulated at the endplate sites (Vigny et al 1976a); in fact, we must not exclude the possibility that the globular forms as well as collagen-tailed forms may be associated in some way with the basal lamina.

Sketelj & Brzin (1979) found that collagenase treatment released only about half of the acetylcholinesterase from the endplate region of the mouse diaphragm. The remaining activity could still be visualized by histochemical staining. It is likely that this collagenase-resistent enzyme corresponds to globular forms, which occur also in aneural regions of the fibers. This is consistent with the finding that papain solubilized not only the collagenase-sensitive enzyme, but also released some activity, at a lower rate, from endplate and endplate-free sections. This papain-released enzyme would thus represent membrane-bound globular forms. It should be recalled that intracellular acetylcholinesterase, particularly associated with the sarco-plasmic reticulum, may represent a significant portion of the muscle's activity.

Globular and Collagen-tailed Forms in Nervous Tissue

Collagenase was found to be inefficient in solubilizing the acetylcholinesterase activity from the rat superior cervical ganglion (Klinar & Brzin 1980). Papain, on the contrary, released the extracellular enzyme, but left the intracellular activity intact. Since the A_{12} form represents less than 5% of the total activity (Gisiger et al 1978), the papain-solubilized activity probably corresponds mostly to globular forms. Although it is difficult to conclude that the A_{12} form is collagenase resistant, in view of its small proportion, the possibility that it is intracellular must be considered.

Collagen-tailed forms are certainly intracellular in motor axons, where they are carried by the fast axonal flow, as discussed below (Di Giamberardino & Couraud 1978, Fernandez et al 1980a). The localization of these forms in nervous tissue is not known and, in particular, it is not certain whether they occur at extracellular sites. Various types of interaction probably exist. For example, recent experiments have shown that the solubilization properties of collagen-tailed forms are quite different in the brain and in the retina of chickens: the A_{12} form, or its modified derivatives, could be solubilized from both tissues by sodium cholate, or by collagenase (Barat et al 1980a,b). This form can also be solubilized in a detergent-containing buffer in the presence of a chelating agent (EDTA) in the case of retina, but not in brain (Barat et al 1980a,b). Whether the detergent interacts directly with the collagen-tailed enzyme, or only facilitates its solubilization, remains an open question.

In the nervous tissue of adult rat and chicken, notably in brain, G_4 constitutes by far the major acetylcholinesterase form (Rieger & Vigny 1976, Marchand et al 1977). We have seen that in mouse and chicken neural cells this G_4 form occurs on the external face of the plasma membrane, whereas the smaller forms, mostly G_1 and G_2, respectively, are intracellular (Lazar & Vigny 1980, Taylor et al 1981). It is therefore likely that in nervous tissue the plasma membrane acetylcholinesterase corresponds to G_4. It is clear, however, that the G_4 form also occurs intracellularly [e.g. it is transported with the fast axonal flow in motor nerves (Di Giamberardino & Couraud 1978, Couraud & Di Giamberardino 1980) and this implies its association with the endoplasmic reticulum]. In neurons, the G_4 form also probably constitutes the bulk of the activity that may be demonstrated histochemically in the cisternae of the endoplasmic reticulum as well as the perinuclear membrane (Somogyi & Chubb 1976, Gisiger et al 1978, Toth & Kreutzberg 1979, Klinar & Brzin 1980).

BIOSYNTHESIS OF ACETYLCHOLINESTERASE AND CHOLINESTERASE IN HIGHER VERTEBRATES

We discuss in this section the biosynthesis of acetylcholinesterase and the relationships between its molecular forms. We examine whether acetylcholinesterase and cholinesterase are metabolically related, or whether they are genetically distinct enzymes.

Biosynthesis and Secretion of Acetylcholinesterase

Walker & Wilson (1976a) observed that in DFP-treated cultures of chick muscle, the acetylcholinestrase activity continued to increase for nearly one hour after the blockade of protein synthesis by cycloheximide. They concluded that the enzyme may be first synthesized as an inactive precursor. The activation of a precursor may perhaps explain why proteolytic treatment has often been observed to increase the activity (Sketelj & Brzin 1977). This has been reported particularly for a solubilized enzyme, the isolated A_{12} form of chicken acetylcholinesterase (Rotundo & Fambrough 1979). This effect does not seem to be related to the removal of diffusion barriers in particulate structures and it probably reflects modification of the catalytic subunits themselves.

In their studies of chick muscle cultures, Wilson and collaborators found that after inhibition by organosphosphates, the acetylcholinesterase activity first reappeared around the nucleus, then progressively spread into the whole cell, and was released into the medium (Wilson & Walker, 1974, Cisson & Wilson 1977). From an electrophoretic analysis they concluded that low molecular weight forms appear first, and suggested that the enzyme undergoes an "orderly process of binding, movement, and assembly" (Wilson & Walker 1974).

It is extremely likely that the endoplasmic reticulum is involved in the synthesis and transport of the enzyme to the cell's outer surface. It should be noted that in rat muscles there is also an elevated acetylcholinesterase activity in the sarcoplasmic reticulum during active synthesis of the enzyme in vivo; i.e. in very young animals, or after treatment with organophosphates (Brzin et al 1980, Grubić et al 1981).

Smilowitz (1980) studied the effect of monovalent ionophores on the release of acetylcholinesterase by cultured chicken muscle cells and concluded that it represents a real secretory phenomenon, and not a sloughing of membrane-bound enzyme. Rotundo & Fambrough (1980a,b) found that the secretory process can be blocked by the antibiotic tunicamycin, which interferes with the glycosylation of proteins. These authors observed that after treating the cells with DFP, active acetylcholinesterase reappeared at the cell membrane and in the medium after a lag of 2 or 3 hr. Their results, however, indicate that the total pool of membrane-bound enzyme is not an intermediate in the secretory process: it has a long half-life (50 hr), whereas the intracellular enzyme turns over rapidly (2–3 hr). The membrane-bound enzyme amounts to one-third of the cellular enzyme, but less than one-tenth of the acetylcholinesterase synthesized becomes incorporated in the plasma membrane. It is remarkable that the released enzyme corresponds to only half of the activity that is lost by the cells, thus raising the possibility that some sites may be inactivated in the process.

Acetylcholinesterase is secreted not only by cultured cells, but apparently also by neurons in vivo, as it occurs in the extracellular space of mammalian central and peripheral nervous tissue (Kreutzberg & Toth 1974, Kreutzberg et al 1975, 1979). By using push-pull canules located in the ventricular space adjacent to the caudate nucleus and substantia nigra of anesthetized rats, Glowinski and his colleagues have been able to analyze the release of acetylcholinesterase at both sites and its modification during stimulation of the nigro-striatal pathway (obtained by applying a depolarizing concentration of potassium chloride to the substantia nigra). They found that this stimulation induces an increase in the release of acetylcholinesterase at the site of the stimulated substantia nigra, as well as postsynaptic modifications in the ipsi and contralateral caudate nuclei, but that it has no effect on the release of cholinesterase (Greenfield et al 1980). Such a neuronal release is probably the source of the soluble acetylcholinesterase found in the cerebrospinal fluid, since its concentration is increased by stimulation of the sciatic nerve (Chubb et al 1974, 1976). Acetylcholinesterase is released mainly as the G_4 form at the neuromuscular junction. It probably originates in part from the nerve terminal (Skau & Brimijoin 1978), where it is transported by the axonal flow (see below), but also from the muscle itself, as indicated by a continuous release of enzyme (over 80% G_4 with a minor G_1 component) from denervated muscles (Carter & Brimijoin 1980). The G_4 form is specifically secreted by rat superior cervical ganglia, maintained in vitro (Gisiger & Vigny 1977); however, neuroblastoma cells in culture release all three globular forms, G_1, G_2, and G_4 (Lazar & Vigny 1980, Kimhi et al 1980).

In dystrophic chickens, a highly increased level of acetylcholinesterase occurs in the serum (15–20 fold the normal level at the age of 4 mo). Temporal correlations indicate that this enzyme is probably released by the muscles, which contain an abnormally high activity (Lyles et al 1980). It is remarkable that although cholinesterase is even more increased in dystrophic chicken muscles over the control value, its level in the plasma remains normal in all cases. This suggests that it is not secreted by the muscles, or that its concentration in the serum is controlled by a specific mechanism, such as removal by the liver. The glycoprotein nature of these enzymes may be relevant to such processes.

Metabolic Relationship Between the Molecular Forms

The problem of possible precursor-product relationships between molecular forms has been approached by analyzing the recovery of acetylcholinesterase after irreversible inactivation by phosphorylating inhibitors such as DFP. In neuroblastoma cell cultures, it has been found that G_1 reappears before G_4 (Rieger et al 1976b), and Lazar & Vigny have been able to find conditions in which G_1 in the absence of protein synthesis appears to be converted into G_4 (M. Lazar and M. Vigny, unpublished results).

We have recently studied the metabolic turnover rate of G_1 and G_4 in T_{28} murine cells by the method of heavy isotope labeling introduced by Fambrough and his colleagues for the study of the acetylcholine receptor (Devreotes et al 1977, Gardner et al 1979). This method avoids the use of protein synthesis inhibitors, or of phosphorylating agents such as DFP, which block the preexisting acetylcholinesterase, but may also affect the biosynthetic processes of the cells (Walker & Wilson 1976a, Cisson & Wilson 1977). The heavy isotope-labeled molecular forms may be separated from the light forms by differential sedimentation. In this way, we found that under equilibrium conditions, the half life of G_1 is approximately 3 hr, whereas that of G_4 is close to 40 hr (the minor form G_2 appears to behave like G_1) (M. Lazar, M. Vigny, and J. Massoulié, in preparation). Since these cells contain nearly equal proportions of G_1 and G_4, this means that less than 10% of G_1 is incorporated into G_4. Considering that G_1 and G_4 are, respectively, intracellular and membrane-bound, these results are strikingly similar to those obtained by Rotundo & Fambrough (1980a,b) for the intracellular and externally accessible pools of enzyme in cultured chick muscle cells, as discussed above. The rapid turnover of G_1 therefore appears correlated with secretion of acetylcholinesterase.

The relative proportions of G_1 and G_2 vary widely in different species: while G_1 is predominant in murine cells, the two forms are approximately equal in human muscle (Carson et al 1979); in chicken tissues the major light form is G_1 for cholinesterase, but G_2 for acetylcholinesterase (Lyles et al 1979, 1980). There are marked differences, therefore, in the relationships between the processes of dimerization, and those of tetramer formation and secretion.

In rat muscle cultures containing the A_{12} form, the recovery of the collagen-tailed enzyme was observed later than that of G_1 and G_4 (Koenig & Vigny 1978a), and the same sequence of reappearance (G_1, G_4, A_{12}) was found in rat superior cervical ganglia either in vivo or in vitro (Gisiger & Vigny 1977). Similarly, the recovery of G_1 and G_4 was found to start very early in rat diaphragm after irreversible inhibition by soman, while that of A_8 and A_{12} was much slower (Sketelj et al 1980). Cells therefore appear to incorporate progressively the catalytic subunits from intracellular forms, probably associated with the endoplasmic reticulum, into a plasma membrane G_4 form, and in some cases into collagen-tailed forms, by active metabolic processes.

It must be emphasized that the different molecular forms may not be metabolically homogeneous. For instance, although the renewal of G_1 in T_{28} cells appears to be satisfactorily described by a first-order process, distinct pools may be precursors of the membrane-bound and secreted forms. In spite of the likelihood of branched or parallel assembly lines, at present it appears possible to interpret the available data by assuming a linear sequence of polymerization, eventually leading to the synthesis of the collagen-tailed forms.

The assembly of the collagen-tailed forms in cells that produce these forms raises interesting problems. By their lack of interaction with nonionic detergents, the collagen-tailed forms are more similar to secreted than to membrane-bound molecules. As in the case of the membrane-bound forms (Rotundo & Fambrough 1980a,b) it is quite possible that they do not derive from the bulk of the intracellular enzyme, but that the two types of insertions correspond to pathways that diverge at an early stage. Are the collagen-tailed molecules assembled intracellularly, or do they result from the binding of secreted or membrane-bound tetramers to collagenic elements that preexist in the extracellular basal lamina? The only indication we have so far is the transport of the collagen-tailed A_{12} form by axonal flow in motor nerves, which clearly points to an intracellular assembly site in the motoneuron cell body. Another question that remains entirely open is the significance of the smaller A_8 and A_4 forms. These forms are generally very minor in proportion, and have been overlooked by most authors. They have not been found to accumulate during the biosynthesis of A_{12}, but this does not preclude their possible role as biosynthetic intermediates. They may also represent the products of proteolytic degradations similar to those observed in vitro. In any case, we must keep in mind that the various molecular forms observed in tissues do not necessarily participate in the physiological hydrolysis of acetylcholine, but may also represent biosynthetic intermediates and/or degradation products of the physiologically active forms. In this respect it is interesting that the level of the A_8 form is high in the developing *Torpedo* electric organ at the stage of maximal accumulation of A_{12}: A_8 is therefore probably a biosynthetic intermediate of the complete molecule. At the time of birth, A_8 is no longer visible, but it reappears in the adult; this tempts us to suggest that it then represents a degradation product of A_{12} (S. Bon, in preparation).

Are Cholinesterase and Acetylcholinesterase Metabolically Related?

Since the discovery that two distinct types of enzymes hydrolyze choline esters, no satisfactory explanation for this duality has been given, and no clear physiological function for the "nonspecific" cholinesterase has been discovered. In fact, the inhibition of this enzyme, even for prolonged periods, does not appear to produce consistent pharmacological effects. However, Koelle and his collaborators have proposed a hypothesis according to which cholinesterase is a precursor of acetylcholinesterase and is also involved in the regulation of its activity (Koelle et al 1976, 1977a,b). This hypothesis is based on studies of the two enzymes in cat autonomic ganglia. In the cat superior cervical ganglion, the activities of cholinesterase and acetylcholinesterase are approximately equal. Davis & Koelle (1978) have shown by histochemical observation at the electron microscopic level that both enzymes are predominantly localized on the postsynaptic neuronal membranes. Decentralization of the cat ganglion induces the disappearance of about half the cholinesterase and of nearly all acetylcholinesterase after two weeks (Koelle et al 1974). This suggests that a presynaptic neurotrophic factor controls the fate of the two enzymes, which therefore appear closely linked. Their most important result, however, is that a maintained selective irreversible inhibition of cholinesterase leads to a significant increase of acetylcholinesterase (about 40%) in the superior cervical ganglion, as well as in other autonomic ganglia. In contrast, such an inhibition reduces the recovery of acetylcholinesterase after its inactivation by the phosphorylating inhibitor, sarin (Koelle et al 1976, 1977a). Conversely, the protection of cholinesterase by a reversible specific inhibitor during sarin treatment increases the recovery of acetylcholinesterase in the ganglia (Koelle et al 1977b).

The observed relationship between the level of cholinesterase and the recovery of acetylcholinesterase after sarin poisoning is clearly consistent with the hypothesis that acetylcholinesterase derives from cholinesterase. The influence of cholinesterase inhibition on the level of acetylcholinesterase in absence of sarin is more difficult to explain: the authors suggest that cholinesterase retroactively inhibits its own production from an inactive precursor, but it is not obvious how such an effect would result in the rise of acetylcholinesterase, unless one postulates the existence of distinct pools of cholinesterase, each involved differently in this regulation and in the production of acetylcholinesterase.

No influence of cholinesterase inhibition on acetylcholinesterase was observed in cat muscle (Koelle et al 1977a) or in rat ganglia (Koelle et al 1979a). Attempts to obtain the interconversion of these enzymes in homogenates were not conclusive (Koelle et al 1979b). Although these results do not directly contradict the hypothesis that acetylcholinesterase derives from cholinesterase, they do indicate that the metabolic relationships that exist in cat ganglia are perhaps not general.

The hypothesis that cholinesterase is a precursor of acetylcholinesterase and not the product of a distinct gene could be most directly approached by analyzing genetic variations in these enzymes. In humans, for example, genetic variants of serum cholinesterase are known, e.g. an "atypical" type of low activity (Kalow & Genest 1957). It would be interesting to analyze whether the muscle or ganglionic enzymes belong to the same type, and whether such genetic variations are correlated with differences in acetylcholinesterase.

In our view, however, the hypothesis appears unlikely for several reasons. First, Brzin et al (1980) have studied the effect of continued specific inhibition of cholinesterase, during the perinatal development in rats, starting 6 d before birth and maintained up to 3 wk after birth. They found that this treatment did not prevent the normal development of the animals, or particularly of the muscle endplate, and had no effect on the activity of acetylcholinesterase: thus the cholinesterase activity did not appear to have any critical relationship with acetylcholinesterase. Indeed, the histochemical distribution of cholinesterase does not coincide systematically with that of acetylcholinesterase, as it is extremely variable among different species (Silver 1974). Second, the two proteins are quite distinct in their sensitivity to thermal reactivation and in their immunoreactivity (Vigny et al 1978b, Koelle et al 1979b). In addition, the mass of each cholinesterase form appears larger than the corresponding acetylcholinesterase form in rat (Vigny et al 1978b), but smaller in chicken (Allemand et al 1981); this does not seem compatible with conversion processes in both species. Finally, it is very difficult to conceive how the parallel sets of acetylcholinesterase and cholinesterase molecular forms could be generated in the precursor product hypothesis, considering the relationships that exist for each enzyme—especially since no hybrid molecules appear to exist in spite of the structural homology between the two series.

In any case, the observations reported by Koelle et al (1976, 1977a,b) for cat autonomic ganglia unambiguously demonstrate that the two enzymes are not controlled independently. It must be pointed out that the effects measured are on the *catalytic activity* of these enzymes. Such effects may be mediated through complex allosteric interactions, or more obviously, through the level of substrate acetylcholine. Because of the very high activity of acetylcholinesterase in the ganglion, however, it is difficult to imagine that the inhibition of cholinesterase alone could be effective in affecting the acetylcholine level, unless it controlled a particular acetylcholine compartment specifically involved in biosynthetic regulations, or unless it controlled the level of another choline ester not hydrolyzed by acetylcholinesterase. This second possibility, however unlikely, would imply a distinct function for the "nonspecific" cholinesterase. It may be recalled in this respect that cholinesterase specifically hydrolyzes iontophoretically applied butyrylcholine, at the frog neuromuscular junction (Pecot-Dechavassine 1968).

THE REGULATION OF ACETYLCHOLINESTERASE AND CHOLINESTERASE MOLECULAR FORMS IN VARIOUS DEVELOPMENTAL AND EXPERIMENTAL CONDITIONS

In this section we focus our discussion on the evolution of the composition of acetylcholinesterase, and in some cases of cholinesterase, during development and in various experimental situations.

Evolution of Acetylcholinesterase Forms in Nerve Cells during Differentiation in Culture and Embryonic Development

When neuroblastoma cells are induced to differentiate in culture, their acetylcholinesterase level rises (Blume et al 1970, Furmanski et al 1971, Schneider 1976). This quantitative variation is correlated in some cases with a marked relative increase in the proportion of G_4 (Vimard et al 1976, Lazar & Vigny 1980). During differentiation of a hybrid cell line between neuroblastoma and sympathetic neuron (T_{28}), for example, the cells preferentially accumulate this external membrane-bound enzyme form. This accumulation takes place even in conditions that do not allow the extension of neurites (Lazar & Vigny 1980). In contrast, the synthesis of the A_{12} form, which is induced by the nerve growth factor in a rat cell line (PC 12) derived from an adrenal chromaffin tumor (pheochromocytoma), appears correlated with such a morphological differentiation (Rieger et al 1980c).

The evolution observed during the differentiation of neuroblastoma cells in vitro mimics the changes that occur in vivo during the maturation of the brain: the G_4 form increases, while the light forms—G_1 in the case of rat (Rieger & Vigny 1976) or G_2 in the case of chicken (Marchand et al 1977, Villafruela et al 1981)—remain at a low steady level.

The A_{12} form is the only collagen-tailed form that has been identified so far in embryonic tissues of birds and mammals. It is present as early as the fifth day of incubation in the chick ciliary ganglion (Kato et al 1980). In chick paravertebral sympathetic ganglia, the major form at early stages (7 d) is G_4, but most of the activity that accumulates before hatching corresponds to G_1, and the A_{12} form does not appear until day 17 (Taylor et al 1980).

Molecular Forms in Peripheral Ganglia: Effects of Decentralization and Axotomy

Peripheral ganglia such as the rat superior cervical ganglia contain both acetylcholinesterase and cholinesterase (Klingman et al 1968). Acetylcholinesterase has been histochemically localized extracellularly (mostly around the presynaptic terminals) and intracellularly (associated with the nuclear envelope and with the endoplasmic reticulum) (Klinar & Brzin 1978, Somogyi & Chubb 1976). Gisiger et al (1978) believe that this endoplasmic reticulum activity corresponds largely to the G_4 form, because of its quantitative predominance. Davis & Koelle (1978) observed in the cat superior cervical ganglion that acetylcholinesterase is present not only in the plasma membranes of the ganglionic axons and their terminals, but also in the membranes of the postsynaptic neurons, where it is accompanied by cholinesterase.

The decentralization of superior cervical ganglia leads to a rapid degeneration of the presynaptic elements. It causes the loss of a large proportion of both acetylcholinesterase and cholinesterase in cat (Koelle et al 1974), as well as in rat (Klinar and Brzin 1978, 1980). In rat, about half of the acetylcholinesterase and a quarter of the cholinesterase have disappeared after two weeks. As indicated by histochemical localization and by solubilization with papain, decentralized ganglia contain essentially no extracellular acetylcholinesterase and about half of the control level of extracellular cholinesterase (Klinar & Brzin 1980).

Gisiger et al (1978) observed that the initial decrease of ganglionic acetylcholinesterase was followed by a partial recovery, starting three days after the operation. During this recovery phase they observed the appearance of the enzyme in glial cells that did not possess it originally. The main forms of acetylcholinesterase in the ganglion—G_1, G_4, and A_{12}—followed the loss and recovery phases to different degrees: most of the increase was due to the G_1 form, which approximately recovered its original activity after two weeks and then remained stable. The A_{12} form appeared to present seasonal fluctuations and showed a more irregular variation, but it is remarkable that its activity increased significantly above the control in a transient manner.

If the acetylcholinesterase activity remaining after decentralization is only intracellular, it would follow that the A_{12} form is itself intracellular, as already suggested by the failure of collagenase to solubilize ganglionic acetylcholinesterase; but here again no conclusion can be safely reached because of its small proportion, which is only 4% of the total activity in control ganglia. The complex variations observed and the existence of acetylcholinesterase in the postsynaptic neuronal membranes of normal ganglia imply that decentralization not only causes the loss of presynaptic enzymes, but induces rearrangements in the postsynaptic neurons, as well as in the glial cells.

Reciprocal pre- and postsynaptic interactions have been demonstrated by Couraud et al (1980b), who followed the level of the molecular forms of acetylcholinesterase in chick ciliary ganglion, after section of the pre- or postganglionic nerves. They found that decentralization decreased the total activity by 35% within 2 d, affecting mainly the A_{12} and G_4 forms (reduced respectively to 25% and 40% of their original level), but left the light forms G_1 and G_2 essentially unchanged. Axotomy did not change the global ganglionic activity, but this apparent lack of effect resulted from a balance between a decrease of A_{12} and G_4 and an increase in G_2. The effects of pre- and postganglionic sections were not additive. Moreover, axotomy and double sections produced marked contralateral effects, both in the total activity and in the porportions of molecular forms. Such experiments therefore cannot define unambiguously pre- and postsynaptic components of acetylcholinesterase or cholinesterase, but they do demonstrate that the levels of the various forms in both ganglionic elements are controlled in a complex manner.

Molecular Forms of Acetylcholinesterase and Cholinesterase in Motor Nerves—Axonal Transport

Several authors have established that a fraction of acetylcholinesterase is transported with the rapid component of axonal flow, in motor nerves of dog (Niemierko & Lubinska 1967, Lubinska & Niemierko 1971), cat (Ranish & Ochs 1972), rabbit (Fonnum et al 1973, Tuček 1975, Brimijoin & Wiermaa 1978), and rat, where histochemically detectable activity was observed within tubules and vesicles (Kása 1968, Kása et al 1973).

Di Giamberardino & Couraud observed that sciatic nerves of rat and chicken contain G_1, G_2, and G_4 forms, together with a low level of A_{12} form (representing less than 5% of the total activity). They found that the rapid accumulation of acetylcholinesterase at the proximal stump of transected nerves is due principally to A_{12} and G_4 (Di Giamberardino & Couraud 1978, Di Giamberardino et al 1979). Recently they reported a quantitative analysis of the movements of the different forms in chicken nerve (Couraud & Di Giamberardino 1980): G_1 and G_2 are transported slowly (anterograde velocity 3.5 mm/d), while 20% of G_4 and all of A_{12} are carried by fast axonal flow (approximately two-thirds forward, at 410 mm/d, and one-third backward, at 150 mm/d). Most of the rapidly transported enzyme is G_4, since this form represents over 80% of the total axonal acetylcholinesterase.

Couraud et al also studied axonal transport in regenerating nerves (Couraud et al 1980a). The amount of transported G_4 was decreased to half of the normal value, and that of A_{12} to only 5%. It is tempting to interpret this as an adaptation to the reduced need for these synaptic components in the growing nerve. The transport of these forms in normal nerves may result from a retrograde induction by the muscle.

It has been reported that axonal transport of acytylcholinesterase is perturbed in certain pathological states. This has been examined in the sciatic nerves of chickens in which a peripheral neuropathy had been induced by injection of acrylamide (Droz et al 1979) and in sciatic and brachial nerves of chickens with a genetically determined muscular dystrophy (Di Giamberardino et al 1979). In the acrylamide neuropathy, the fast orthograde transport of A_{12} was severely decreased, although its level in the nerve was increased five-fold over the normal. The G_4 form was less affected. The reduced transport of the A_{12} form could be correlated with degenerative processes in the distal branches of the nerves and with a decrease of this form in the muscle. In the dystrophic chickens, both anterograde and retrograde transports were found to be quantitatively and qualitatively normal, although the muscles contained an abnormally elevated level of all forms of acetylcholinesterase (Lyles et al 1979).

Brimijoin and collaborators have shown that in rabbit nerves, the rapidly transported enzyme fraction is protected from the nonpermeating inhibitor phospholine, whereas an immobile fraction is sensitive to this compound (Brimijoin & Wiermaa 1978, Brimijoin et al 1978). The immobile, externally accessible fraction probably corresponds to the nontransported G_4 form, localized in the axonal membrane. It is very likely that the slowly transported G_1 and G_2 forms are axoplasmic and that the rapidly moving G_4 and A_{12} forms are associated with the endoplasmic reticulum (Droz et al 1979, Couraud & Di Giamberardino, 1980, Rambourg & Droz 1980). It is legitimate to identify the various pools of enzyme with distinct molecular forms because the amount of each form was found to be globally invariant in an isolated chicken nerve segment during a 4 hour period, in spite of an extensive spatial redistribution; this indicates that the transport processes occur without interconversion of the molecular forms, at least during this range of time (Couraud & Di Giamberardino 1980).

The fate of the transported enzymes at the nerve terminal is not clear: a fraction of these molecules is carried back by retrograde flow, and it is likely that a fraction of G_4 is released, as mentioned above. Whether collagen-tailed A_{12} molecules are also released and allowed to stick to the endplate basal lamina is not known; however, no correlation appears to exist between the level of A_{12} at the endplates and its axonal transport. For example, dystrophic and normal chickens present a large difference in the level of this enzyme at the endplate, but show no change in axonal transport (Di Giamberardino et al 1979). This is even clearer in the case of ectopic reinnervation in rat muscles, where acetylcholinesterase, including A_{12}, reappears at the original endplate in the absence of nerve (Weinberg & Hall 1979).

Couraud & Di Giamberardino (1980) also studied the cholinesterase content of chicken nerve. They found that the activity of this enzyme amounted to 15% of the acetylcholinesterase activity and that the enzyme showed molecular forms that were homologous to those of acetylcholinesterase, including the A_{12} form. Cholinesterase did not behave like acetyl-cholinesterase, however, since very little accumulated at the proximal extremity of transected nerves; therefore, this enzyme is probably localized mostly in the Schwann cells, in agreement with histochemical data (Koelle 1951).

Molecular Forms in Skeletal Muscles During Development

Skeletal muscles of very young embryos contain globular forms (G_1 and G_4); A_{12} appears only at the stage at which the first neuromuscular contacts are established, i.e. at 7 d in ovo in the hindlimb of chicken embryos (Kato et al 1980) and at 14 d of fetal development in the posterior leg muscles of rat embryos (Vigny et al 1976a, Koenig & Vigny 1978a).

Koenig & Vigny (1978a) found that muscle cells in culture, obtained from the posterior limb of 13–14 day-old rat embryos, did not produce the A_{12} form. This enzyme was synthesized, however, in cultures obtained from 18 day-old embryos in which neuromuscular contacts had already taken place in vivo. Quite spectacularly, the muscle cells from the younger embryos could be induced to synthesize A_{12} acetylcholinesterase in vitro when co-cultured with motoneurons which themselves do not produce the A_{12} form. These experiments indicate that neuromuscular interactions specifically control the synthesis of collagen-tailed forms in muscles. We discuss this point at length in another section.

The synthesis of A_{12} acetylcholinesterase in nerve-free cultures, as observed in the case of 18 day-old embryonic muscles (Koenig & Vigny 1978a) and in a transformed muscle cell line (Sugiyama 1977), contrasts with the fact that adult denervated muscles do not maintain this form. This illustrates the different potentialities of adult and embryonic or transformed cells. In the 18 d muscle cultures, there is no histochemically detectable concentration of acetylcholinesterase activity, which may indicate that there are no specific anchoring sites on the myotubes. It is not known whether the collagen-tailed form is localized at the outer surface of the cells in this case, or whether it remains intracellular.

This situation may be similar to that which exists in the muscles of new born rats where the A_{12} form is present along the whole fibers (Sketelj & Brzin 1980, Koenig et al 1980). In adult rat, however, the A_{12} form is restricted to the sites of the end-plates (Hall 1973, Vigny et al 1976a).

Cholinesterase levels are quite low in adult rat muscle, and in particular collagen-tailed forms appear to be absent (Vigny et al 1978b). The level of this enzyme is, however, much higher during early postnatal development (Brzin & Kiauta 1979) and the A_{12} form exists in significant proportion in the muscles of newborn rats (Brzin et al 1980). Cholinesterase therefore regresses quantitatively and qualitatively after birth.

In chicken muscles, cholinesterase levels are also relatively high at an
early stage and drop to a very low value after hatching (Lyles et al 1980).
Both cholinesterase and acetylcholinesterase levels remain abnormally high
in the fast twitch muscles of dystrophic chickens, which in some respects
maintain embryonic characteristics.

The Nervous Control of Acetylcholinesterase in Skeletal Muscles: the Regulation of the Collagen-tailed Forms and Their Physiological Significance

It is well known that motor nerves control the acetylcholinesterase content
of muscles (Guth 1968). Denervation produces widely different effects on
the acetylcholinesterase content of muscles from different species. In rat
muscles, the most spectacular effect of denervation is the disappearance of
the collagen-tailed 17 S (A_{12}) form after about 2 wk (Hall 1973, Vigny et
al 1976a, Fernandez et al 1979b). Denervation, however, also affects the
globular forms, and in fact the earliest modification (2 d) observed in the
extensor digitorum longus (a white muscle) was a marked decrease in G_1,
accompanied by a transient increase in G_4 (the level of both forms later
decreased below the control values) (McLaughlin & Bosmann 1976). Carter
& Brimijoin (1980) showed that the release of acetylcholinesterase by the
diaphragm (removed and incubated in organ baths) was greatly increased
soon after denervation. This release later decreased below control in absolute terms, but remained higher in proportion to the muscle's enzyme
content. Carter & Brimijoin suggested that although the secreted enzyme
appeared mostly as G_4, it could originate from the muscle's G_1 form. This
hypothesis would be in agreement with the rapid turnover rate of G_1 (see
above).

In chicken, denervation leads to the loss of the 20 S (A_{12}) form (Vigny
et al 1976b, Sketelj et al 1978) in spite of the large increase in total acetylcholinesterase activity (Linkhart & Wilson 1975a). This shows clearly that
the biosynthesis of the collagen-tailed forms is submitted to distinct regulatory factors.

Denervation of the semimembranous leg muscle of the rabbit induces a
large increase in activity, which corresponds to markedly different variations in the white and red parts of the muscle. In the white section, the G_1
form remains predominant, and the A forms essentially disappear after one
month; however, in the red conoidal section, both A and G forms are
elevated and the proportion of A_{12} in fact increases from about 10% to more
than 25%, a 25-fold increase in specific activity (F. Bacou, P. Vigneron, and
J. Massoulié, in preparation). These findings illustrate the large differences
that may exist among the different types of muscles with respect to acetylcholinesterase metabolism (we have verified that the level of acetylcholine

receptors is elevated in both red and white muscles) and demonstrates that the biosynthesis of the collagen-tailed forms does not depend upon innervation of the muscle in all cases.

The persistence of acetylcholinesterase within the basal lamina, in the experiments of McMahan and his colleagues (26 d denervation, and two weeks of muscle degeneration) (Marshall et al 1977, McMahan et al 1978), suggests that the A forms do not readily disappear from denervated frog muscles.

In dystrophic chickens the content of cholinesterase and acetylcholinesterase in fast twitch muscles is greatly elevated over the normal level (Wilson et al 1968, Wilson et al 1973a). This situation recalls that observed in denervated muscles, except that dystrophic muscles also contain an increased amount of collagen-tailed A_{12} forms, which disappears after denervation (Sketelj et al 1978). In dystrophic and in normal chickens, the collagen-tailed forms of cholinesterase as well as acetylcholinesterase disappear after denervation; this indicates that the enzymes are regulated in a parallel manner (Silman et al 1979).

The A forms reappear after reinnervation of rat muscles at original or ectopic sites (Vigny et al 1976a). As discussed above, whereas globular forms exist very early in embryonic muscle, the synthesis of the A forms seems to be triggered by the establishment of nerve-muscle contact. The nature of the nerve appears critical: Inestrosa et al (1979) have shown that while the A_{12} acetylcholinesterase form is normally absent in a smooth muscle, e.g. the nictitating membrane of the cat, it can be induced by applying a dual innervation with a motor nerve.

Lyles & Barnard (1980) reported that during the growth of chickens, the A_{12} and A_8 forms of acetylcholinesterase and cholinesterase disappear from the multiply innervated, tonic *anterior latissimus dorsi* (ALD) muscle, whereas they always become predominant in focally innervated, fast-twitch muscles such as the *posterior latissimus dorsi* (PLD) or the pectoral muscle. Thus, the exclusively tonic type of activity of the ALD muscle appears neither to necessitate the presence of A forms nor to support their synthesis.

In the frog some acetylcholinesterase activity remains histochemically detectable at the endplate in denervated muscles after ten weeks, but this enzyme appears inefficient in hydrolyzing focally applied acetylcholine (Pecot-Dechavassine 1968). Cangiano et al (1980) paralyzed rat soleus muscles by applying a compression cuff to the sciatic nerve and obtained a denervation-like decrease in their acetylcholinesterase content. They found that after two weeks, the functional enzyme acting on nerve-released acetylcholine was greatly reduced, as indicated by the decay time of miniature endplate potentials, although there was little change in the histochemical staining of the endplate activity. These experiments show that there is no simple correspondance between histochemically detectable, and physiologically active enzyme at the endplate. Molecular forms of acetylcholinesterase unfortunately have not yet been examined in this type of experiment. Assuming, however, that denervation or paralysis induced a selective decrease of the collagen-tailed forms, these observations would be consistent with the hypothesis that the collagen-tailed forms represent the physiologically active species, at least in certain muscles. The experiments of Rubin et al (1980) in chicken nerve-muscle cultures indeed suggest that there may be a correlation between the presence of the A_{12} form and a reduction in the half-life of the synaptic potentials; however, the respective roles of A and G forms in the neuromuscular junction have not been unambiguously evaluated.

Factors Involved in the Regulation and in the Localization of Acetylcholinesterase

In view of the complex effects exerted by the nerve on the biosynthesis, stability, and localization of the collagen-tailed forms in muscles, it is clear that several regulatory factors must be involved.

ROLE OF PROTEASES IN THE DEGRADATION OF ENDPLATE ACETYLCHOLINESTERASE   Proteases released by the muscle cells appear to be important in the modeling of neuromuscular junctions and in their stability (Vrbová et al 1978). In particular, the decrease of endplate acetylcholinesterase that follows denervation probably results from proteolysis, since protease inhibitors have recently been found to maintain the normal level of activity of rat neuromuscular endplates for 24 hr after the operation (Fernandez & Duell 1980). Therefore, the disappearance of the A forms from denervated rat muscles is probably actively regulated.

EVIDENCE FOR THE MUSCULAR SYNTHESIS OF ACETYLCHOLINESTERASE A FORMS   The capacity of muscle cells to synthesize the collagen-tailed A_{12} form in the absence of nervous elements has been demonstrated unambiguously in the case of a murine clone (Sugiyama 1977) and also under certain conditions in cultures of embryonic rat (Koenig & Vigny 1978a) and chicken muscles (Kato et al 1980).

As noted above, although acetylcholinesterase is transported by the axonal flow in motor nerves, the endplate enzyme, particularly A_{12}, is probably of predominantly muscular origin. This is particularly obvious in ectopic reinnervation experiments. Guth & Zalewski (1963) observed that ectopic reinnervation restored the acetylcholinesterase activity at the original endplates of a previously denervated rat muscle, i.e. at a site distinct from the new nerve muscle contact. Weinberg & Hall (1979) later demonstrated that this induced muscular synthesis includes the A_{12} form (together with a small proportion of A_8). It should be emphasized that these collagen-tailed forms did not appear in muscle regions that contained neither old endplates nor newly formed ectopic endplates.

POSSIBLE ROLE OF THE BASAL LAMINA   The ectopic reinnervation experiments not only demonstrate that the muscle can synthesize the collagen-tailed forms, they also show that, in adult rat muscle, the sites of neuromuscular contact selectively maintain the capacity of accumulating these forms, either by allowing their local production or by stabilizing them, when the muscle is reinnervated at another site. Weinberg & Hall (1979) suggested that modifications of the extracellular basal lamina constitute this permanent trace left by the nerve on the muscle surface. The role of the extracellular basal lamina as an organizing structure during muscle regeneration has long been recognized (Vracko & Benditt 1972); however, is now clear that it does not simply behave as a passive "scaffold." In particular, the basal lamina of the endplate regions has distinctive staining characteristics (Zacks et al 1973), possesses specific antigenic components (Sanes & Hall 1979), and can induce specific pre- and postsynaptic differentiation in regenerating frog muscle (Marshall et al 1977, Sanes et al 1978, Burden et al 1979).

IS ACETYLCHOLINE ITSELF A REGULATORY FACTOR? INFLUENCE OF THE ACTIVITY OF ACETYLCHOLINESTERASE ON ITS OWN SYNTHESIS   The synthesis of acetylcholinesterase may be increased by acetylcholine or by acetyl-β-methylcholine in cultures of chicken embryonic muscle (Goodwin & Sizer 1965, Oh & Johnson 1972, Wilson et al 1973b) and spinal cord (Turbow & Burkhalter 1968). These effects are blocked by curare, which suggests that acetylcholine receptors are involved (Walker & Wilson 1976a, 1978). It is clear that they do not reproduce the influence exerted by the nerve, which decreases the total activity of acetylcholinesterase in chicken muscles. It has not been analyzed whether A and G forms are differently affected in these experiments.

The effect of acetylcholine on the level of acetylcholinesterase may be related to the fact that the activity of the enzyme appears to control its own synthesis. Wilson and collaborators found, for example, that after exposure of cultures of embryonic chicken muscle to irreversible inhibitors (organophosphates such as DFP or paroaxon), the initial rate of acetylcholinesterase recovery and its release into the medium were faster if the inhibition was more complete (Walker & Wilson 1976a, Cisson & Wilson 1977). This effect appeared somewhat selective, although the global protein synthesis of the cells was stimulated by the organophosphates. A similar phenomenon was noted by Gisiger & Vigny (1977) in the rat superior cervical ganglion. Brzin et al (1980) observed that exposure to organophosphates stimulates the synthesis of acetylcholinesterase in the sarcoplasmic reticulum of the muscle cell and also in the Schwann cell that is adjacent to the neuromuscular junction.

If the level of acetylcholinesterase activity exerts its influence by controlling the level of acetylcholine, as suggested in connection with cholinesterase, it must be assumed that acetylcholine is present in muscle cultures. The possibility of acetylcholine synthesis within muscle cells has indeed been demonstrated recently (Molenaar & Polak 1980).

ROLE OF THE MUSCULAR ACTIVITY   The paralysis of muscles by tetrodotoxin induces a denervation-like decrease in acetylcholinesterase (Butler et al 1978), which illustrates the importance of muscular activity. A similar result was reported by Cangiano et al (1980) in rat soleus paralyzed by nerve compression. Conversely, the changes that occur after denervation can be partially prevented by electrical stimulation in rat (Lømo & Slater 1980) and in chicken (Linkhart & Wilson 1975b, Weidoff & Wilson 1977). In culture, electrical stimulation of chicken muscle cells has also been shown to reduce the synthesis of acetylcholinesterase (Walker & Wilson 1975, 1976b). Although the molecular forms have not been analyzed in these experiments, it is likely that the A forms are particularly affected. The synthesis of A forms at the old endplates of denervated rat muscle after reinnervation at an ectopic site illustrates that in addition to its local influence, the nerve exerts a nonlocal effect on muscle, which is probably due to the muscular activity (Weinberg & Hall 1979).

The paralysis of chicken embryos by a curare-like drug reduces both the histochemical staining of endplates for acetylcholinesterase and the level of the A_{12} form (Bourgeois et al 1978, Betz et al 1980). The biosynthesis of the A_{12} form in cultures of rat muscle cells from 18 day-old embryos appears correlated with their spontaneous contractions, since both are reversibly blocked by tetrodotoxin (Koenig & Vigny 1979, Rieger et al 1980b, Koenig et al 1980).

Kato et al (1980) observed that the level of A_{12} was not increased in chicken muscle cultures by the presence of ciliary ganglion neurons, although their presence did induce muscular contractions. It is possible that in the nerve-free cultures spontaneous activity was sufficient to promote the synthesis of the collagen-tailed forms.

Contractile activity itself is probably not necessary. This is suggested because the regulation of acetylcholinesterase is normal in cultured muscles of muscular dysgenic mice, which possess normal membrane properties but do not contract (Powell et al 1979). Ionic fluxes through the membrane probably play an essential role in this regulation: it has been shown, for example, that ouabain induces a denervation-like increase in acetylcholinesterase activity (and in acetylcholine receptor level) in chick muscle cultures (Weidoff et al 1979).

Rubin et al (1980) showed in their cultures of chicken muscle, that the synthesis of the A_{12} form was induced in the presence of motor neurons, which develop synaptic contacts with the muscle fibers. Curare was found to block this synthesis, but synthesis could be restored by addition of dibutyryl-cyclic guanosine monophosphate. This suggests that the regulatory influence of the muscle's electrical activity is exerted through the cellular level of cyclic guanosine monophosphate (cGMP); however, these experiments showed that cGMP was ineffective in the absence of nerve, which indicates that neural factors are also necessary in these conditions.

INFLUENCE OF NEURAL SUBSTANCES   Fernandez et al (1979b) and Ranish et al (1980) found that the length of the nerve stump that remains attached to the muscle after denervation determines the time of onset of decrease of the enzyme at rat endplates. This is particularly clear in the case of the A_{12} form (Fernandez et al 1979b). These experiments demonstrate the existence of an activity-independent nervous control of this endplate enzyme that involves transported substances. Previous experiments showed that the blockade of axonal flow in the motor nerve by colchicine induced a transitory denervation-like decrease of acetylcholinesterase activity at the endplates of cat muscles (Fernandez & Inestrosa 1976, Inestrosa et al 1977). Oh & Markelonis (1978) have indeed reported that a protein fraction from nerves increased the enzyme content in cultures of chicken muscle. Younkin and collaborators also found that the level of acetylcholinesterase of rat muscles in organ culture can be elevated by nerve tissue extracts or by substances released from stimulated muscle, which are transported by the axonal flow in motor nerves (Younkin et al 1978, Davey et al 1979).

Various factors, both neural and muscular, therefore participate in the complex regulation of the acetylcholinesterase collagen-tailed forms at the neuromuscular junction, and their contributions appear to depend largely on the physiological state of the muscle cells.

CONCLUSION

A few points emerge very clearly from this discussion of the polymorphism of cholinesterase and acetylcholinesterase, yet many questions remain unsolved. Both enzymes exist in a number of molecular forms, including collagen-tailed molecules as well as tailless or globular molecules, and obvious homologies exist in general between the different forms observed in all vertebrate species. We believe cholinesterase and acetylcholinesterase thus correspond to two isozymic sets of molecules. We define molecular forms by their hydrodynamic parameters, i.e. mostly by sedimentation analysis. This analysis is not sensitive to modifications such as intrasubunit cleavage, to the loss of small peptide fragments, or to differences in the carbohydrate moieties of these glycoproteins. It is clear that the molecular forms defined in this way are heterogeneous and may be further subdivided, e.g. by electrophoretic analysis. With some reservations concerning degradation processes and solubilization artifacts, the different molecular forms correspond to distinct pools of enzyme at the cellular level.

The forms have different abilities to undergo inter molecular interactions and this probably determines their localization in the cell, e.g. in plasma membranes (hydrophobic interactions) or in basal laminae (probably ionic interactions). The first case seems to correspond to globular forms, and the second to collagen-tailed forms. The globular forms also occur as secreted or intracellular soluble enzymes.

The presence of the collagen-tailed enzymes at the neuromuscular junction is precisely controlled. Their biosynthesis appears usually dependent upon the muscle's innervation, but this is not universal and they may in some cases actually increase after denervation, both in proportion and in absolute level. It is likely that these forms are important to specific tasks. It is far from certain, however, that they are the only physiologically active enzyme. In general, they cannot be considered as an exclusively endplate component.

Apart from the problems of the physiological meaning of the different forms, and of the molecular interactions that allow various types of insertions in different tissues and physiological conditions, many questions remain open. The detailed biochemical structure of the molecular forms is still unknown, notably with regard to the existence of hydrophobic regions. Also, the carbohydrate moieties have not been extensively investigated although they possibly play an important role in the determination of the fate of the molecules. Finally, the biosynthesis of the different forms and its regulation offer fascinating problems in molecular biology. In particular, the participation of elements distinct from the catalytic subunits, either as structural components or as catalysts for the assembly processes, remains to be elucidated.

ACKNOWLEDGMENTS

We thank Dr. Marc Vigny for critical discussions, Mr. Pierre Allemand for his help with experiments, and Mrs. Jacqueline Pons and Mrs. Solange Duchâtel for the preparation of the manuscript.

Literature Cited

Adamson, E. D., Ayers, S. E., Deussen, Z. A., Graham, C. F. 1975. Analysis of the forms of acetylcholinesterase from adult mouse brain. *Biochem. J.* 147: 205–14

Adamson, E. D. 1977. Acetylcholinesterase in mouse brain erythrocytes and muscle. *J. Neurochem.* 28:605–15

Allemand, P., Bon, S., Massoulié, J., Vigny, M. 1981. The quaternary structure of chicken acetylcholinesterase and butyrylcholinesterase; effect of collagenase and trypsin. *J. Neurochem.* 36:860–67

Alles, G. A., Hawes, R. C. 1940. Cholinesterases in the blood of man. *J. Biol. Chem.* 133:375–90

Anglister, L., Silman, I. 1978. Molecular structure of elongated forms of electric eel acetylcholinesterase. *J. Mol. Biol.* 125:293–311

Anglister, L., Silman, I. 1979. Limited digestion with pepsin as a tool for studying the molecular structure of elongated forms of electric eel acetylcholinesterase. *Presented at 7th Meet. Int. Soc. Neurochem. Jerusalem,* p. 196. (Abstr.)

Augustinsson, K. B. 1959a. Electrophoresis studies on blood plasma esterases. I. Mammalian plasmata. *Acta Chem. Scand.* 13:571–92

Augustinsson, K. B. 1959b. Electrophoresis studies on blood plasma esterases. II. Avian reptilian, amphibian and piscine plasmata. III. Conclusions. *Acta Chem. Scand.* 13:1081–1105

Augustinsson, K. B., Nachmansohn, D. 1949. Distinction between acetylcholinesterase and other choline estersplitting enzymes. *Science* 110:98–99

Barnard, E. A., Wieckowski, J., Chiu, T. H. 1971. Cholinergic receptor molecules and cholinesterase molecules at skeletal muscle junctions. *Nature* 234:207–9

Barnard, E. A., Chiu, T. H., Jedrzejcyzk, J., Porter, C. W., Wieckowski, J. 1973. Acetylcholine receptors and cholinesterase molecules of vertebrate skeletal muscles and their nerve junctions. In *Drug Receptors,* ed. H. P. Rang, pp. 225–40. London: Macmillan

Barat, A., Escudero, E., Gomez-Barriocanal, J., Ramirez, G. 1980a. Solubilization of 20 S acetylcholinesterase from chick retina. *Biochem. Biophys. Res. Commun.* 96:1421–26

Barat, A., Escudero, E., Gomez-Barriocanal, J., Ramirez, G. 1980b. Solubilization of 20 S acetylcholinesterase from the chick central nervous system. *Neuroscience Lett.* 20:205–10

Barnett, P., Rosenberry, T. L. 1979. Functional identity of catalytic subunits of acetylcholinesterase. *Biochem. Biophys. Acta* 567:154–60

Betz, H., Bourgeois, J. P., Changeux, J. P. 1980. Evolution of cholinergic proteins in developing slow and fast skeletal muscles in chick embryo. *J. Physiol.* 302:197–218

Betz, W., Sakmann, B. 1973. Effect of proteolytic enzymes on function and structure of frog neuromuscular junctions. *J. Physiol.* 230:673–88

Blume, A., Gilbert, F., Wilson, S., Farber, J., Rosenberg, R., Nirenberg, M. 1970. Regulation of acetylcholinesterase in neuroblastoma cells. *Proc. Natl. Acad. Sci. USA* 67:786–92

Bon, S., Massoulié, J. 1976a. An active monomeric form of *Electrophorus electricus* acetylcholinesterase. *FEBS Lett.* 67:99–103

Bon, S., Massoulié, J. 1976b. Molecular forms of *Electrophorus* acetylcholinesterase; the catalytic subunits: fragmentation, intra and inter-subunit disulfide bonds. *FEBS Lett.* 71:273–78

Bon, S., Massoulié, J. 1978. Collagenase sensitivity and aggregation properties of *Electrophorus* acetylcholinesterase. *Eur. J. Biochem.* 89:89–94

Bon, S., Massoulié, J. 1980. Collagen-tailed and hydrophobic components of acetylcholinesterase in *Torpedo marmorata* electric organ. *Proc. Natl. Acad. Sci. USA* 77:4464–68

Bon, S., Vigny, M., Massoulié, J. 1979. Asymmetric and globular forms of acetylcholinesterase in mammals and birds. *Proc. Natl. Acad. Sci. USA* 76:2546–50

Bon, S., Cartaud, J., Massoulié, J. 1978. The dependence of acetylcholinesterase aggregation at low ionic strength upon a polyanionic component. *Eur. J. Biochem.* 85:1–14

Bon, S., Huet, M., Lemonnier, M., Rieger, F., Massoulié, J. 1976. Molecular forms of *Electrophorus* acetylcholinesterase; Molecular weight and composition. *Eur. J. Biochem.* 68:523–30

Bon, S., Rieger, F., Massoulié, J. 1973. Propriétés des formes allongées de l'acétylcholinestérase en solution; rayon de Stokes, densité et masse. *Eur J. Biochem.* 35:372–79

Bourgeois, J. P., Betz, H., Changeux, J. P. 1978. Effet de la paralysie chronique de l'embryon de Poulet par le flaxédil sur le développement de la jonction neuro-musculaire. *Compt. Rend. Acad. Sci. Paris* 286:773–78

Brimijoin, S., Skau, K., Wiermaa, M. J. 1978. On the origin and fate of external acetylcholinesterase in peripheral nerve. *J. Physiol.* 285:143–58

Brimijoin, S., Wiermaa, M. J. 1978. Rapid orthograde and retrograde axonal transport of acetylcholinesterase as characterized by the stop-flow technique. *J. Physiol.* 285:129–42

Brzin, M., Pucihar, S. 1976. Iodide, thiocyanate and cyanide ions as capturing reagents in one-step copper-thiocholine methods for cytochemical localization of cholinesterase activity. *Histochemistry* 48:283–92

Brzin, M., Kiauta, T. 1979. Cholinesterases of the motor end-plate of the rat diaphragm. *Prog. Brain Res.* 49:313–22

Brzin, M., Sketelj, J., Grubic, Z., Kiauta, T. 1980. Cholinesterases of neuromuscular junction. *Neurochem. Intl.* 2:149–59

Burden, S. J., Sargent, P. B., McMahan, U. J. 1979. Acetylcholine receptors in regenerating muscle accumulate at original synaptic sites in the absence of the nerve. *J. Cell Biol.* 82:412–25

Butler, I. J., Drachman, D. B., Goldberg, A. M. 1978. The effect of disease on cholinergic enzymes. *J. Physiol.* 274:593–600

Cangiano, A., Lømo, T., Lutzemberger, L., Sveen, O. 1980. Effects of chronic nerve conduction block on formation of neuromuscular junctions and junctional AchE in the rat. *Acta Physiol. Scand.* 109:283–96

Carson, S., Bon, S., Vigny, M., Massoulié, J., Fardeau, M. 1979. Distribution of acetylcholinesterase molecular forms in neural and non-neural sections of human muscle. *FEBS Lett.* 96:348–52

Cartaud, J., Rieger, F., Bon, S., Massoulié, J. 1975. Fine structure of electric eel acetylcholinesterase. *Brain Res.* 88:127–30

Cartaud, J., Bon, S., Massoulié, J. 1978. *Electrophorus* acetylcholinesterase; biochemical and electron microscope characterization of low ionic strength aggregates. *J. Cell Biol.* 77:315–22

Carter, J. L., Brimijoin, S. 1980. Effects of acute and chronic denervation on release of acetylcholinesterase and its molecular forms in rat diaphragms. *J. Neurochem.* 36:1018–25

Chang, C. H., Blume, A. J. 1976. Heterogeneity of acetylcholinesterase in neuroblastoma. *J. Neurochem.* 27:1427–35

Changeux, J. P. 1966. Responses of acetylcholinesterase from *Torpedo marmorata* to salts and curarizing agents. *Mol. Pharmacol.* 2:369–92

Chubb, I. W., Goodman, S., Smith, A. D. 1974. Increased concentration of an isozyme of acetylcholinesterase in rabbit cerebrospinal fluid after peripheral stimulation. *J. Physiol.* 242:118–20

Chubb, I. W., Goodman, S., Smith, A. D. 1976. Is acetylcholinesterase secreted from central neurons into the cerebrospinal fluid? *Neuroscience* 1:57–62

Cisson, C. M., Wilson, B. W. 1977. Recovery of acetylcholinesterase in cultured chick embryo muscle treated with paraoxon. *Biochem. Pharmacol.* 26:1955–60

Couraud, J.-Y., Di Giamberardino, L. 1980. Axonal transport of the molecular forms of acetylcholinesterase in chick sciatic nerve. *J. Neurochem.* 35:1053–66

Couraud, J.-Y., Di Giamberardino, L., Hässig, R. 1980a. Reduction of axonal flow rates of acetylcholinesterase forms in regnerating chick sciatic nerve. In *Synaptic constituents in Health and Diocese,* ed. M. Brzin, D., Sket, H. Bachelard, p. 574. Oxford: Pergamon (Abstr.)

Couraud, J.-Y., Koenig, H. L., Di Giamberardino, L. 1980b. Acetylcholinesterase molecular forms in chick ciliary ganglion: Pre- and postsynaptic distribution derived from denervation, axotomy and double section. *J. Neurochem.* 34:1209–18

Couteaux, R. 1960. Motor endplate structure. In *Structure and Function of Muscle,* ed. G. H. Bourne, 1:337–80. New York: Academic

Couteaux, R., Taxi, J. 1952. Recherches histochimiques sur la distribution des activités cholinestérasiques au niveau de la synapse myoneurale. *Arch. Anat. Microsc. Morphol. Exp.* 41:352–92

Dale, H. H. 1914. The action of certain esters of choline and their relation to muscarine. *J. Pharmacol. Exp. Therap.* 6:147–90

Davey, B., Younkin, L. H., Younkin, S. G. 1979. Neural control of skeletal muscule cholinesterase—a study using organ cultured rat muscle. *J. Physiol.* 289:501–15

Davis, R., Koelle, G. B. 1978. Electron microscope localization of acetylcholinesterase and butyrylcholinesterase in the superior cervical ganglion of the cat. I. Normal ganglion. *J. Cell Biol.* 78:785–809

Devreotes, P. N., Gardner, J. M., Fambrough, D. M. 1977. Kinetics of biosynthesis of acetylcholine receptor and subsequent incorporation into plasma membrane of cultured chick skeletal muscle. *Cell.* 10:365–73

Di Giamberardino, L., Couraud, J.-Y. 1978. Rapid accumulation of high molecular weight acetylcholinesterase in transected sciatic nerve. *Nature* 271:170–72

Di Giamberardino, L., Couraud, J.-Y., Barnard, E. A. 1979. Normal axonal transport of acetylcholinesterase forms in peripheral nerves of dystrophic chickens. *Brain Res.* 160:196–202

Droz, B., Koenig, H., Di Giamberardino, L., Couraud, J.-Y., Chrétien, M., Souyri, F. 1979. The importance of axonal transport and endoplasmic reticulum for the function of cholinergic synapse in normal and pathological conditions. *Prog. Brain Res.* 49:23–44

Dudai, Y., Herzberg, M., Silman, I. 1973. Molecular structures of acetylcholinesterase from electric organ tissue of the electric eel. *Proc. Natl. Acad. Sci. USA* 70:2473–76

Dudai, Y., Silman, I. 1973. The effect of Ca^{2+} on interaction of acetylcholinesterase with subcellular fractions of electric organ tissue from the electric eel. *FEBS Lett.* 30:49–52

Dudai, Y., Silman, I. 1974. The molecular weight and subunit structure of acetylcholinesterase preparations from the electric organ of the electric eel. *Biochem. Biophys. Res. Commun.* 59:117–24

Emmerling, M. R., Johnson, C. D., Mosher, D. F., Lipton, B. H., Lilien, J. E. 1981. Crosslinking and binding of fibronectin with asymmetric acetylcholinesterase. *Biochemistry* 20:3242–46

Eränkö, O., Teräväinen, H. 1967. Cholinesterases and eserine resistant esterases in degenerating and regenerating motor endplates of the rat. *J. Neurochem.* 14:947–54

Fernandez, H. L., Duell, M. J. 1980. Protease inhibitors reduce effects of denervation on muscle endplate acetylcholinesterase. *J. Neurochem.* 35:1166–71

Fernandez, H. L., Inestrosa, N. C. 1976. Role of axoplasmic transport in neurotrophic regulation of muscle endplate acetylcholinesterase. *Nature* 262:55–56

Fernandez, H. L., Duell, M. J., Festoff, B. W. 1979a. Cellular distribution of 16 S acetylcholinesterase. *J. Neurochem.* 32:581–85

Fernandez, H. L., Duell, M. J., Festoff, B. W. 1979b. Neurotrophic control of 16 S acetylcholinesterase at the vertebrate neuromuscular junction. *J. Neurobiol.* 10:441–54

Fernandez, H. L., Duell, M. J., Festoff, B. W. 1980a. Bidirectional axonal transport of 16 S acetylcholinesterase in rat sciatic nerve. *J. Neurobiol.* 11:31–39

Fernandez, H. L., Patterson, M. R., Duell, M. J. 1980b. Neurotrophic control of 16 S acetylcholinesterase from mammalian skeletal muscle in organ culture. *J. Neurobiol.* 11:557–70

Fertuck, H. C., Salpeter, M. M. 1976. Quantitation of junctional and extrajunctional acetylcholinesterase receptors by electron microscope autoradiography after ^{125}Iα-bungarotoxin binding at mouse neuromuscular junctions. *J. Cell Biol.* 69:144–58

Fonnum, F., Fritzell, M., Sjöstrand, J. 1973. Transport, turnover and distribution of choline acetyltransferase and acetylcholinesterase in the vagus and hypoglossal nerves of the rabbit. *J. Neurochem.* 21:1109–20

Frenkel, E. J., Roelofsen, B., Brodbeck, U., Van Deenen, L. L. M., Ott, P. 1980. Lipid-protein interactions in human erthyrocyte-membrane acetylcholinesterase. Modulation of enzyme activity by lipids. *Eur. J. Biochem.* 109:377–82

Furmanski, P., Silverman, D. J., Lubin, M. 1971. Expression of differentiated functions in mouse neuroblastoma mediated by dibutyryl cyclic adenosine monophosphate. *Nature* 232:413–15

Gardner, J. M., Fambrough, D. M. 1979. Acetylcholine receptor degradation measured by density labeling: Effects of cholinergic ligands and evidence against recycling. *Cell* 16:661–74

Gisiger, V., Vigny, M. 1977. A specific form of acetylcholinesterase is secreted by rat sympathetic ganglia. *FEBS Lett.* 84:253–56

Gisiger, V., Vigny, M., Gautron, J., Rieger, F. 1978. Acetylcholinesterase of rat sympathetic ganglion: Molecular forms, localization and effects of denervation. *J. Neurochem.* 30:501–16

Goodwin, B. C., Sizer, I. W. 1965. Effect of spinal cord and substrate on acetylcholinesterase in chick embryonic skeletal muscle. *Dev. Biol.* 11:136–53

Grafius, M. A., Millar, D. B. 1965. Reversible aggregation of acetylcholinesterase. *Biochim. Biophys. Acta* 110:540–47

Grafius, M. A., Millar, D. B. 1967. Reversible aggregation of acetylcholinesterase. Interdependance of pH and ionic strength. *Biochemistry* 6:1034–46

Grassi, J., Vigny, M., Massoulié, J. 1981. Molecular forms of acetylcholinesterase in bovine caudate nucleus and superior cervical ganglion: solubility properties and hydrophobic character. *J. Neurochem.* In press

Greenfield, S., Chéramy, A., Leviel, V., Glowinski, J. 1980. *In vivo* release of acetylcholinesterase in cat substantia nigra and caudate nucleus. *Nature* 284: 355–57

Grossman, H., Liefländer, M. 1979. Acetylcholinesterase aus Rindererythrozyten. Reinigung und Eigenschaften des mit und ohne Triton X 100 solubilisierten Enzyms. *Z. Naturforsch. Teil C* 34: 721–25

Guillon, G., Massoulié, J. 1976. Multiplicité des formes moléculaires de l'acétylcholinestérase et acclimatation thermique chez le *Carassius auratus. Biochimie* 58:465–71

Guth, L. 1968. "Trophic" influences of nerve on muscle. *Physiol. Rev.* 48:655–87

Guth, L., Zalewski, A. A. 1963. Disposition of cholinesterase following implantations of nerve into innervated and denervated muscle. *Exp. Neurol.* 7: 316–26

Hall, E. R., Brodbeck, U. 1978. Human erythrocyte membrane-acetylcholinesterase. Incorporation into the lipid bilayer structure of liposomes. *Eur. J. Biochem.* 89:159–67

Hall, Z. W. 1973. Multiple forms of acetylcholinesterase and their distribution in endplate and non-endplate regions of rat diaphragm muscle. *J. Neurol.* 4:343–61

Hall, Z. W., Kelly, R. B. 1971. Enzymatic detachment of endplate acetylcholinesterase from muscle. *Nature New Biol.* 232:62

Heuser, J. 1980. 3 D visualization of membrane and cytoplasmic specializations at the frog neuromuscular junction. In *Ontogenesis and Functional Mechanisms of Peripheral Synapses,* ed. J. Taxi, pp. 139–55. Amsterdam: Elsevier-North Holland

Hollunger, E. G., Niklasson, B. H. 1973. The release and molecular state of mammalian brain acetylcholinesterase. *J. Neurochem.* 20:821–36

Inestrosa, N. C., Ramirez, B. U., Fernandez, H. L. 1977. Effect of denervation and of axoplasmic transport blockage on the *in vitro* release of muscle endplate acetylcholinesterase. *J. Neurochem.* 28: 941–45

Inestrosa, N. C., Mendez, B., Luco, J. V. 1979. Acetylcholinesterase like that of skeletal muscle in smooth muscle reinnervated by a motor nerve. *Nature* 280:504–6

Jedrzejczyk, J., Silman, I., Lyles, J. M., Barnard, E. A. 1981. Molecular forms of the cholinesterases inside and outside muscle endplates. *Biosci. Rep.* 1:45–51

Johnson, C. D., Smith, S. P., Russell, R. L. 1977. *Electrophorus electricus* acetylcholinesterase: Separation and selective modification by collagenase. *J. Neurochem.* 28:617–24

Kalow, W., Genest, K. 1957. A method for the detection of atypical forms of human serum cholinesterase: Determination of dibucaine numbers. *Can. J. Biochem. Physiol.* 35:339

Kása, P. 1968. Acetylcholinesterase transport in the central and peripheral nervous tissue: The role of tubules in the enzyme transport. *Nature* 218:1265–67

Kása, P., Mann, S. P., Karcsu, S., Tóth, L., Jordan, S. 1973. Transport of choline acetyltransferase and acetylcholinesterase in the rat sciatic nerve: A biochemical and electron histochemical study. *J. Neurochem.* 21:431–36

Kato, A. C., Vrachliotis, A., Fulpius, B., Dunant, Y. 1980. Molecular forms of acetylcholinesterase in chick muscle and ciliary ganglion embryonic tissues and cultured cells. *Dev. Biol.* 76:222–28

Kaufman, K., Silman, I. 1980. Interaction of electric eel acetylcholinesterase with natural and synthetic lipids. *Neurochem. Intl.* 2:205–7

Kimhi, Y., Mahler, A., Saya, D. 1980. Acetylcholinesterase in mouse neuroblastoma cells; intracellular and released enzyme. *J. Neurochem.* 34:554–59

Klinar, B., Brzin, M. 1978. Cytochemical demonstration of the postnatal development of cholinesterase in the sympathetic ganglion of the rat. *Neuroscience* 3:1129–34

Klinar, B., Brzin, M. 1980. The separation of extra and intracellular cholinesterase of the rat superior cervical ganglion by mild proteolytic treatment. A quantitative, histochemical and electron microscope cytochemical study. *Cell Mol. Biol.* 26:459–67

Klingman, G. I., Klingman, J. D., Poliszczuk, A. 1968. Acetyl- and pseudocholinesterase activities in sympathetic ganglia of rats. *J. Neurochem.* 15: 1121–30

Koelle, G. B. 1951. The elimination of enzymatic diffusion artefacts in the histochemical localization of cholinesterases and a survey of their cellular distributions. *J. Pharmacol. Exp. Ther.* 103: 153–71

Koelle, G. B., ed. 1963. *Cholinesterases*

and Anticholinesterase Agents. Berlin: Springer

Koelle, G. B., Friedenwald, J. S. 1949. A histochemical method for localizing cholinesterase activity. Proc. Soc. Exp. Biol. Med. 70:617–22

Koelle, G. B., Davis, R., Koelle, W. A. 1974. Effects of aldehyde fixation and of preganglionic denervation on acetylcholinesterase and butyrylcholinesterase of cat autonomic ganglia. J. Histochem. Cytochem. 22:244–51

Koelle, G. B., Koelle, W. A., Smyrl, E. G. 1977a. Effects of inactivation of butyrylcholinesterase on steady-state and regenerating levels of ganglionic acetylcholinesterase. J. Neurochem. 28:313–19

Koelle, G. B., Rickard, K. K., Smyrl, E. G. 1979a. Steady state and regenerating levels of acetylcholinesterase in the superior cervical ganglion of the rat following selective inactivation of propionylcholinesterase. J. Neurochem. 33:1159–64

Koelle, G. B., Rickard, K. K., Ruch, G. A. 1979b. Interrelationships between ganglionic acetylcholinesterase and nonspecific cholinesterase of the cat and rat. Proc. Natl. Acad. Sci. USA 76:6012–16

Koelle, W. A., Hossaini, K. S., Akbarzadeh, P., Koelle, G. B. 1970. Histochemical evidence and consequences of the occurrence of isoenzymes of acetylcholinesterase. J. Histochem. Cytochem. 18:812–19

Koelle, W. A., Koelle, G. B., Smyrl, E. G. 1976. Effects of persistent selective suppression of ganglionic butyrylcholinesterase on steady-state and regenerating levels of acetylcholinesterase: Implications regarding function of butyrylcholinesterase and regulation of protein synthesis. Proc. Natl. Acad. Sci. USA 73:2936–38

Koelle, W. A., Smyrl, E. G., Ruch, G. A., Siddons, V. E., Koelle, G. B. 1977b. Effects of protection of butyrylcholinesterase on regeneration of ganglionic acetylcholinesterase. J. Neurochem. 28:307–11

Koenig, J., Vigny, M. 1978a. Neural induction of the 16 S acetylcholinesterase in muscle cell cultures. Nature 271:75–77

Koenig, J., Vigny, M. 1978b. Formes moléculaires d'acétylcholinestérase dans le muscle lent et le muscle rapide du poulet. Compt. Rend. Soc. Biol. 172:1069–74

Koenig, J., Vigny, M. 1979. Influence of neurones and contractile activity on acetylcholinesterase and acetylcholine receptors in muscle cells. Prog. Brain Res. 49:484 (Abstr.)

Koenig, J., Bournaud, R., Rieger, F. 1980. Acetylcholinesterase and synaptic formation in striated muscle. See Heuser 1980, pp. 313–25

Kreutzberg, G. W., Toth, L. 1974. Dendritic secretion: A way for the neuron to communicate with the vasculature. Naturwissenschaften 61:37

Kreutzberg, G. W., Toth, L., Kaiya, H. 1975. Acetylcholinesterase as a marker for dendritic transport and dendritic secretion. Adv. Neurol. 12:269–81

Kreutzberg, G. W., Kaiya, H., Toth, L. 1979. Distribution and origin of acetylcholinesterase activity in the capillaries of the brain. Histochem. 61:111–22

Lazar, M., Vigny, M. 1980. Modulation of the distribution of acetylcholinesterase molecular forms in a murine neuroblastoma x sympathetic ganglion cell hybrid cell line. J. Neurochem. 35:1067–79

Lee, S., Camp, S., Taylor, P. 1981. Structural characterization of the asymmetric (17 S + 13 S) species of acetylcholinesterase from Torpedo: Analysis of subunit composition. J. Biol. Chem. In press

Linkhart, T. A., Wilson, B. W. 1975a. Acetylcholinesterase in singly and multiply innervated muscles of normal and dystrophic chickens. II. Effects of denervation. J. Exp. Zool. 193:191–200

Linkhart, T. A., Wilson, B. W. 1975b. Role of muscle contraction in trophic regulation of chick muscle acetylcholinesterase activity. Exp. Neurol. 48:557–68

Lockridge, O., Eckerson, H. W., La Du, B. N. 1979. Interchain disulfide bonds and subunit organisation in human serum cholinesterase. J. Biol. Chem. 254:8324–30

Loewi, O., Navratil, E. 1926. Über humorale Übertragbarkeit der Herznervenwirkung XI über den mechanismus der Vaguswirkung von Physostigmin und Ergotamin. Pflügers Arch. 214:689–96

Lømo, T., Slater, C. R. 1980. Control of junctional acetylcholinesterase by neural and muscular influences in the rat. J. Physiol 303:191–202

Lubinska, L., Niemierko, S. 1971. Velocity and intensity of bidirectional migration of acetylcholinesterase in transected nerves. Brain Res. 27:329–42

Lwebuga-Mukasa, J. S., Lappi, S., Taylor, P. 1976. Molecular forms of acetylcholinesterase from Torpedo californica: Their relationship to synaptic membranes. Biochem. 15:1425–34

Lyles, J. M., Barnard, E. A. 1980. Disappearance of the "endplate" form of acetylcholinesterase from a slow tonic muscle. *FEBS Lett.* 109:9–12

Lyles, J. M., Silman, I., Barnard, E. A. 1979. Developmental changes in levels and forms of cholinesterases in muscles of normal and dystrophic chickens. *J. Neurochem.* 33:727–38

Lyles, J. M., Barnard, E. A., Silman, I. 1980. Changes in the levels and forms of cholinesterases in the blood plasma of normal and dystrophic chickens. *J. Neurochem.* 34:978–87

Marchand, A., Chapouthier, G., Massoulié, J. 1977. Developmental aspect of acetylcholinesterase activity in chick brain. *FEBS Lett.* 78:233–36

Marnay, A. 1937. Cholinestérase dans l'organe électrique de la Torpille. *Compt. Rend. Soc. Biol.* 126:573–74

Marnay, A., Nachmansohn, D. 1937. Cholinestérase dans le muscle strié. *Compt. Rend. Soc. Biol.* 124:942–44

Marnay, A., Nachmansohn, D. 1938. Cholinesterase in voluntary muscle. *J. Physiol.* 92:37–47

Marshall, L. M., Sanes, J. R., McMahan, U. J. 1977. Reinnervation of original synaptic sites on muscle fiber basement membrane after disruption of the muscle cells. *Proc. Natl. Acad. Sci. USA.* 74:3073–77

Massoulié, J., Rieger, F. 1969. L'acétylcholinestérase des organes électriques de poissons (Torpille et Gymnote). *Eur. J. Biochem.* 11:441–55

Massoulié, J., Rieger, F., Bon, S. 1970a. Relations entre les complexes moléculaires de l'acétylcholinestérase. *Compt. Rend. Acad. Sci. Paris* 270:1837–40

Massoulié, J., Rieger, F., Tsuji, S. 1970b. Solubilisation de l'acétylcholinesterase des organes électriques de gymnote: Action de la trypsine. *Eur. J. Biochem.* 14:430–39

Massoulié, J., Rieger, F., Bon, S. 1971. Espèces acétylcholinesterasiques globulaires et allongées des organes électriques de Poissons. *Eur. J. Biochem.* 21:542–51

Massoulié, J., Bon, S., Vigny, M. 1980. The polymorphism of cholinesterase in vertebrates. *Neurochem. Int.* 2:161–84

Matthews-Bellinger, J., Salpeter, M. M. 1978. Distribution of acetylcholine receptors at frog neuromuscular junctions with a discussion of some physiological implications. *J. Physiol.* 279:197–213

Mays, C., Rosenberry, T. L. 1981. Characterization of pepsin-resistant collagen-like tail subunit fragments of 18 S and 14 S acetylcholinesterase from *Electrophorus electricus. Biochemistry* 20:2810–17

McCann, W. F. X., Rosenberry, T. L. 1977. Identification of discrete disulfide linked oligomers which distinguish 18 S from 14 S acetylcholinesterase. *Arch. Biochem. Biophys.* 183:347–52

McLaughlin, J., Bosmann, H. B. 1976. Molecular species of acetylcholinesterase in denervated rat skeletal muscle. *Exp. Neurol.* 52:263–71

McMahan, U. J., Sanes, J. R., Marshall, L. M. 1978. Cholinesterase is associated with the basal lamina at the neuromuscular junction. *Nature* 271:172–74

Mendel, B., Mundell, D. B., Rudney, H. 1943. Studies on cholinesterase. A specific test for true cholinesterase and pseudocholinesterase. *Biochem. J.* 37:473–76

Mendel, B., Rudney, H. 1943. Studies on cholinesterase. Cholinesterase and pseudocholinesterase. *Biochem. J.* 37:59–63

Millar, D. B., Christopher, J. P., Burrough, D. O. 1978. Evidence that eel acetylcholinesterase is not an integral membrane protein. *Biophys. Chem.* 9:9–14

Molenaar, P. C., Polak, R. L. 1980. Acetylcholine synthetizing enzymes in frog skeletal muscle. *J. Neurochem.* 35:1021–25

Morrod, P. J., Marshall, A. G., Clark, D. G. 1975. Structural stability and composition of acetylcholinesterase purified by affinity chromatography from fresh electroplax tissue of *Electrophorus electricus. Biochem. Biophys. Res. Commun.* 63:335–42

Nachmansohn, D., Lederer, E. 1939. Sur la biochimie de la cholinestérase. Préparation de l'enzyme, rôle de groupements SH. *Bull. Soc. Chim. Biol.* 21:797–808

Nachmansohn, D., Rothenberg, M. A. 1945. Studies on cholinesterase I on the specificity of the enzyme in nerve tissue. *J. Biol. Chem.* 158:653–66

Niday, E., Wang, C. S., Alaupovic, P. 1977. Studies on the characterization of human erythrocyte acetylcholinesterase and its interaction with antibodies. *Biochim. Biophys. Acta* 459:180–93

Niemierko, S., Lubinska, L. 1967. Two fractions of axonal acetylcholinesterase exhibiting different behavior in severed nerves. *J. Neurochem.* 14:761–67

Oh, T. H., Johnson, D. D. 1972. Effects of acetyl-β-methyl-choline on development of acetylcholinesterase and butyrylcholinesterase activities in cultured chick embryonic skeletal muscle. *Exp. Neurol.* 37:360–70

Oh, T. H., Markelonis, G. J. 1978. Neurotrophic protein regulates muscle acetylcholinesterase in culture. *Science* 200: 337–38

Ott, P., Brodbeck, U. 1978. Multiple molecular forms of acetylcholinesterase from human erythrocytes membranes; interconversion and subunit composition of forms separated by density gradient centrifugation in zonal rotor. *Eur. J. Biochem.* 88:119–25

Ott, P., Jenny, B., Brodbeck, U. 1975. Multiple molecular forms of purified human erythrocyte acetylcholinesterase. *Eur. J. Biochem.* 57:469–80

Pécot-Dechavassine, M. 1968. Evolution de l'activité des cholinestérases et de leur capacité fonctionnelle au niveau des jonctions neuromusculaires et musculotendineuses de la Grenouille après section du nerf moteur. *Arch. Int. Pharmacodyn. Thér.* 176:118–33

Powell, J. A., Friedman, B. A., Cossi, A. 1979. Tissue culture study of murine muscular dysgenesis: Role of spontaneous action potential generation in the regulation of muscle maturation. *Ann. NY Acad. Sci.* 217:550–70

Rambourg, A., Droz, B. 1980. Smooth endoplasmic reticulum and axonal transport. *J. Neurochem.* 35:16–25

Ranish, N., Ochs, S. 1972. Fast axoplasmic transport of acetylcholinesterase in mammalian nerve fibres. *J. Neurochem.* 19:2641–49

Ranish, N. A., Dettbarn, W. D., Wecker, L. 1980. Nerve stump length-dependent loss of acetylcholinesterase activity in endplate regions of rat diaphragm. *Brain Res.* 191:379–86

Rieger, F., Bon, S., Massoulié, J., Cartaud, J. 1973. Observation par microscopie électronique des formes allongées et globulaires de l'acétylcholinestérase de Gymnote *(Electrophorus electricus). Eur. J. Biochem.* 34:539–47

Rieger, F., Bon, S., Massoulié, J., Cartaud, J., Picard, B., Benda, P. 1976a. *Torpedo marmorata* acetylcholinesterase; a comparison with the *Electrophorus electricus* enzyme: Molecular forms, subunits, electron microscopy, immunological relationship. *Eur. J. Biochem.* 68:513–21

Rieger, F., Faivre-Bauman, A., Benda, P., Vigny, M. 1976b. Molecular forms of acetylcholinesterase: Their de novo synthesis in mouse neuroblastoma cells. *J. Neurochem.* 27:1059–63

Rieger, F., Chételat, R., Nicolet, M., Kamal, L., Poullet, M. 1980a. Presence of tailed, asymmetric forms of acetylcholinesterase in the central nervous system of vertebrates. *FEBS Lett.* 121:169–74

Rieger, F., Koenig, J., Vigny, M. 1980b. Spontaneous contractile activity and the presence of the 16 S form of acetylcholinesterase in rat muscle cells in culture. Reversible suppressive action of tetrodotoxin. *Dev. Biol.* 76:358–65

Rieger, F., Shelanski, M. L., Greene, L. A. 1980c. The effects of nerve growth factor on acetylcholinesterase and its multiple forms in cultures of rat PC 12 pheochromocytoma cells: Increased total specific activity and appearance of the 16 S molecular form. *Devl Biol.* 76:238–43

Rieger, F., Vigny, M. 1976. Solubilization and physiochemical characterization of rat brain acetylcholinesterase: Development and maturation of its molecular forms. *J. Neurochem.* 27:121–29

Römer-Luthi, C. R., Hajdu, J., Brodbeck, U. 1979. Molecular forms of purified human erythrocytes membrane acetylcholinesterase investigated by crosslinking with diimidates. *Hoppe Seylers Z. Physiol. Chem.* 360:929–34

Römer-Luthi, C. R., Ott, P., Brodbeck, U. 1980. Reconstitution of human erythrocyte membrane acetylcholinesterase in phospholipid vesicles. Analysis of the molecular forms of cross-linking studies. *Biochim. Biophys. Acta* 601: 123–33

Rosenberry, T. L. 1975. Acetylcholinesterase. *Adv. Enzymol.* 43:103–218

Rosenberry, T. L., Chen, Y. T., Bock, E. 1974. Structures of 11 S acetylcholinesterase, subunit composition. *Biochemistry* 13:3068–79

Rosenberry, T. L., Richardson, J. M. 1977. Structure of 18 S and 14 S acetylcholinesterase. Identification of collagen-like subunits that are linked by disulfide bonds to catalytic subunits. *Biochemistry* 16:3550–58

Rosenberry, T. L., Barnett, P., Mays, C. 1980. The collagen-like subunits of acetylcholinesterase from the eel *Electrophorus electricus. Neurochem. Intl.* 2:135–48

Rotundo, R. L., Fambrough, D. M. 1979. Molecular forms of chicken embryo acetylcholinesterase in vitro and in vivo, isolation and characterization. *J. Biol. Chem.* 254:4790–99

Rotundo, R. L., Fambrough, D. M. 1980a. Synthesis, transport and fate of acetylcholinesterase in cultured chick embryo muscle cells. *Cell* 22:583–94

Rotundo, R. L., Fambrough, D. M. 1980b. Secretion of acetylcholinesterase: Relation to acetylcholine receptor metabolism. *Cell* 22:595–602

Rubin, L. L., Schuetze, S. M., Wiell, C. L., Fischbach, G. D. 1980. Regulation of acetylcholinesterase appearance of neuromuscular junctions in vitro. *Nature* 283:264–67

Salpeter, M. M., Plattner, H., Rogers, A. W. 1972. Quantitative assay of esterases in endplates of mouse diaphragm by electron microscope autoradiography. *J. Histochem. Cytochem.* 20:1059–68

Sanes, J. R., Marshall, L. M., McMahan, U. J. 1978. Reinnervation of muscle fiber basal lamina after removal of myofibers. Differentiation of regenerating axons at original synaptic sites. *J. Cell Biol.* 78:176–98

Sanes, J. R., Hall, Z. W. 1979. Antibodies that bind specifically to synaptic sites on muscle fiber basal lamina. *J. Cell Biol.* 83:357–70

Scarsella, G., Toschi, G., Chiappinelli, V. A., Giacobini, E. 1978. Molecular forms of acetylcholinesterase in the ciliary ganglion and iris of the chick. Developmental changes and effects of axotomy. *Dev. Neurosci.* 1:133–41

Schneider, F. H. 1976. Effects of sodium butyrate on mouse neuroblastoma cells in culture. *Biochem. Pharmacol.* 25:2309–17

Silman, I., Dudai, Y. 1975. Molecular structure and catalytic activity of membrane-bound acetylcholinesterase from electric organ tissue of electric eel. *Croat. Chim. Acta* 47:181–200

Silman, I., Lyles, J. M., Barnard, E. 1978. Intrinsic forms of acetylcholinesterase in skeletal muscle. *FEBS Lett.* 94:166–70

Silman, I., Di Giamberardino, L., Lyles, J., Couraud, J.-Y., Barnard, E. A. 1979. Parallel regulation of acetylcholinesterase and pseudocholinesterase in normal denervated and dystrophic chicken skeletal muscle. *Nature* 280:160–62

Silver, A. 1974. *The Biology of Cholinesterases.* Amsterdam: North Holland

Skau, K. A., Brimijoin, S. 1978. Release of acetylcholinesterase from rat hemidiaphragm preparations stimulated through the phrenic nerve. *Nature* 275:224–26

Skau, K. A., Brimijoin, S. 1980. Multiple molecular forms of acetylcholinesterase in rat vagus nerve, smooth muscle and heart. *J. Neurochem.* 35:1151–54

Sketelj, J., Brzin, M. 1977. Increase in the apparent AchE activity in the mouse diaphragm induced by proteolytic treatment. *J. Neurochem.* 29:109–14

Sketelj, J., Brzin, M. 1979. Attachment of acetylcholinesterase to structures of the motor endplate. *Histochemistry* 61:239–48

Sketelj, J., Brzin, M. 1980. 16 S acetylcholinesterase in endplate-free regions of developing rat diaphragm. *Neurochem. Res.* 5:655–50

Sketelj, J., McNamee, M. G., Wilson, B. W. 1978. Effect of denervation on the molecular forms of acetylcholinesterase in normal and dystrophic chicken muscles. *Exp. Neurol.* 60:624–29

Sketelj, J., Grubic, Z., Klinar, B., Brzin, M. 1980. Regeneration of acetylcholinesterase in the diaphragm, brain and plasma of the rat after irreversible inhibition by soman. Cytochemical study and molecular forms in the endplate region of the diaphragm. See Couraud et al 1980a, p. 586 (Abstr.)

Smilowitz, H. 1980. Monovalent ionophores inhibit acetylcholinesterase release from cultured chick embryo skeletal muscle cells. *Mol. Pharmacol.* 16:202–14

Somogyi, P., Chubb, I. W. 1976. The recovery of acetylcholinesterase activity in the superior cervical ganglion of the rat following the inhibition by diisopropylphosphofluoridate: By a chemical and cytochemical study. *Neuroscience* 1:413–21

Stedman, Edgar, Stedman, Ellen. 1935. The relative choline esterase activities of serum and corpuscles from the blood of certain species. *Biochem. J.* 29:2107–11

Stedman, Edgar, Stedman, Ellen, Easson, L. H. 1932. Choline-esterase. An enzyme present in the blood serum of the horse. *Biochem. J.* 26:2056–66

Sugiyama, H. 1977. Multiple forms of acetylcholinesterase in clonal muscle cells. *FEBS Lett.* 84:257–50

Taylor, P. B., Rieger, F., Greene, L. A. 1980. Development of the multiple molecular forms of acetylcholinesterase in chick paravertebral sympathetic ganglia: An in vitro and in vivo study. *Brain Res.* 182:383–96

Taylor, P. B., Rieger, F., Shelanski, M. L., Greene, L. A. 1981. Cellular localization of the multiple molecular forms of acetylcholinesterase in cultured neuronal cells. *J. Biol. Chem.* 256:3827–30

Tennyson, V. M., Kremzner, L. T., Brzin, M. 1977. Electron microscopic-cytochemical and histochemical studies of acetylcholinesterase in denervated muscle of rabbits. *J. Neuropathol. Exp. Neurol.* 36:245–75

Toth, L., Kreutzberg, G. W. 1979. Acetyl-cholinesterase in neurons of the rat cerebral cortex. *Acta Anat.* 103:125–29

Tsuji, S., Rieger, F., Peltre, G., Massoulié, J., Benda, P. 1972. Acetylcholinesterase of muscle, spinal chord and brain of the electric eel *Electrophorus electricus. J. Neurochem.* 19:989–97

Tuček, S. 1975. Transport of choline acetyl-transferase and acetylcholinesterase in the central stump and isolated segment of a peripheral nerve. *Brain Res.* 86:259–70

Turbow, M. M., Burkhalter, A. 1968. Acetyl-cholinesterase activity in the chick em-bryo spinal cords. *Dev. Biol.* 17:233–44

Vigny, M., Di Giamberardino, L., Couraud, J.-Y., Rieger, F., Koenig, J. 1976a. Molecular forms of chicken acetyl-cholinesterase: Effect of denervation. *FEBS Lett.* 69:277–80

Vigny, M., Koenig, J., Rieger, F. 1976b. The motor endplate specific form of acetyl-cholinesterase: Appearance during em-bryogenesis and reinnvervation of rat muscle. *J. Neurochem.* 27:1347–53

Vigny, M., Bon, S., Massoulié, J., Leterrier, F. 1978a. Active site catalytic efficiency of acetylcholinesterase molecular forms in *Electrophorus, Torpedo,* rat and chicken. *Eur. J. Biochem.* 85:317–23

Vigny, M., Gisiger, V., Massoulié, J. 1978b. "Nonspecific" cholinesterase and ace-tylcholinesterase in rat tissues: Molecu-lar forms, structural and catalytic prop-erties, and significance of the two en-zyme systems. *Proc. Natl. Acad. Sci. USA* 75:2588–92

Vigny, M., Bon, S., Massoulié, J., Gisiger, V. 1979. The subunit structure of mam-malian acetylcholinesterase: Catalytic subunits, dissociating effect of proteol-ysis and disulfide reduction of the polymeric forms. *J. Neurochem.* 33: 559–65

Villafruela, M. J., Barat, A., Villa, S., Ramirez, G. 1980. Molecular forms of acetylcholinesterase in the chick visual system: Collagenase-released 21.5 S and 16.5 S species. *FEBS Lett.* 110:91–95

Villafruela, M. J., Barat, A., Manrique, F., Villa, S., Ramirez, G. 1981. Molecular forms of acetylcholinesterase in the de-veloping chick visual system. *Dev. Neu-rosci.* 4:25–36

Vimard, C., Jeantet, C., Netter, Y., Gros, F. 1976. Changes in the sedimentation properties of acetylcholinesterase dur-ing neuroblastoma differentiation. *Bio-chimie* 58:473–78

Viratelle, O. M., Bernhard, S. A. 1980. Major component of acetylcholinesterase in

Torpedo electroplax is not basal lamina associated. *Biochemistry* 19:4999–5007

Vracko, R., Benditt, E. P. 1972. Basal lamina: The scaffold for orderly cell re-placement. Observations on regenera-tion of injured skeletal muscle fibres and capillaries. *J. Cell Biol.* 55:406–19

Vrbová, G., Gordon, T., Jones, R. 1978. *Nerve Muscle Interaction.* London: Wiley

Walker, C. R., Wilson, B. W. 1975. Control of acetylcholinesterase by contractile activity of cultured muscle cells. *Nature* 256:215–16

Walker, C. R., Wilson, B. W. 1976a. Regula-tion of acetylcholinesterase in chick muscle cultures after treatment with diisopropylphosphorofluoridate: Ribo-nucleic acid and protein synthesis. *Neuroscience* 1:509–13

Walker, C. R., Wilson, B. W. 1976b. Regula-tion of acetylcholinesterase in cultured muscle by chemical agents and elec-trical stimulation. *Neuroscience* 1: 191–96

Walker, K. B., Wilson, B. W. 1978. Regula-tion of acetylcholinesterase in cultured chick embryo spinal chord neurons. *FEBS Lett.* 93:81–85

Watkins, M. S., Hitt, A. S., Bulger, J. E. 1977. the binding of 18 S acetyl-cholinesterase to sphingomyelin and the role of the collagen-like tail. *Biochem. Biophys. Res. Commun.* 79:640–47

Webb, G. 1978. Acetylcholinesterase: Char-acterization of native and proteolyti-cally derived forms and identification of structural protein components. *Can. J. Biochem.* 56:1124–32

Webb, G., Clark, D. G. 1978. Acetylcholines-terase: Differential affinity chromato-graphic purification of 11 S and 18 S plus 14 S forms; the importance of mul-tiple site interactions and salt concen-tration. *Arch. Bioch. Biophys.* 191: 278–88

Weidoff, P. M. Jr., Wilson, B. W. 1977. Influ-ence of muscle activity in trophic regu-lation of acetylcholinesterase activity in dystrophic chicken. *Exp. Neurol.* 57: 1–12

Weidoff, P. M., McNamee, M. G., Wilson, B. W. 1979. Modulation of cholinergic proteins and RNA by ouabain in chick muscle cultures. *FEBS Lett.* 100: 389–93

Weinberg, C. G., Hall, Z. W. 1979. Junc-tional form of acetylcholinesterase re-stored at nerve-free endplate. *Dev. Biol.* 68:631–35

Wiedmer, T., Di Francesco, C., Brodbeck, U. 1979. Effects of amphiphiles on structure and activity of human erythrocyte membrane acetylcholinesterase. *Eur. J. Biochem.* 102:59–64

Wilson, B. W., Montgomery, M. A., Asmundson, R. V. 1968. Cholinesterase activity and inherited muscular dystrophy of the chicken. *Proc. Soc. Exp. Biol. Med.* 129:199–206

Wilson, B. W., Linkhart, S. G., Nieberg, P. A. 1973a. Acetylcholinesterase in singly and multiply innervated muscles of normal and dystrophic chickens. *J. Exp. Zool* 186:187–92

Wilson, B. W., Nieberg, P. S., Walker, C. R., Linkhart, T. A., Fry, D. M. 1973b. Production and release of acetylcholinesterase by cultured chick embryo muscle. *Dev. Biol.* 33:285–99

Wilson, B. W., Walker, C. R. 1974. Regulation of newly synthetized acetyl-cholinesterase in muscle cultures treated with diisopropylfluorophosphate. *Proc. Natl. Acad. Sci. USA* 71:3194–98

Witzemann, V. 1980. Characterisation of multiple forms of acetylcholinesterase in electric organ of *Torpedo marmorata. Neurosci. Lett.* 20:277–82

Younkin, S. G., Brett, R. S., Davey, B., Younkin, L. H. 1978. Substances moved by axonal transport and released by nerve stimulation have an innervation-like effect on muscle. *Science* 200:1292–95

Zacks, S. I., Saito, A., Sheff, M. F. 1973. Cytochemical properties of the external lamina of myofibers and neuromuscular junctions. In *The Striated Muscle.* C. M. Pearson, F. K. Mostofi, pp. 123–43. Baltimore: William & Wilkins

Reference added in proof:

Grubić, Z., Sketelj, J., Klinar, B., Brzin, M. 1981. Recovery of acetylcholinesterase in the diaphragm, brain and plasma of the rat after irreversible inhibition by Soman: A study of cytochemical localization and molecular forms of the enzyme in the motor endplate. *J. Neurochem.* 37:909–16

Ann. Rev. Neurosci. 1982. 5:107–20

INTRACELLULAR PERFUSION

P. G. Kostyuk

Department of General Physiology of the Nervous System, A. A. Bogomoletz
Institute of Physiology, Kiev, Ukrainian SSR, 252601 GSP, USSR

INTRODUCTION

Intracellular perfusion is of great importance for effective elucidation of membrane mechanisms underlying cell excitability, since it allows the control and change at will of the physico-chemical conditions near the internal surface of the cell membrane. Until quite recently, application of this method was confined to squid giant axons, which are very suitable for mechanical extrusion of the axoplasm and for its replacement by an artificial saline solution of desirable composition (Oikawa et al 1961, Baker et al 1962). Fundamental data dealing with the nature of excitability obtained from giant squid axons are well known. The search for methods of controlled substitution of intracellular medium applicable to a greater variety of cells, especially to nerve cells, has led to substantial discoveries about the mechanisms underlying the basic physiological processes occurring in cells.

PRINCIPLES OF INTRACELLULAR PERFUSION

A method for controlled substitution of the ionic composition of the nerve cell cytoplasm was first suggested by Krishtal & Pidoplichko (1975). Its main idea is to use isolated cells and to make a hole several tens of micrometers diameter in their surface membrane. The isolated cell is placed in saline solution corresponding to the extracellular medium, whereas the hole in the membrane is brought in contact with the solution representing the desirable intracellular ionic medium. The replacement of intracellular ions through the hole by ions of this solution can be completed in several tens of seconds. As a matter of fact, this process can be regarded as intracellular dialysis.

To perform this method successfully the following measures should be undertaken: (*a*) a reliable separation of the working part of the cell mem-

107

brane facing the external solution from the system supplying the cell with internal solution; (*b*) a reliable prevention of the resealing of the pore in the membrane. There are various technical ways to meet these demands. The most convenient is the fixation of the investigated cell in a conical pore made in a plastic film separating two perfusion chambers (Figure 1). Insertion of the cell into the pore is facilitated by a suction effect due to a hydrostatic pressure difference in upper and lower chambers; rupture of the cell membrane in the smaller opening of the pore can be easily performed by a negative hydrostatic pressure pulse in the lower chamber. The absence of fixed charges on the plastic film ensures a good adhesion of the external surface of the cell membrane with the wall of the pore; in addition, the wall of the pore is covered with a neutral glue (Kostyuk et al 1975, Kostyuk & Krishtal 1977a). Other authors suggested the use of a glass micropipette as a pore for the fixation of the isolated cell. In this case the cell can be located either outside the pipette, in which it is partly sucked (Lee et al 1978), or inside the pipette; here a part of the membrane facing the pipette opening is ruptured (Takahashi & Yoshii 1978). Since surface charges of the glass prevent a reliable fixation of the cell in the glass micropipette, the glass is first neutralized by chemical pretreatment (for example, with protamine).

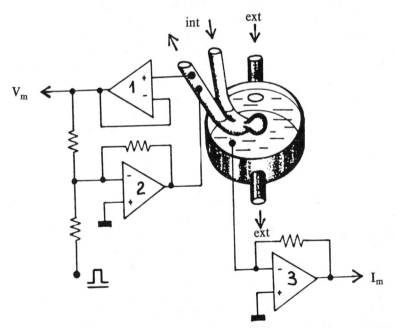

Figure 1   Experimental set-up for cell dialysis. 1: input amplifier; 2: voltage-clamp amplifier; 3: current amplifier; ext, int-extra- and intracellular solutions.

A further development of the method was a transition from intracellular dialysis to perfusion of the isolated cell. For this purpose two holes are made on opposite sides of the cell membrane, through which the cell interior is connected to two perfusing systems (Figure 2). The difference in hydrostatic pressures in these systems ensures a controlled flow of the perfusing solution through the cell (perfusion proper); in this case the replacement of intracellular solutions is substantially accelerated (Krishtal 1978).

A fundamental advantage created by the usage of intracellular dialysis or perfusion is the creation of ideal electrical conditions for voltage clamping and measuring transmembrane ionic currents under conditions in which the ionic gradients of the inside and the outside of the cell can be controlled. There is no need to use intracellular microelectrodes: intracellular perfusion allows a practically complete separation of the circuits for passing the feedback current and recording the membrane potential, which reduces the series resistance to a minimum and makes the accuracy of measurements much more precise.

Whether structural changes occur inside the cell during intracellular dialysis or perfusion has been extensively investigated. Nerve cells were fixed during the dialysis and were found to have their cell cytoplasm and intracellular organelles preserved (as contrasted to the perfused axons, where following the mechanical extrusion of the axoplasm, only about 4% of the latter remain attached to the inner surface of the membrane). The only evidence for structural changes in perfused cells is a substantial vacuolization of the cytoplasmic matrix (probably, due to an abrupt distension of

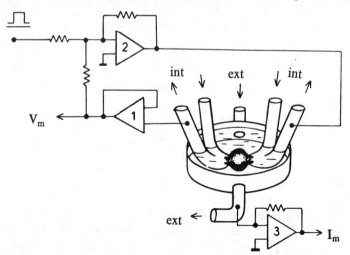

Figure 2 Experimental set-up for cell perfusion. 1: input amplifier; 2: voltage-clamp amplifier; 3: current amplifier; ext, int- extra- and intracellular solutions.

the endoplasmic reticulum, as well as of mitochondria). This vacuolization seems to be the main factor contributing to the high velocity of substitution of the intracellular ions by ions from the perfusing solution (Pogorelaya et al 1980).

The intracellular perfusion technique has yielded its most important results in the study of the mechanisms of electrical excitability of the somatic membrane of nerve cells. This technique was originally used for the analysis of ionic currents in the somatic membrane of giant mollusc neurons, which, due to their large dimensions, were the most suitable objects for experimentation. The method was subsequently applied successfully to other types of neurons.

PERFUSION IN INVERTEBRATE NEURONS

The replacement of intracellular K^+ ions during intracellular perfusion can be performed by $Tris^+$ ions. In this case the fast and delayed potassium currents (in isolated mollusc neurons) are completely eliminated and it becomes possible to record undistorted calcium currents, which are practically absent in the axonal membrane (Kostyuk et al 1975, Kostyuk & Krishtal 1977a,b, Lee et al 1978). These currents have been detected in all investigated neurons, and detailed characteristics for the corresponding specific electrically operated ionic channels have been obtained. It was shown that these channels pass both Ca^{2+} ions and other divalent cations in the following order: Ba>Sr>Ca>Mg; the corresponding relative permeabilities are: 2.8:2.6:1.0:0.2 (Doroshenko et al 1978b). The current-voltage characteristics of calcium current are nonlinear to a considerable extent: they reach exponentially a zero value at high positive testing shifts of the membrane potential (Doroshenko et al 1978a,b). Upon increasing the concentration of carrier ions in the external solution, ionic currents through calcium channels of the somatic membrane show a clear saturation that indicates the presence of a binding site at the outer mouth of each channel. The dissociation constants for such binding of divalent cations estimated for *Helix* cell membranes are correspondingly $K_{Ca} = 5.4$ mM, $K_{Sr} = 10$ mM and $K_{Ba} = 15$ mM (Akaike et al 1978a,b, Valeyev 1979). Due to the presence of this binding site, cations having a higher capability of binding exert a blocking effect on inward currents carried by less effectively binding cations, which confirms the competitive nature of such interaction. The dissociation constants estimated for the ion-channel complexes are as follows: $K_{Ni} = K_{Co} = 0.74$ mM and $K_{Mn} = 0.36$ mM (Ponomaryov et al 1980). Cd^{2+} ($K_{Cd} = 0.07$ mM) (Krishtal 1976) turned out to be the most effective blocker of calcium channels.

The intracellular perfusion technique allowed controlled changes of ionized calcium content inside the cell (by introduction of solutions containing Ca-EGTA buffer). Increasing the internal free calcium concentration to 5×10^{-8} M causes a decrease of the calcium inward current by 50%; a further increase of the intracellular calcium produces a complete block of this current (Kostyuk & Krishtal 1977b, Doroshenko & Tsyndrenko 1978). Due to this peculiarity of calcium channels it is impossible to reverse the calcium current by changing the electrochemical gradient for the penetrating ions; these channels prove to be perfectly rectifying.

The presence of effective binding of penetrating ions by at least two functional groups (at the external and internal mouth of the channel) can be used for modeling the energy profile in these channels, using Eyring's theory of absolute reaction rates. The ion penetration through the channel can be described as a series of successive binding of the ion with functional groups (potential wells) separated by energy barriers. On the basis of such an approach, using the diagram technique, models of the energy profile of the calcium channel have been proposed (Naruševičius & Rapoport 1979, Kostyuk et al 1980b). These models describe within a wide range the concentration and potential dependence of calcium currents in the nerve cell membrane.

The calcium current has been shown to depend specifically on the anionic content of the intracellular medium; introduction of fluoride anions into the cell produced an irreversible block of calcium current (Kostyuk et al 1975, Kostyuk & Krishtal 1977b).

The analysis of kinetic characteristics of the calcium currents has been successfully performed on perfused mollusc neurons. The activation kinetics of the calcium currents can be satisfactorily described by the Hodgkin-Huxley model by use of the square power of the Hodgkin-Huxley m-variable (Kostyuk & Krishtal 1977a, Kostyuk et al 1979b). The inactivation kinetics of calcium current are more complex and can be described as a sum of at least two exponential processes—the fast and the slow ones. The fast decline of the calcium current may be an artifact due to activation of a nonspecific outward current that is preserved after the replacement of intracellular K^+ ions and carried by $Tris^+$ ions. The slow component probably reflects a true inactivation of the calcium current (Doroshenko et al 1978b). Since the penetration of Ca^{2+} ions into the cell may exert an autoblocking effect on calcium channels, it can result in a decrease of the calcium current; in this case the degree of inactivation should depend not on the membrane potential but on the concentration of Ca^{2+} ions penetrated into the cell (Tillotson 1979). At the same time, the presence of potential dependent inactivation of calcium channels can be observed under stationary conditions during prolonged displacements of the holding potential. The function $h_\infty(V)$, like the similar function for sodium channels, has an S-shaped form; however, it is displaced to more positive potentials (Kostyuk & Krishtal 1977a).

The elucidation of the membrane mechanisms underlying the kinetic characteristics of electrically operated sodium channels has made substantial progress since the discovery of the intramembrane charge displacement that accompanies the activation of the ionic conductance of the membrane (the so-called gating currents). Successful separation and analysis of gating currents related to the activation of sodium conductance led to hopes for similar success in respect to the calcium conductance. In fact, Adams & Gage (1976) by analyzing charge displacements in the membrane of *Aplysia* neurons found a component that could correspond to the activation of calcium channels because of its slow time course. Kostyuk, Krishtal & Pidoplichko (1977) used the intracellular perfusion of the cell with fluoride-containing solutions for a reliable separation into two components of asymmetric displacement current in the somatic membrane; i.e. the fast and slow currents, the latter of which is irreversibly abolished by introduction of fluoride anions into the cell. It was shown that the fluoride-sensitive component of this current corresponds well in its potential and kinetic characteristics to the expected gating current of calcium channels (Kostyuk et al 1979a, 1981). If one assumes that such a charge displacement represents the Boltzmann distribution of charged particles between two steady-state positions, the effective valency for displaced charge can be determined; it equals 3. The data obtained are in good conformity with the model, according to which two gating particles should be simultaneously displaced in order to open the calcium channel.

A stationary decrease of the holding potential produces a decrease of the gating current, which occurs in parallel with the inactivation of calcium conductance, thus confirming the idea about the "immobilization" of gating particles as the base for inactivation of ionic channels.

In summary, the intracellular perfusion method established that the gating mechanism of calcium channels is in principle similar to the gating mechanism of sodium channels and differs only in some quantitative characteristics.

Finally, the application of the intracellular perfusion technique made it possible to elaborate a method for an effective analysis of the fluctuations of ionic currents in the somatic membrane—fluctuations reflecting the discrete nature of molecular structures that create such currents. The fluctuations of transmembrane current were recorded from a micropatch of the membrane (an area of about 20 μm^2) that was sucked into a micropore of the plastic membrane (Kostyuk et al 1978, 1980a, Krishtal et al 1981). In this way it became possible to record the fluctuations of the order of 10^{-11} A suitably for statistical analysis. It was also possible to measure the ionic current fluctuations at a large part of the perfused cell membrane (Akaike et al 1978a,b); however, the analysis of these fluctuations was less extensive.

Statistical analysis of such fluctuations has shown that they can be regarded as a result of the opening and closing of unitary ionic channels that have a constant conductance. The unitary conductance of the calcium channel was found to be about 0.5 pS, i.e. one order less than the conductance of the unitary sodium channel. Judging by this value, the single calcium channel can pass at most 3×10^5 Ca^{2+} ions (or 6×10^5 Ba^{2+} ions) per second.

The spectrum density of the membrane noise produced by calcium channels can be approximated by the Lorentz function characteristic of the processes with a single relaxation time constant. The mean value of the relaxation time constant for the calcium channel of the membrane of *Helix* neuron, determined by cut-off frequency, was found to be equal to 0.7 ± 0.2 msec at 22°C (Kostyuk et al 1980a, Krishtal et al 1981). The relaxation time of the calcium channel determined from the spectral characteristics of the current fluctuations was only slightly dependent on the testing potential. This finding is not in conformity with the predictions of the model that the displacement of the gating particle directly transfers the channel from a nonconducting state to a conducting one. Such a discrepancy obviously indicates the existence of intermediate stages in the course of activation of calcium conductance. The displacement of charged gating particles is a necessary but insufficient condition in this process; it only creates conditions for the realization of the next stage—the transition from open to closed state, which may be independent of potential.

The investigations of calcium currents described above demonstrate that in many nerve cells the density of calcium channels and the corresponding calcium inward current are sufficient to produce a regenerative response in the form of a propagating action potential. At the same time, in the majority of investigated membranes the inward current produced by the sodium conductance is strong enough to generate an active response, and the calcium conductance causes only a slight modification of this active response. This circumstance indicates a predominant role of the calcium currents in the maintenance of some other cellular functions; one of these may be a recurrent influence of Ca^{2+} ions on potassium conductance of the cell membrane (Meech 1974; and others). Such an influence has been observed in many excitable membranes; it is thought to be functionally important. However, a direct mechanism of the action of intracellular Ca^{2+} ions on potassium conductance remained unclear, because of the lack of precise control of internal Ca^{2+} content and the interaction of Ca^{2+} with different intracellular structures. With the intracellular perfusion technique, a controlled change in the level of free Ca^{2+} ions inside the cell was possible. Using this method, it was found that an increase of free calcium to 10^{-7} M was accompanied by a clear potentiation of outward currents in the membrane of nerve cells (Doroshenko et al 1979). Only those potassium channels that were resistant to the blocking action of tetraethylammonium (TEA) and which had a lowered selectivity and special kinetic features (the absence of a noticeable time-dependent inactivation) were affected. The channels producing the TEA-sensitive fast and delayed potassium currents were not affected by intracellular Ca^{2+} ions; moreover, their increase had a depressive effect on the corresponding components of potassium conductance (Kostyuk & Krishtal 1977b). The effect of Ca^{2+} ions was highly specific and could not be reproduced by Sr^{2+} ions in much higher concentrations. The study of the potential dependence of outward currents induced by increased internal concentration of free Ca^{2+} ions demonstrated that the Ca^{2+} dependent K^+ channel is not a special kind of chemically operated channel, but an electrically operated channel that apparently is transferred from a nonactive state into a state ready for activation in the presence of Ca^{2+}.

PERFUSION IN VERTEBRATE NEURONS

Experiments made on intracellularly dialyzed or perfused nerve cells of vertebrates (dorsal root ganglion neurons of frog or rat) originally failed to detect the presence of a specific calcium inward current (Veselovsky et al 1977a,b, 1979). Sometimes this current could be observed at the beginning of cell perfusion, but then it decreased rapidly, while the sodium and potassium currents remained stable. It was suggested that this effect was due to washing out from the cell of some cytoplasmic factor (or factors) necessary for normal functioning of calcium channels. It should be noted that in mollusc neurons the calcium conductance decreases much more rapidly in the course of cell perfusion than potassium or sodium conductance; probably, in this case a loss of the corresponding factor from the cell also takes place (Kostyuk et al 1981). The most probable cytoplasmic process related to the maintenance of ionic conductance of the surface membrane is the phosphorylation of corresponding proteins by a cAMP dependent protein kinase (Nathanson 1977, Greengard 1979). An attempt was made therefore to prevent the distortion of calcium conductance by the introduction into the cell of substances capable of restoring the activity of this system (cAMP, ATP, Mg^{2+} ions). Indeed, this procedure not only prevented the decline of calcium inward current during perfusion, but it restored to a considerable extent the diminished calcium conductance (Veselovsky & Fedulova 1980). This restoration is especially prominent in neurons of newborn rats; in adult animals it is less pronounced.

The use of this technique allowed the separation and detailed investigation of specific calcium currents in the somatic membrane of rat dorsal root ganglion neurons. These currents were proportional to the extracellular concentration of Ca^{2+} ions at concentrations between 2 and 14 mM; at higher concentrations a saturation effect was observed, thus indicating the presence of a binding site at the external mouth of the calcium channels. The replacement of Ca^{2+} by Ba^{2+} ions caused about a two-fold increase in the amplitude of the inward current. Introduction of Co^{2+}, Mn^{2+} ions, verapamil, and its derivatives into the extracellular medium produced a competitive block of the calcium current. Activation kinetics of this current could be approximated, like that in mollusc neurons, by a modified Hodgkin-Huxley equation using the square power of the m-variable. The calcium current inactivation develops very slowly; a half-value of steady state inactivation can be achieved at a holding potential of about –60 mV. It should be noted that the currents described here have been recorded in all cells investigated. The conclusion (Matsuda et al 1978) that calcium currents exist only in a small percentage ($\sim$3%) of dorsal root ganglion neurons, based on the recordings of action potentials in sodium-free solutions seems to be related to the low sensitivity of this indirect method for detection of calcium currents.

An interesting property of some dorsal root ganglion neurons in rats is the presence, together with the usual TTX-sensitive inward sodium current, of a special TTX-resistant current that differs from the TTX-sensitive current in its potential-dependent and kinetic characteristics. The presence of this current was suggested from the investigation of the peculiarities of action potentials in nondialyzed neurons (Yoshida et al 1978). The use of intracellular dialysis made it possible to separate and to study this current in detail. The activation and inactivation time constants of the slow TTX-resistant sodium inward current were about one order higher than those for TTX-sensitive current; in addition, it was highly sensitive to all extracellular agents known to block the calcium channels in other excitable membranes (Co^{2+}, Mn^{2+}, Cd^{2+} ions; D-600, and its derivatives). However, the selectivity of the TTX-resistant sodium channels does not differ much from the selectivity of the TTX-sensitive channels. Thus, these ion-conducting structures combine the properties of sodium (selectivity filter) and calcium (gating mechanism, pharmacological sensitivity) channels (Veselovsky et al 1980).

The possibility of artificial transformation of calcium ionic channels into sodium channels that preserve the pharmacological properties of calcium channels has been the subject of several experiments with intracellular perfusion (Kostyuk & Krishtal 1977b, Krishtal 1978). The introduction of Ca-chelating agents (EGTA or EDTA) into the external medium, which is deprived of divalent cations, induces in the somatic membrane an intense sodium conductance. However, the kinetic characteristics of this induced sodium current correspond to those for calcium currents; it can be blocked by verapamil and other specific blockers of calcium channels. The study of the fluctuations of these currents has shown that the conductance of the corresponding unitary ionic channel comes close to the conductance of usual sodium channels (Krishtal et al 1981). Apparently, in this case we also observe in a single channel a combination of the gating mechanism peculiar to the calcium channels and the selectivity filter of the "sodium" channel. Since the reorganization of the selective properties of calcium channels can be easily performed by extracellular chemical influences, one may suggest that the corresponding functional group is located near the external mouth of these channels.

PERFUSION IN EMBRYONIC CELLS

Extensive experimentation with microelectrode recordings of action potentials in sodium-free solutions has revealed that calcium conductance is one of the primary properties of the cell membrane and is well pronounced already in oocytes. The application of the voltage clamp technique to the egg cell membrane of tunicates, sea urchins, and rats (Okamoto et al 1976a,b, 1977) has shown that in all cases a fast and a slow inward current are present; the fast current is carried by Na^+ ions, and the slow one by Ca^{2+} ions. By using a modified method of intracellular dialysis, Takahashi & Yoshii (1978) were able to analyze the calcium currents in the membrane of tunicate egg cells. In this case the calcium current, like that in the somatic membrane of differentiated nerve cells, has no clear-cut reversal potential and can be blocked by the introduction of fluoride anions into the cell. This current can be obtained in a pure form by lowering the holding potential to –34 mV, at which point the sodium current becomes completely inactivated.

FURTHER PROSPECTS OF THE APPLICATION OF THE INTRACELLULAR PERFUSION TECHNIQUE

So far the intracellular perfusion technique has been used primarily for the study of the mechanisms of electrical excitability of the neuronal membrane. We can expect further substantial progress from the application of intracellular perfusion to such problems as dependence of the functioning of different types of ionic channels on intracellular metabolic systems; duration of the existence of functioning channels in the membrane; and clarification of the processes responsible for the synthesis, transport, and incorporation of the corresponding protein structures in the surface membrane. For these purposes nerve cells in tissue culture, as well as peripheral cell lines that possess electrical excitability (for instance, neuroblastoma cells), may prove highly suitable. As preliminary experiments have shown, neuroblastoma cells are particularly suitable for intracellular perfusion (Veselovsky et al 1977b, Veselovsky & Tsyndrenko 1979, Tsyndrenko 1980); the changes in ionic channel properties and their distribution can be successfully investigated both at different stages of mitosis and during induction or suppression of morphogenesis.

Intracellular perfusion can also be applied to other excitable cells, in particular to isolated cardiac muscle fibers (Undrovinas et al 1980a). In cardiac muscle cells it is possible to make an effective separation of the total ionic current into its components and to study their characteristics, in particular the effect of extra- and intracellular application of pharmacological drugs (Undrovinas et al 1980b). The intracellular perfusion technique may also be useful for understanding the functioning of chemically operated channels (Bregestovski & Iljin 1980) and the action of cyclic nucleotides (Akopyan et al 1979b).

Perfused cells may also be suitable for studying the active transport of ions through the cell membrane. Dialyzed mollusc neurons (Akopyan et al 1979a) and mollusc muscle fibers (Nelson & Blaustein 1980) preserve the capability of transferring sodium and potassium ions against the concentration gradients if ATP is introduced into the cell; a "pump" current reflecting the electrogenic character of this transfer can also be recorded. The possibility of full control over the ionic medium on both sides of the membrane, as well as of an easy introduction of any substances that may interact with the mechanisms of active transport, opens the way for a precise solution of the problems of stochiometry and potential dependence of this process and its control from the membrane receptor systems. Recently, results obtained by the intracellular perfusion technique have been reported regarding the effect of serotonin on active transport in the neuronal membrane (Piskunova & Chemeris 1979).

CONCLUDING REMARKS

The discovery that the intracellular ionic medium of isolated cells can be rapidly and effectively replaced by any other desirable medium has opened new possibilities for the analysis of ionic mechanisms of active cellular responses. The use of intracellular dialysis and perfusion techniques is limited only by cell dimensions (it is difficult for cells less than 20–30 μm in diameter); however, it is quite possible that in the future these techniques will also be applicable to smaller cells. Cell processes do not present great difficulties—if such processes are torn off in the process of isolation, the openings in the surface membrane in most cases become rapidly resealed. The promising results obtained through the intracellular dialysis or perfusion techniques are due to the practically ideal conditions for the measurement of currents, the ability to separate completely the total transmembrane currents into individual specific components, and the precise control of the electrochemical gradients responsible for the production of each component.

Literature Cited

Adams, D. J., Gage, P. W. 1976. Gating currents associated with sodium and calcium currents in an *Aplysia* neurone. *Science* 192:783–84

Akaike, N., Fishman, H. M., Lee, K. S., Moore, L. E., Brown, A. M. 1978a. The units of calcium conduction in *Helix* neurone. *Nature* 274:379–82

Akaike, N., Lee, K. S., Brown, A. M. 1978b. The calcium current of *Helix* neuron. *J. Gen. Physiol.* 71:509–31

Akaike, N., Lee, K. S., Fishman, H. M., Moore, L. E., Brown, A. M. 1978c. Calcium binding and its relation to calcium current noise in snail neurons. *Biophys. J.* 21:208a

Akopyan, A. R., Airapetyan, S. N., Chemeris, N. K. 1979a. Dialyzed neurone as a new model for study of the electrogenic sodium pump. *Dokl. Akad. Nauk Arm. SSR (Yerevan)* 68:45–49

Akopyan, A. R., Piskunova, G. M., Chemeris, N. K. 1979b. Dialyzed neurone as a model for study of cyclic nucleotide effect on membrane electrical parameters. *Dokl. Akad. Nauk SSSR Moscow* 246:997–99

Baker, P. F., Hodgkin, A. L., Shaw, T. I. 1962. Replacement of the axoplasm of giant nerve fibres with artificial solutions. *J. Physiol.* 164:330–54

Bregestovski, P. D., Iljin, V. N. 1980. Effect of "calcium antagonist" D-600 on the postsynaptic membrane. *J. Physiol. Paris* 76:515–22

Doroshenko, P. A., Kostyuk, P. G., Tsyndrenko, A. Y. 1978a. Studies of reversal potential for slow component of inward current in snail neuron membrane. *Neurophysiology Kiev* 10:206–8

Doroshenko, P. A., Kostyuk, P. G., Tsyndrenko, A. Y. 1978b. Separation of potassium and calcium channels in the somatic membrane of a nerve cell. *Neurophysiology Kiev* 10:645–53

Doroshenko, P. A., Kostyuk, P. G., Tsyndrenko, A. Y. 1979. Studies of TEA-resistive inward current in the somatic membrane of the perfused nerve cells. *Neurophysiology Kiev* 11:460–68

Doroshenko, P. A., Tsyndrenko, A. Y. 1978. Effect of intracellular calcium on the calcium inward current. *Neurophysiology Kiev* 10:203–5

Greengard, P. 1979. Cyclic nucleotides, phosphorilated proteins, and the nervous system. *Fed. Proc.* 38:2208–17

Kostyuk, P. G., Krishtal, O. A. 1977a. Separation of sodium and calcium currents in the somatic membrane of mollusc neurones. *J. Physiol.* 270:545–68

Kostyuk, P. G., Krishtal, O. A. 1977b. Effects of calcium and calcium chelating agents on the inward and outward currents in the membrane of mollusc neurones. *J. Physiol.* 270:569–80

Kostyuk, P. G., Krishtal, O. A., Pidoplichko, V. I. 1975. Intracellular dialysis of nerve cells: Effect of intracellular fluoride and phosphate on the inward current. *Nature* 257:691–93

Kostyuk, P. G., Krishtal, O. A., Pidoplichko, V. I. 1977. Asymmetrical displacement currents in nerve cell membrane and effect of internal fluoride. *Nature* 267:70–72

Kostyuk, P. G., Krishtal, O. A., Pidoplichko, V. I. 1978. The estimation of single calcium channel conductance using the current fluctuation method and EGTA effect. *Dokl. Akad. Nauk SSSR Moscow* 238:478–81

Kostyuk, P. G., Krishtal, O. A., Pidoplichko, V. I. 1981. Calcium inward current and related charge movements in the membrane of snail neurones. *J. Physiol.* 310:403–21

Kostyuk, P. G., Krishtal, O. A., Pidoplichko, V. I., Shakhovalov, Y. A. 1979a. Gating mechanism of calcium channels— steady state characteristics and kinetics. *Dokl. Akad. Nauk SSSR Moscow* 249:1470–73

Kostyuk, P. G., Krishtal, O. A., Pidoplichko, V. I., Shakhovalov, Y. A. 1979b. Kinetics of calcium inward current activation. *J. Gen. Physiol.* 73:675–77

Kostyuk, P. G., Krishtal, O. A., Pidoplichko, V. I., Shakhovalov, Y. A. 1980a. A spectral analysis of conductance fluctuations of calcium channels in a nerve cell membrane. *Dokl. Akad. Nauk SSSR Moscow* 250:219–22

Kostyuk, P. G., Mironov, S. L., Doroshenkov, P. A. 1980b. An energy profile of a calcium channel in a mollusc neuronal membrane. *Dokl. Akad. Nauk SSSR Moscow* 253:978–81

Krishtal, O. A. 1976. Blocking effect of cadmium ions on a calcium inward current in a nerve cell membrane. *Dokl. Akad. Nauk SSSR Moscow* 231:1003–5

Krishtal, O. A. 1978. Modification of calcium channels in a nerve cell using EGTA. *Dokl. Akad. Nauk SSSR Moscow* 238:482–85

Krishtal, O. A., Pidoplichko, V. I. 1975. Intracellular perfusion of snail giant neurones. *Neurophysiology Kiev* 7:327–29

Krishtal, O. A., Pidoplichko, V. I., Shakhovalov Y. A. 1981. Conductance of the calcium channel in the membrane of snail neurones. *J. Physiol.* 310:423–34

Lee, K. S., Akaike, N., Brown, A. M. 1978. Properties of internally perfused, voltage-clamped, isolated nerve cell bodies. *J. Gen. Physiol.* 71:489–507

Matsuda, Y., Yoshida, S., Yonezawa, T. 1978. Tetrodotoxin sensitivity and Ca component of action potentials of mouse dorsal root ganglion cells cultured in vitro. *Brain Res.* 154:69–82

Meech, R. W. 1974. The sensitivity of *Helix aspersa* neurones to injected calcium ions. *J. Physiol.* 237:259–77

Naruševičius, E. V., Rapoport, M. S. 1979. The influence of membrane potential and extracellular concentration of Sr and Ca ions upon the inward current in dialyzed neurones of *Helix pomatia*. *Dokl. Akad. Nauk SSSR Moscow* 246:217–19

Nathanson, J. L. 1977. Cyclic nucleotides and nervous system function. *Physiol. Rev.* 57:157–256

Nelson, N. T., Blaustein, M. P. 1980. Properties of sodium pump in internally per-

fused barnacle muscles fibres. *J. Gen. Physiol.* 75:183–206

Oikawa, T., Spyropoulos, C. S., Tasaki, I., Teorell, T. 1961. Methods for perfusing the giant axon of *Loligo pealii*. *Acta Physiol. Scand.* 52:195–96

Okamoto, H., Takahashi, K., Yamashita, N. 1977. Ionic currents through the membrane of the mammalian oocyte and their comparison with those in the tunicate and sea urchin. *J. Physiol.* 267:465–95

Okamoto, H., Takahashi, K., Yoshii, M. 1976a. Membrane currents of the tunicate egg under the voltage-clamp condition. *J. Physiol.* 254:607–38

Okamoto, H., Takahashi, K., Yoshii, M. 1976b. Two components of the calcium current in the egg cell membrane of the tunicate. *J. Physiol.* 255:527–61

Piskunova, G. M., Chemeris, N. K. 1979. Study of the serotonine electrogenic effect on the dialyzed soma of a *Helix pomatia* neurone. *Dokl. Akad. Nauk SSSR Moscow* 248:763–66

Pogorelaya, N. K., Skibo, G. G., Troitskaya, N. K. 1980. Structural features of isolated and perfused neurones in the *Helix pomatia*. *Neurophysiology Kiev* 12:297–302

Ponomaryov, V. N., Naruševičius, E. V., Chemeris, N. K. 1980. Blocking effect of Ni^{2+}, Co^{2+}, Mn^{2+} and Mg^{2+} ions on the value of inward current through calcium channels of *Limnaea* neurones. *Neurophysiology Kiev* 12:221–23.

Takahashi, K., Yoshii, M. 1978. Effects of internal free calcium upon the sodium and calcium channels in the tunicate egg analysed by the internal perfusion technique. *J. Physiol.* 279:519–49

Tillotson, D. 1979. Inactivation of Ca conductance dependent on entry of Ca ions in molluscan neurons. *Proc. Natl. Acad. Sci. USA* 76:1497–1500

Tsyndrenko, A. Y. 1980. Comparative analysis of potassium channel characteristics in a membrane of spinal ganglion neurones and neuroblastoma cells. *Neurophysiology Kiev* 12:208–10

Undrovinas, A. I., Yushmanova, A. V., Hering, S., Rosenshtraukh, L. V. 1980a. Voltage clamp method on single cardiac cells from adult rat heart. *Experientia* 36:572–73

Undrovinas, A. I., Yushmanova, A. V., Rosenshtraukh, L. V. 1980b. Changes in sodium current under the action of ethmozine and lidocain inside and outside the membrane of single rat myocardial cells. *Bull. Eksp. Biol. Med. Moscow* 90:178–80

Valeyev, A. E. 1979. Selectivity of calcium channels of a somatic membrane in *Helix* neurones for calcium, strontium and barium ions. *Neurophysiology Kiev* 11:371–74

Veselovsky, N. S., Fedulova, S. A. 1980. Revealing of calcium channels in a somatic membrane of spinal ganglion neurones of rats under an intracellular dialysis with cAMP. *Dokl. Akad. Nauk SSSR Moscow* 253:1493–95

Veselovsky, N. S., Kostyuk, P. G., Krishtal, O. A., Pidoplichko, V. I. 1977a. Separation of ionic currents responsible for a generation of the action potential in isolated neurones of frog spinal ganglia. *Neurophysiology Kiev* 9:638–40

Veselovsky, N. S., Kostyuk, P. G., Krishtal, O. A., Naumov, A. P., Pidoplichko, V. I. 1977b. Transmembrane ionic currents in neuroblastoma cells. *Neurophysiology Kiev* 9:641–43

Veselovsky, N. S., Kostyuk, P. G., Tsyndrenko, A. Y. 1979. Separation of ion currents responsible for a generation of action potential in a somatic membrane of spinal ganglion neurones in newborn rats. *Dokl. Akad. Nauk SSSR Moscow* 249:1466–69

Veselovsky, N. S., Kostyuk, P. G., Tsyndrenko, A. Y. 1980. "Slow" sodium channels in a somatic membrane of spinal ganglion neurones in newborn rats. *Dokl. Akad. Nauk SSSR Moscow* 250:216–18

Veselovsky, N. S., Tsyndrenko, A. Y. 1979. Comparative analysis of sodium channel characteristics in the membrane of rat spinal ganglion neurones and neuroblastoma cells. *Neurophysiology Kiev* 11:607–9

Yoshida, S., Matsuda, Y., Samejima, A. 1978. Tetrodotoxin-resistant sodium and calcium components of action potentials in dorsal root ganglion cells of the adult mouse. *J. Neurophysiol.* 41:1096–106

Ann. Rev. Neurosci. 1982. 5:121–70
Copyright © 1982 by Annual Reviews Inc. All rights reserved

ELECTRORECEPTION

Theodore Holmes Bullock

Neurobiology Unit, Scripps Institution of Oceanography, Department
of Neurosciences, School of Medicine, University of California at San Diego,
La Jolla, California 92093

A system of receptors and central processes mediating the specialized detection and analysis of feeble electric fields in the aquatic milieu of fishes was suggested on behavioral evidence by Lissmann in 1958 (see also Lissmann & Machin 1958, Machin & Lissmann 1960). Confirmation of a new class of specialized receptors by direct physiological experiment, and justification for designating them "electroreceptors" by showing their adequate stimuli to be biologically significant electric currents, followed shortly thereafter (Bullock et al 1961, Fessard & Szabo 1961, Szabo 1962, 1963, Hagiwara et al 1962, Hagiwara & Morita 1963, Dijkgraaf & Kalmijn 1962, 1963, Kalmijn 1966, 1971, Szabo & Fessard 1965). It soon became evident that there are, in fact, several kinds of specialized receptors with distinct, nonclassical modes of encoding intensity. Since then, the study of these sense organs, their central physiology, and behavioral significance, has grown to occupy several scores of workers and hundreds of papers. Several reviews on general or particular aspects have appeared (Lissmann 1963, Bennett 1965, 1967, 1970, 1971, 1978, Akoev et al 1971, Bullock 1973, 1974, 1979a, Szabo 1974a,b, Fessard & Szabo 1974, Murray 1974, Kalmijn 1974, 1978a,c, Scheich & Bullock 1974, Heiligenberg 1977, Bell 1979, Bullock et al 1979, Viancour 1979c). The present review, while selective, perforce overlaps somewhat with the others; it emphasizes recent additions to our knowledge of anatomy and physiology and to such behavioral insight as bears upon ideas of neural mechanisms. Nomenclature and distinctions among the several kinds of receptors has been confusing. A systematization is provided (Table 1), based partly on new findings.

121

0147-006X/82/0301-0121$02.00

Table 1   Types of electroreceptors[a]

I. AMPULLARY ORGANS; LOW FREQUENCY SENSITIVE RECEPTORS

(Ampulla with patent canal to the surface and receptor cells in epithelial lining; passband ca. 0.1–50 Hz; afferent nerve fibers are frequency coders; not synchronized with EOD except partially in mormyriforms)

A. Ampullae of Lorenzini (afferent nerve fibers = "L" units)

In elasmobranchs, holocephalans, chondrosteans, and possibly including organs of Fahrenholz of dipnoans and polyteriforms, also unknown organs of lampreys
Cathodal ampulla, depolarizing apical membrane causes excitation
Hair cell receptors with kinocilium but no microvilli
Long canal in marine species; short duct microampullae in fresh water species

B. Ampullae of teleosts ("F" units)

In siluriforms, gymnotiforms[b], and mormyriforms[c]
Anodal ampulla, depolarizing basal membrane causes excitation
Hair cell receptors without kinocilium but with microvilli
Long canal in marine species (sometimes, inappropriately called Lorenzinian), "small pit organs," with short duct in fresh water species

II. TUBEROUS ORGANS; HIGH FREQUENCY SENSITIVE RECEPTORS

(Cavity occupied by sense cells projecting from basal attachment, usually without patent duct; sense cells with microvilli protruding or lining an invagination, or without them; pass band ca. 50–2000 Hz; afferent fibers synchronized with EOD and encoding intensity in various ways not appropriately called frequency coders)

A. Pulse marker units ("M" units) in gymnotiform pulse species[d]

Large axon; myelinated preterminal branches
High threshold in some species, low in others
Single spike follows EOD with fixed latency

B. Phase coders ("T" units) in gymnotiform wave species[d]

Medium sized axon; myelinated preterminals
Single spike follows each EOD in physiological intensity range; shifts latency with intensity, with small jitter
More sensitive (absolute) than P units

C. Burst duration coders ("B" units) in gymnotiform pulse species[e]

Small axon, unmyelinated preterminals, many sense cells per organ
Burst of spikes follows EOD, graded in number at fixed interval, little latency shift

D. Probability coders ("P" units) in gymnotiform wave species

Morphology may be similar to preceding
Single spike follows some EODs probability graded with intensity; smaller, more jittery latency shift and more phasic than "T" units in same fish

E. Knollenorgane ("K" units) in mormyrids

"Large pore" organs, large axon; myelinated preterminals
More sensitive (absolute) than D units
Spontaneous receptor potential, rhythmic, usually spike-like, up to several kHz under some conditions
Useful for detecting social signals by EOD waveform

Table 1   *(Continued)*

II. TUBEROUS ORGANS; HIGH FREQUENCY SENSITIVE RECEPTORS *(continued)*

  F. Mormyromasts; burst dispersion coders ("D" units) of mormyrids

  "Medium pore" organs; small axon with unmyelinated preterminals
  Two forms of sense cells, complex innervation
  Threshold can be high, up to 10 mv
  Receptor potential not spontaneous; damped ringing
  Burst of spikes graded in latency, number and spacing (likely to be a heterogeneous category)

  G. Gymnarchomasts type I ("S" units) in gymnarchids

  Like preceding but single nerve fiber, with unmyelinated preterminals
  Pair of unequal sense cells, or more
  Like "T" units but high spontaneous firing rate, less tendency to burst below 1:1 range, small dynamic range of phase coding

  H. Gymnarchomasts type II ("O" units) in gymnarchids

  Single nerve fiber with unmyelinated preterminals
  Twelve or more sensory units per organ; sense cells with long microvilli at bottom of invagination
  Like "P" units but spontaneous with more difficulty following 1:1, small dynamic range of phasic part of probability coding; high intensity suppression.

[a] Revised from Bullock 1974, Table 3, p. 9.
[b] Type I of Szabo 1974a, type 1 of Lissmann & Mullinger 1968.
[c] "Small pore" mormyromast type I of Szabo 1974a, gymnarchid type A of Szabo 1974a.
[d] Part of Szabo's 1974a common tuberous, type I.
[e] Part of Szabo's 1974a common tuberous, type II.

DISTRIBUTION AND EVOLUTION OF ELECTRORECEPTION

Lissmann (1958) speculated on the evolution of electroreception at a time when receptors were only inferred. The picture has changed greatly in the meantime, due particularly to recent findings. Electroreceptors apparently have been invented at least three times among the vertebrates. First they appeared in primitive vertebrates—ancestral to all the nonteleost fishes (except possibly hagfish)—only to be lost by the immediate ancestors of holosteans and teleosts (Northcutt et al 1980). They were reinvented at least twice among teleosts, and now appear to be ordinal characteristics of Gymnotiformes [recently elevated to the status of an order by Mago-Leccia (1978) and Mago-Leccia & Zaret (1978)], Siluriformes (which might possibly have shared a common ancestor not shared by cypriniforms), and the Mormyriformes, which are only distantly related.

In the nonteleosts, a single major class of electric sense organs is probably general: the ampullae of Lorenzini and similar organs. In a few teleosts a new type of ampullary organ, best called the "ampullae of teleosts", is found and in some of these—the gymnotiform and mormyriform teleosts, which are electric fish—still another major class of sense organs is found as well, the tuberous organs. Ampullae of Lorenzini or similar organs are now known in almost all the nonteleost fishes: Agnatha, Petromyzoniformes (lampreys), Holocephala (chimaeras), Elasmobranchii (sharks, skates, and rays), Dipnoi (lungfish), Crossopterygii (*Latimeria,*) Chondrostei (sturgeon, paddlefish), Polypteriformes (brachiopterygians); but not in Holostei (gars and bowfin), which probably secondarily lost them, as also did the ancestors of teleosts (Kleerekoper & Sibakin 1956, Pfeiffer, 1968, Kalmijn 1971, 1974, Jørgensen et al 1972, Roth 1973b, Bullock 1979b, Teeter et al 1980, Fields & Lange 1980, Jørgensen 1980, Bodznick & Northcutt 1981; D. A. Bodznick, R. G. Northcutt, and T. H. Bullock, in preparation). Receptors are accompanied by special brain structures. Characteristic and diagnostic of the ampullae of Lorenzini is a well developed *dorsal nucleus* of the lateral line lobe, which is found in all the same groups and segregates electrosensory input from common lateral line and eighth nerve input in adjacent nuclei (Boord & Campbell 1977, McCormick 1978, 1981a,b, Bodznick & Northcutt 1980, 1981, Northcutt et al 1980, Northcutt 1980, 1981b). Among living representatives of these nonteleost groups, only two families in one of them (Elasmobranchii: Rajidae, Torpedinidae) evolved electric organs.

The ancestral teleosts, like some 30 of the 33 orders of living teleosts, presumably lacked electroreception. However, we have direct evidence only for representatives of some nine of the 30 (named below).

This raises the question of definition: When can an animal be said to possess electroreception? Fortunately, there is not a continuum between taxa that clearly possess and those that lack this sense. Those considered to be electroreceptive exhibit the following characteristics:

1. Adaptive behavior to naturally available electric fields—either transient pulses produced by their own and their conspecific's specialized system of electric organs and brain structures controlling them, or low frequency fields (0.1–50 Hz) produced by organisms unspecialized for such emission or by inanimate sources.
2. Sensitivity in the latter two situations, extending to thresholds usually below 10 μV/cm (in freshwater species in water of ca. 20 k·ohm·cm resistivity) and as low as 0.005 μV/cm in marine species in sea water (Kalmijn 1981 and to be submitted).
3. In the case of response to their own or their conspecific's brief electric organ discharge (EOD), the absolute threshold may be as high as 10 mV/cm, but its specialization is indicated by consistent responses to a natural EOD exceeding this value or/and by high increment sensitivity.
4. By extension from species where adaptive behavior has been shown, we

may cautiously use physiological measures from the brain, nerves, or sense organs. Electroreceptors and relevant regions of the brain may be identified with the above criteria.

On such grounds, classical galvanotaxis (Kalmijn 1974) seems to have nothing to do with electroreception. The generally high thresholds for the "first reaction" in experiments on galvanotaxis may be taken as suggestive though inconclusive evidence that electroreception is lacking in the families Cyprinidae, Cottidae, Mugilidae, Sparidae, Serranidae. Evoked potential study confirmed this for goldfish (T. H. Bullock, unpublished). Recent, careful attempts to find evoked potentials in likely parts of the brain, probably the best method so far for this purpose, strongly suggest the absence of specialized, high sensitivity electroreception in *Notopterus* (Notopteridae), *Anguilla* (Anguillidae), *Oncorhynchus* (Salmonidae), and *Astroscopus* (Uranoscopidae) (D. A. Bodznick, R. G. Northcutt, and T. H. Bullock, in preparation). Curiously, *Xenomystus* (Notopteridae) has been found to possess a lateral line lobe specialized like the electrosensory lobe of catfish (M. R. Braford, to be submitted) and to be extremely sensitive by this method (responses to < 5 μv/cm; T. H. Bullock, unpublished).

It is notable that so far there is little indication of intermediate or equivocal species. *Anguilla* may be one. Although no evoked potentials have been found with stimuli up to several millivolts per centimeter (D. A. Bodznick, R. G. Northcutt, and T. H. Bullock, in preparation) and eels were indifferent to stimuli that attracted sharks and catfish (Kalmijn 1978c, Kalmijn et al 1976), behavioral responses extended down to about 0.5 mV/cm (Enger et al 1976a, Berge 1979; contra McCleave et al 1971, Rommel & McCleave 1972, Richardson et al 1976). Eels have been reported to be sensitive to magnetic fields as weak as the earth's (Branover et al 1971, Gleizer & Khodorkovsky 1971, Rommel & McCleave 1972). This claim is not thrown into question by the electric field measurement; magnetic detection can be direct (Blakemore 1975, Kalmijn & Blakemore 1978, Frankel et al 1979), not necessarily involving movement-induced voltage gradient sensitivity as the indirect magnetic detection does in sharks and rays (Kalmijn 1978a,c, 1979, 1981). Direct detection of the magnetic field of the earth may be widespread (Blakemore 1975, Teague et al 1979, Gilson & Kalmijn 1980, Blakemore et al 1980, Kalmijn 1981, and A. J. Kalmijn, to be submitted). Many other cases of magnetic detection are reported which might prove to be direct but where the evidence is not yet in (Phillips 1977 in salamanders; Lindauer & Martin 1972 in bees; Brown 1971 and Gottlieb & Caldwell 1967 in snails and planaria; Keeton et al 1974 in pigeons; Wiltschko 1968 and Wiltschko & Wiltschko 1976 in song birds, among others).

The stargazer (*Astroscopus*), apparently lacking specialized electroreceptors (T. H. Bullock, unpublished) or electroreceptive nuclei (R. G. North-cutt, unpublished), is a special case because it is an electric fish, albeit exceptional in a family that is mostly nonelectric—the only case where the electric organ is not a family character. The function of its EOD has been problematical: although temporally associated with prey capture, it would seem to be too low in power even to disorient a prey and too late in time to be a sensory guide (Pickens & McFarland 1964). Seasonal changes and sex differences in intensity, duration, and form of the EOD are reported (Mikhailenko 1971); two types of EOD are given, short and long, and of different peak voltage. Mikhailenko (1971, 1973) suggested that it has a social communicating function, especially in the mating season, as well as a role in deterring predators. He also found synchronous acoustic pulses and suggested that the electric and the acoustic event are generated by the same mechanism. Our failure to find high sensitivity electroreception, while certainly not definitive, suggests that ordinary lateral line or other receptors with thresholds in the (untested) 10+ mV/cm range may serve any elec-troreception that is needed.

Apparently all members of three orders have electroreceptors (Table 1) —Siluriformes (Bullock 1979b), Gymnotiformes, and Mormyriformes. The first two are closely related but the last named is systematically distant from them. Siluriforms have only ampullary receptors of the teleost type—per-haps best regarded as not homologous to those of nonteleosts. Siluriforms are mostly without electric organs. Gymnotiform and mormyriform fishes, all of which are electric, have ampullae of the teleost type and also tuberous electrosense organs. Because of the remoteness of the relationship of these two orders and the absence of electroreceptors in the vast majority of teleosts, the position is tenable that the receptors should not be assumed to be homologous among these orders. Counting the nonteleosts as one, it is likely that electroreceptors were separate inventions at least three times, from common lateral line origins. The term "ampullary" implies only a partial analogy, not homology.

Electroreception is unknown in invertebrates or in vertebrates outside of the fishes. However, the suggestion that it may occur in some caecilian amphibians (Lissmann 1958) appears to be supported by histological evi-dence from larval ichthyophids (Hetherington & Wake 1979).

RECEPTION BY AMPULLARY SYSTEMS

Behavior: Orientation and Object Detection

Evidence summarized in earlier reviews, cited above, explains the current view that a broad category of electroreceptors called collectively "ampul-lary receptors," on histological grounds, shares certain properties and roles in the wide range of taxa that possess them, including elasmobranchs and most nonteleosts, siluriforms, gymnotiforms, and mormyriforms. Ampullary receptors have proven to be heterogeneous in mechanism, structure, and central projection (Table 1). At present two quite distinct classes are recognized: ampullae of Lorenzini (in nonteleost fishes) and ampullae of teleosts. Limited to detecting low frequencies (ca. 0.1–50 Hz), they are believed to be stimulated by incidental voltage gradients from animate sources other than electric organs, both from the self and from other organisms, and from inanimate sources.

Best established is the detection by sharks, rays, and skates of prey and of voltage gradients induced by motion in magnetic fields like the earth's (Kalmijn 1974, 1978a–c, 1981). New evidence (Butsuk & Bessonov 1981) adds to previous information that common teleosts leak enough current from skin, gills, mouth, etc to cause millivolt potentials close to the body relative to a remote reference; e.g. adequate for a shark to detect at 50 cm. Limited evidence of social communication by EOD has been reported for skates (*Raja:* Mikhailenko 1971), stargazers (*Uranoscopus,* receptors unknown; Mikhailenko 1971), and electric catfish (*Malapterurus;* Kastoun 1971, Belbenoit et al 1979). Detection of the fish's own respiratory or locomotor movements and of inert nonconducting objects, presumably by the distortion of the fish's own field or of standing fields from foreign sources, is also known (Dijkgraaf & Kalmijn 1966, Akoev et al 1976a, Roth 1969, Fields & Lange 1980). These behavioral and electroreceptor achievements, as well as heart rate decelerations, involve feeble electric fields that are best expressed in terms of their voltage gradient in the water; gradients down to 0.005 μV/cm are detected behaviorally by elasmobranchs. Voltage gradient was preferred by Kalmijn (1974) because the fish as a whole is a high resistance device, acting more like a voltmeter than an ammeter, both in salt and in fresh water species. Today we can go further, based on the accumulated literature on voltage sensitive currents and gates in neurons; it seems likely that transmembrane voltage is more relevant than current, especially the current density in the water or in the tissues of the fish at a gross level. A. J. Kalmijn (unpublished) finds that catfish maintain a behavioral threshold of 1 μV/cm in water from 2 to 20 k·ohm·cm resistivity, over which range the current density changes by a factor of 10. He has shown in field as well as laboratory tests that sharks use low frequency or DC fields of the prey to locate it when within about the predator's body length—ca. 50 cm in small sharks. Siluriforms (catfish) and gymnotiforms can also locate prey in this way (Roth 1972, Kalmijn 1974). With relative motion Kalmijn has also demonstrated useful sensitivity to both local electromagnetic fields and large homogenous fields like that of the earth and even much weaker in guiding the orientation of sharks and rays. The effective field depends on the direction in which the animal is moving, so the information is available for an electromagnetic compass sense; experimentally, elasmobranchs show the ability both to detect and to analyze such information. This is true both for the large fields due to ocean currents and for those due to the animal's swimming. In principle, by comparing the results of some intentional swerves, the shark can know its heading relative to the earth and that relative to the axis of the ocean stream, its velocity, and even its latitude (Kalmijn 1981).

Extending considerably the findings of Roth (1968, 1969, 1972), Peters & Bretschneider (1972), and Peters & van Wijland (1974), A. J. Kalmijn et al (1976) showed that catfish in fresh water can also orient in strictly uniform electric fields as low as 1.0 μV/cm. Testing with a gradient of 10 μV/cm, the fish can select the correct direction from among 12 arranged in a circle, using only the angle in the uniform field (V. Kalmijn et al 1976). In this, as in comparable physiological measurements, it appears that ampullary organs in marine species have lower thresholds, in terms of voltage gradient, than sensitive fresh water species by a factor of up to 200.

Because of its general interest, something must be said about earthquake prediction by catfish. This is so far the best documented case of animal responses prior to the major shock. The series of papers from Japan in the early 1930s that established a statistically significant behavior change some hours before an earthquake are reviewed by Kalmijn (1974). The Japanese authors gave some evidence that the adequate stimulus was electric potentials from the earth. Only later quantitative measurements of catfish sensitivity and of the magnitude of potentials anticipating the main quake (which are not universal accompaniments of earthquakes and are perhaps peculiar to certain regions) showed the signals to be well above threshold and the original observations therefore quite plausible.

Anatomy: Receptors and Central Projections

Szabo (1974a) reviewed the anatomy of ampullary receptors and added new observations. More than 20 papers have dealt with light and electron microscopy of one or more species of the following groups: Holocephali, Elasmobranchii (both marine and fresh water), Dipnoi, Chondrostei, Polypteriformes (brachiopterygians), Siluriformes (both fresh water and marine), Gymnotiformes, and Mormyriformes. Hetherington & Wake (1979) confirm the suggestion of Lissmann (1958) that organs histologically like ampullary organs of fish occur in certain species of the caecilian amphibians (Gymnophiona, Ichthyophiidae; see also Taylor 1970). Receptors have not yet been described in lampreys, though electroreception is physiologically demonstrated. Except for the suggestion of electric sense organs in the head in *Latimeria* (Hetherington & Bemis 1979), and confirmation of the Lorenzinian character of the ampullary receptors in sturgeons (Jør-gensen 1980, Teeter et al 1980), in chimaeras (Fields & Lange 1980), in freshwater rays (Szamier & Bennett 1980), and in siluriforms (Gelinek 1978), almost nothing new has been published since Szabo's review. For the present purposes the main fact is that ampullary organs are quite hetero-geneous (Table 1) even within each of the two unrelated types—the Loren-zinian in nonteleosts and the teleost type—from the gross anatomy of the canal and the ampullae, to the ultrastructure of the receptor hair cells, their cilia and their innervation. Each of the two types can have long canals (up to several cm) or short canals (ca. 0.1 mm); the former is found in marine, the latter in freshwater species. Bennett (1971) and Kalmijn (1974) pointed out that long canals together with the skin resistance improve sensitivity in a high conductance medium, but are not needed in freshwater. Hair cells of Lorenzinian ampullae typically have one kinocilium, whereas those in teleost ampullae do not, but instead have microvilli; but there is variation within each category (Derbin 1974). One may expect new distinctions between subtypes, which are perhaps inappropriately lumped together to-day. It is time for a new, broadly based study, comparing not only the major taxa but genera and families within them, different organs within the array on one animal, ontogenetic stages, and stages of regeneration. Peters et al (1974) have provided an account of the distribution of ampullary organs over the body of the catfish together with relevant aspects of the skin resistance that affect the axis of greater sensitivity.

    In contrast to the peripheral system, our knowledge of the central projec-tions of ampullary receptors is primarily based, not on older work, but on recent studies, dependent largely on experimental anatomical techniques (Andrianov & Volkova 1976, Andrianov & Enin 1977, Knudsen 1977, Northcutt 1981a, Bodznick & Northcutt 1980, 1981).

    In elasmobranchs and other nonteleost fish that possess ampullae of Lorenzini, or their equivalents, the afferent axons from those receptors end in a nucleus called the dorsal nucleus of the octavolateral area of the medulla (Boord & Campbell 1977, Koester & Boord 1978, Bodznick & Northcutt 1980, 1981; see also Paul et al 1977, Paul & Roberts 1977a,b, McCormick 1978, 1981a,b, Northcutt 1980, 1981a,b). Some authors call it the "anterior lateral line lobe" but this is not felicitous since the part of the brain in teleosts in which electroreceptor afferents end is called the "poste-rior lateral line lobe" by some authors. I here follow the most recent analyses (McCormick 1978, 1981a,b, Northcutt 1981a), which largely agree with Pearson (1936) but reject the usage of Larsell (1967). The common lateral line mechanoreceptors go to a part of the medial nucleus ("posterior lateral line lobe" in elasmobranchs, "anterior lateral line lobe" in teleosts, in the usage of Paul & Roberts 1977a and Maler 1974). Eighth nerve afferents end in their own nuclei in most teleost species studied to date (McCormick 1978, 1981a,b, Northcutt 1981b) but seem to converge with lateral line afferents in the medial nucleus in mormyriforms (Bell 1981), so that the octavolateral modalities are typically well segregated. The dorsal nucleus can be said, on the basis of recent comprehensive surveys (McCormick 1978, 1981a,b, Northcutt 1981a,b; D. A. Bodznick, R. G. Northcutt, and T. H. Bullock, in preparation) of all the major groups of fishes, including the use of correlated anatomical and physiological criteria, to be coextensive among nonteleost taxa with electroreception. It is the best anatomical indicator of the presence of that sensory modality—better than skin histology or other brain structures. Its internal structure and cell types have been described, mainly from elasmobranchs by light microscope methods (Pearson 1936, Paul & Roberts 1977a, McCormick 1978, 1981a,b. Northcutt 1981a). Paul et al (1977) contributed an EM study and found seven types of terminals; they emphasized the similarity between the molecular layer of this lobe and that of the cerebellum.

In the three teleost orders that possess ampullary organs, afferents enter the brain via the anterior lateral line nerve, as in elasmobranchs (McCready & Boord 1976) and in mormyriforms to some extent via the posterior lateral line nerve and terminate, not in a dorsal nucleus, which is lacking in teleosts, but, to use Pearson's terminology, in a portion of the medial nucleus—the pars dorsalis, which is enlarged in these orders (Maler, 1973, Bell & Russell 1978b, Maler et al 1981, Bell 1981, Bell et al 1981). The medial nucleus seems to have become specialized in electroreceptive teleosts, independently in each group, or at least twice (mormyriforms and others). Its pars dorsalis receives the electric, its pars medialis the mechanoreceptive afferents. Following the usage of most recent workers, hereafter I call the nucleus medialis pars dorsalis the PLLL (posterior lateral line lobe) (Maler 1974 and others). The structure and the cells of the medial nucleus have been described in catfish (Pearson 1936, Andrianov & Volkova 1976). Entering myelinated afferents from the ampullary or "small pit organs" terminate in the more highly differentiated pars dorsalis, dorsalis, which is distinct from the pars medialis ("anterior lateral line lobe"), which receives lateral line mechanoreceptor afferents. There is only slight overlap of areas of termination of ampullary afferents and eighth nerve afferents in catfish. In mormyrids, where in addition there is a new submodality of electroreceptive input, the tuberous afferents, it appears that the ampullary input is generally segregated from the tuberous in the ventrolateral part of the medial nuclear complex (Bell & Russell 1976, 1978b).

Because the sensory nucleus is different for ampullary receptors in nonteleosts and in electroreceptive teleosts, and there is no compelling similarity of structure of the sense organs, or adequate knowledge of their origins, the ampullary systems of nonteleosts cannot seriously be homologized with those of teleosts. Indeed, it is not clear whether the receptors and nuclei should be considered homologous in siluriforms, gymnotiforms, and mormyriforms, except in the sense that they are all homologous to the lateral line system. As independent developments, cropping up in separate lineages, they exemplify convergent evolution and a mixture of similarities and differences in structure and function.

Projections from the dorsal nucleus in elasmobranchs are mainly to the torus semicircularis (TS) of the midbrain, but also to the optic tectum and to the reticular formation, including a part that may be equivalent to an olivary nucleus (McCormick 1978, 1981a,b, Boord & Northcutt 1979, Northcutt 1981a). Higher order projections are little known, but there is peroxidase evidence for axons from the TS to the cerebellar corpus and the tegmental reticular formation as well as physiological evidence of projections to a circumstribed part of the telencephalon in sharks (Platt et al 1974, Bullock 1979b).

In teleosts the projections from the medullary nuclei (PLLL) are best known in mormyriforms and gymnotiforms, but the ampullary projections are only beginning to be distinguished from the tuberous. Tuberous system projections are treated below. The chief targets of the medullary nuclear output are the TS, nucleus praeeminentialis, nucleus lemniscus lateralis, and reticular formation. The TS in gymnotiforms concentrates its ampullary input in laminae 8b and/or 8d and 9 (Carr et al 1981). The TS in catfish has been shown by Knudsen (1976a,b, 1977), using both anatomical and physiologic techniques, to segregate electric from both common lateral line and acoustic input into three subdivisions: respectively, lateral, central, and medial. The electric or ampullary division is itself differentiated, topographically mapping the body surface. Additional physiological analysis is treated below.

From the TS, in the catfish where we do not have the complication of a tuberous system, projections have been traced to tectum, reticular formation, nucleus praeeminentialis, pretectum, and a specific part of the thalamus. The electroreceptive part of the thalamus is separate from the common lateral line mechanoreceptive part (Finger 1980, Finger & Bullock 1982). From ampullary laminae of the TS in gymnotiforms, projections go to a lateral thalamic region bilaterally, to the tectum ipsilaterally, to the nucleus praeeminentialis ipsilaterally, and to the reticular formation (Carr et al 1981). Electric sense receiving areas of the telencephalon are to be expected but have not yet been reported.

Physiology

RECEPTORS   More than a dozen papers have accumulated since the reviews of Fessard (1974), Fessard & Szabo (1974), Kalmijn (1974), Murray (1974), Viancour (1979c), and Bell (1979). A series from Leningrad deals with rays and catfish. Broun et al (1972) confirmed Kalmijn (1966, 1971, Dijkgraaf & Kalmijn 1963, 1966) with respect to respiratory and cardiac responses to weak electric fields in elasmobranchs and attribution of the sensitivity to the ampullae of Lorenzini. They reported a sensitivity, as expected, to a changing magnetic field, not only behaviorally but in impulse rate of single ampullary fibers (Brown & Ilyinsky 1978). This impulse response depends on (a) the rate of change of the magnetic field or of movement of the fish or of the water in a steady field, (b) the orientation of the canal (actually, probably that of the straight line between the ampulla and pore to the outside) relative to the induced currents, and (c) the length of the canal. Akoev & Ilyinsky (1973) and Akoev et al (1976a,b) claimed that in skates there are not only ampullary afferent fibers that increase their impulse discharge when the pore, hence the free end of the hair cells, is made cathodal (and at the "off" of anodal current), as known from Murray (1960, 1962, 1965, 1974) and Obara & Bennett (1972), but also other fibers that are excited by anodal current (and cathode "off"). Heretofore the latter type of fiber has been known only in teleosts (catfish, confirmed by Zhadan et al 1976), and it is believed that the polarity for excitation is a consistent difference between teleosts and elasmobranchs. (Known to all of these authors was the excitation to the reverse polarity at high stimulus intensities.) The claim that sensitivities to both polarities exist in afferent fibers in skates cannot be regarded as established without assuring, for example, that the fibers could not have supplied adjacent ampullae whose pores were far apart; and that the stimulating anode was not by chance just outside the skin over an ampulla, or too near a gill slit. The Russian authors supported the idea, which has been cautiously suggested before, of object detection by distortion of incidental low frequency (ca. 1–10 Hz) electric fields. Ampullary afferents alter their ongoing impuse rate for objects, both nonconducting and conducting. Akoev and co-workers invoked the animal's own "periodic respiratory potential field." They also proposed that *Torpedo* discharges its electric organ in a low voltage mode for the purpose of orientation and detection of prey; this is found to alter the activity of the electroreceptors. Several differences in spontaneous activity between the ampullae of Lorenzini in vitro, as usually studied, and in vivo, as Akoev et al studied them, are regarded as significant. Adaptation is attributed to the receptor cells, rather than to the nerve fiber terminals. It is emphasized that maintained stimuli can elicit in some receptors a phasic-tonic response, i.e. adaptation is not complete (Zhadan et al 1976). Latency, stimulus duration and interval effects, temperature, drug, and ionic and salinity effects are described in skates and catfish (Akoev et al 1976a,b, Akoev et al 1980a,b, Brown et al 1980). Warming, which was known to slow down afferent firing rate phasically (Sand 1938), is said to increase input resistance of canal and ampulla and to cause a potential shift in the ampulla-negative direction (Broun et al 1980). This direction should excite and the authors discussed but did not reconcile the discrepancy with Clusin & Bennett (1977a,b), beyond invoking the possibility of involvement of the basal membrane.

A good deal is known about the mechanism at the receptor cell, the locus of its receptor potential, the calcium dependence and calcium conductance increase, sodium independence, oscillatory character, and other properties. These have been elucidated by Roth (1971), Obara & Bennett (1972), Obara & Oomura (1972, 1973), Obara (1974), Akutsu & Obara (1974), Peters et al (1975), Zhadan & Zhadan (1976), Bauswein (1977), Clusin & Bennett (1977a,b, 1979), Bennett & Clusin (1978, 1979), and Obara & Sugawara (1979) and have been reviewed by Obara (1976), Viancour (1979b,c), and Schouten & Bretschneider (1980). Briefly, the elasmobranch receptor potential is a depolarizing, inward Ca current at the *apical* (i.e. lumenal, ciliated) membrane, followed by a late outward K current, in the presence of a standing current outwards across that membrane to the ampullary jelly (Clusin & Bennett, 1977a,b). Hence, the Lorenzinian ampulla is excited when a cathode is at the pore of the skin where the canal opens. The catfish receptor potential is a depolarizing, inward Ca current at the *basal* membrane. Hence, the teleost ampulla is excited by an anode at the pore. Okitsu et al (1978) have analyzed the canal jelly in the marine catfish, *Plotosus* (whose ampullae have a long duct resembling the canal of the Lorenzinian ampullae of marine elasmobranchs, whereas fresh water teleost ampullae have short ducts—as do those of fresh water rays, sturgeons, bichirs, lung fish, and others). They found evidence for a steady bias current flowing inwards through the apical and outwards through the basal membranes of the receptor cells, arising from a high potassium in the jelly, as in *Xenopus* (Russell & Sellick 1976) and mammals (Sellick & Johnstone 1975).

The synaptic transmission between hair cell receptor and afferent nerve fiber has been studied (Bennett 1965, 1971, Roth 1973a, 1978, Obara 1976, Obara et al 1980, Higuchi et al 1980, Umekita et al 1980, Teeter & Bennett 1981). It conforms to chemical transmission, and evidence indicates that glutamate is an agonist or mimic of the transmitter but is not the actual transmitter. Roth (1971, 1972, 1973a), Peters & Buwalda (1972), Dunning (1973), Bromm et al (1976), McCreery (1977a,b), Bretschneider et al (1980) and J. DeWeille (to be submitted) have provided some details on the properties of the response of afferent units, including their best frequency (ca. 10 Hz).

Ampullary receptors have been little studied in electric fish. As in catfish, they are low frequency band pass filters (Suga 1967), but can be noticeably influenced by the EOD in mormyriforms (Bell & Russell 1978a), presumably by a DC component.

CENTRAL NERVOUS SYSTEM   A good many authors have studied the responses of lower and higher levels of the brain to electroreceptor input, using evoked potential and single unit measures. Much of this literature has been reviewed by Andrianov & Enin (1977) and Bell (1979). In the present context it is useful to survey first the responses of the medulla to ampullary input, chiefly in elasmobranchs and siluriforms, then to deal in turn with other levels of the brain.

Nonteleost ampullary receptor input to the octavolateral area of the **medulla** was first examined by Il'inskii et al (1972) in rays (*Raja*), comparing evoked potentials to short pulse stimulation of various branches of the cranial nerves that supply ampullae of Lorenzini or lateral line canals. They found differences in the evoked potentials to the different nerves both in the form and in the focus of maximal activity within the octavolateral region. Platt et al (1974) also stimulated electroreceptive nerves and mechanoreceptive nerves in the elasmobranch (*Torpedo*) and found in addition differences in the facilitation or antifacilitation to a train of stimuli. Recording from single units in the elasmobranch (*Trygon*), Andrianov et al (1974) and Andrianov & Broun (1976) showed that most units are spontaneously active and modulated in rate, up or down, according to the polarity of current in the water or a change in the magnetic flux. Two main populations of units were distinguished in respect to polarity: those excited by applying an anode or a south pole to the ventral side and those excited by a cathode or north pole applied to the ventral surface; both are inhibited by the opposite stimuli. Four main patterns of response were observed: phasic, phasic-tonic, tonic sustained (persisting after cessation of the stimulus), and bursting. Paul & Roberts (1977b,c) also recorded from the octavolateral area and distinguished field potentials from the dorsal nucleus, where afferent axons terminate, and from the overlying molecular layer. They also distinguished incoming primary fibers from second order cells in the nucleus and from efferent neurons. It is likely that they were mainly dealing with mechanoreceptive units.

In teleosts (*Ictalurus*) Andrianov & Ilyinsky (1973) and Andrianov & Volkova (1976) studied second order units and found much the same as in skates. They noted a bimodal sensitivity, to both electrical and mechanical stimulation. Roth (1975) described the properties of units (in *Kryptopterus*) located deep to the cerebellar surface in its posterolateral corner: most likely they were in the medial nucleus of the octavolateral medulla. He distinguished three main types of second order units and compared them with primary efferents. Type *a* are much less active spontaneously and, at 1 to 10 Hz sinusoidal stimulation, 10 to 30 times more sensitive than primary afferents. Type *b* have no spontaneous activity and are most responsive to stimuli below 1 Hz. Type *c* has a regular spontaneous activity and best frequency of 10 to 20 Hz. McCreery (1977a) also examined the second order neurons (*Ictalurus*) and found that those excited by anodal (inward) current through the hair cells, like all the first order afferents, receive monosynaptic excitation from those afferents. The second order units excited by cathodal (outward) current receive disynaptic inhibitory input from the afferents. He also noted that the most effective stimulus frequency for second order units is lower than for primary afferents. J. DeWeille (to be submitted) also found this, using a white noise cross-correlation technique: the first order axons have their best frequency at ca. 11 Hz, while the system including first and second order neurons, choosing second order cells whose receptive field embraces the first order unit studied, has its best frequency at ca. 7 Hz. First order Wiener kernels adequately describe afferent axon responses but second order predictions are considerably better than first order for second order units, which suggests that receptor properties are followed by an integrative filter and nonlinearity.

Receptive fields of second order units are from a few to many times larger than those of first order afferents, sometimes with two disjunct maxima, usually more rostral for caudally moving stimuli and vice versa (McCreery 1977b). The size of the receptive fields is thought to represent a compromise between enhanced sensitivity due to spatial averaging and loss of spatial resolution, perhaps matching the spatial frequency of the analyzer properties with that of the normal stimuli. The temporal frequency properties of the neural filter are likewise thought to be matched to those of the natural stimuli.

It is to be presumed that there will be found a topographic mapping of the body surface in the dorsal nucleus of elasmobranchs and their allies and in the medial nucleus of siluriforms. Such a map has been found in the cortex of the ventrolateral zone of the PLLL of mormyriforms, which is the region receiving ampullary afferent terminals (Bell & Russell 1978b). In this region the same authors (see Bell 1979) found two cell types much like McCreery's in the catfish.

In and under the **tectum opticum** of elasmobranchs, ampullary input causes a conspicuous and moderately complex series of evoked waves (Platt et al 1974, Bullock 1979b, 1981, Bullock & Corwin 1979). Alterations in the waveform show sensitivities to electric field orientation, sign of the change of current, position of a dipole, and serial number in the first few cycles of a train, as well as interval between electric events in a series. Separate component peaks of the evoked wave complex have different dependencies upon these factors. To sinusoidal stimuli the "best" response by a certain criterion is at 20 to 30 Hz in a shark (*Carcharhinus*), 10 to 15 Hz in a freshwater ray (*Potamotrygon*). The lowest threshold, with moderate averaging, was 0.015 μV/cm (= 0.8 nA/cm^2) in the marine shark, <50 μV/cm (= 0.7 nA/cm^2) in the freshwater ray. The response depends on the locus of recording, the type of electrode, and the state of the brain. There is evidence of topographic segregation of some stimulus parameters but mapping has not been undertaken in elasmobranchs. The best loci for electric evoked potentials are distinct from those for acoustic, although they are contiguous and appear to overlap. No interaction between them has been found. There appear to be differences between species in respect to dynamics of responses and preferred stimuli, that correlate with behavior.

In siluriform fishes, evoked potentials give similar evidence and, from their generality among all 14 genera tested in 9 families, allow us to believe that this sense is common to the very large and diversified order (ca. 30 families) of catfishes.

Detailed studies of the **torus semicircularis** (TS) of catfish showed that a circumscribed, lateral part is responsive to ampullary input; a medial part is acoustic and the intermediate region responds to lateral line input (Knudsen 1976a,b, 1977, 1978). In the electroreceptive part of the TS tactile, acoustic and optic stimuli also influence unit activity. Five types of units based on responses to DC step stimuli are (a) phasic units with a high resting discharge rate, (b) phasic units with a low resting discharge, (c) tonic units, (d) phasic-tonic units, and (e) weakly responding erratic units. Unit latency varies widely, from 8 to 130 ms, usually 8 to 28 ms, as a function of stimulus intensity. Within each type there is a sepctrum of best frequencies of sinusoidal stimuli, from <0.1 to ca. 25 Hz, and of sensitivites at the best frequency, from 0.8 μV/cm to 15 μV/cm (8×10^{-11} to 15×10^{-10}A/cm^2). Four types of units based on phase locking to sinusoidal stimuli are distinguishable: monophasic multispiking, monophasic oligospiking, biphasic multispiking, and biphasic oligospiking. Another difference is that increase of stimulus intensity can advance or retard or do nothing to the phase. Some units show after-discharge following cessation of a high intensity stimulus; most do not. Some show a frequency specific inhibition and this is typically a frequency above the best excitatory frequency. Unit receptive fields show a somatotopic organization of electroreceptive input, primarily contralateral, with the main axis of the body represented by an anterior-posterior axis in the TS. Besides longitudinal somatotopy there is a systematic dorso-ventral distribution of preferred electric field orientation. The evoked potential usually shows somewhat different orientation preference than the units isolated in the same region. There is also a tendency for a systematic distribution of best frequencies of sinusoidal fields; dorsal units are tuned to 0.05–10 Hz, ventral units to 4–15 Hz. The more dorsal unit spikes are also wider (1 ms) than the ventral (0.5 ms).

The **cerebellum** is unusually large in elasmobranchs, gymnotiforms, mor-

myriforms, and most siluriforms. In mormyriforms it is also highly specialized histologically. The suspicion has long been voiced, but still cannot be
settled, that this is an enlargement and elaboration associated with the
development of electroreception. Studies have shown an important representation of this modality, but not enough as yet to explain the unusual size
or specialization.

Ampullary input in sharks and rays gives rise to evoked potentials in the
lobus caudalis (auricle) and corpus cerebelli (Platt et al 1974, Bullock
1979b). The waveforms, latencies, and dynamics cannot be given within the
space available here. In catfish, electroreceptor input excites or silences
units in the lateral part of lobus caudalis, not in the medial part or in the
corpus cerebelli (Tong & Bullock 1982). In mormyriforms, ampullary
input goes to a large medial area of the valvula (Russell & Bell 1978, Finger
et al 1981).

Ampullary input causes localized evoked activity in the **telencephalon** in
elasmobranchs (Platt et al 1974, Bullock 1979b) in a region in the medial
third, middle third antero-posteriorly and central third dorsoventrally. It
appears to overlap with the acoustic response area but the foci of maximal
response are separate. Something is known of the dynamics of response but
no studies on units or best stimuli have been done as yet, nor have there
been any but preliminary explorations in siluriforms or electric teleosts. The
principal finding so far is therefore that even in the telencephalon this
modality is still distinct. That puts it in agreement with each of the other
modalities so far examined in fishes—not converging into a common integrative sensorium in the midbrain, as was once believed, but projecting
independently to the forebrain, as in mammals. How integration between
modalities takes place remains to be discovered.

RECEPTION BY TUBEROUS SYSTEMS

A special class of receptors called "tuberous," from their gross histology,
is adapted to detect the brief, feeble pulses of current discharged by the
electric organs of electric fish of the two orders Gymnotiformes and Mormyriformes. The receptors detect (*a*) the small differences in local intensity
caused by field distortions associated with objects and (*b*) the frequency
modulations associated with social communications. The former role can
be spoken of as active and the latter as passive electroreception. I deal briefly
first with the behavioral findings of recent years.

Behavioral Evidence

OBJECT LOCATION   Active electroreception for object localization and
discrimination was first demonstrated by Lissmann & Machin in 1958. It

has been repeatedly confirmed since then, most recently by Schlegel (1975), Heiligenberg (1975a, 1976), Meyer et al (1976), Feng (1977), Bombardieri & Feng (1977a), Matsubara & Heiligenberg (1978), and Bauer (1979). J. H. Meyer (to be submitted) adds behavioral evidence of discrimination of complex impedance, for which physiological evidence of receptor discrimination had been reported (Scheich & Bullock 1974, Feng & Bullock 1977; see below); that is, the EOD/electroreceptor system can discriminate not only local discontinuities in the water in conductance but also small parallel capacitances.

Most studies of electrolocation use objects and motion but Meyer et al (1976) and Feng (1977) measured postural tilt of a fish hovering above a tilted floor that had to be detected electrically. The fish's vertical axis approached 90° to the floor in both longitudinal and transverse planes. Although there is abundant empirical data on sizes and distances of objects detected, such as rods, plates, and fish dummies, to my mind we have not yet found the best measure of the performance of this system. For example small *Brienomyrus niger* under certain conditions (water conductivity etc.) react to a 4 mm plastic rod at a distance of 30–40 mm (Heiligenberg 1976); *Apteronotus albifrons* react to a 3 mm metal rod at 60 mm (Bombardieri & Feng 1977a). However, it is not clear what the world looks like through this new sense and what the best stimuli and the limiting cases are, e.g. object or aperture sizes, shapes, movements, or complex impedance. Heiligenberg (1975b) experimented with a digital simulation of a fish, its EOD field and distortion due to objects; he showed the importance of body geometry and resistances of body interior and of skin but did not test a wide range of object spacings, shapes, and motions. Hoshimiya et al (1980) simulated the EOD field in a wave species. Comparison of animal performance and simulation experiments raises one's confidence in the simulation approach; however, major assumptions have to be made about the distribution of resistances in the skin and sense organs. Bastian (1981a) has recently made the first direct study of the signals available as a result of objects that distort the EOD, by measuring the changes in EOD voltage across the skin, between a reference electrode in the stomach and a semimicroelectrode just touching the skin. He gives the amplitude modulation and the spectral change as functions of object size, shape, conductivity, distance, and movement velocity. It is therefore possible now to specify what amplitude change sensitivity would be necessary to detect an object of given position and character. Bastian goes on to record the first and second order afferent unit responses to specified amplitude modulation (see below, Physiology).

A new clue in active electrolocation has been discovered by Chichibu (1981), who has found that the mormyrid *Gnathonemus* fires its left and its right electric organs alternately. The small distance between them and the resulting asymmetry of their EOD fields could provide enhanced depth perception. A separate clue, which has been hinted at for years, is due to imperfect synchrony of discharge of the electrocytes in the electric organ. This makes the EOD waveform different as recorded at different points along the body or at different distances; hence the distortion by objects will be different, according to their distance, from what it would be if the electrocytes were synchronous. Hoshimiya et al (1980) have examined this using a computer simulation approach to the EOD of the high frequency wave fish, *Apteronotus*. Under the section on the physiology of receptors, below, evidence for regional differences in their tuning curves is discussed.

SOCIAL COMMUNICATION   Weakly electric fish offer unparalleled op-portunities for quantitative work leading to inferences about the neural operations that must be taking place in social behavior. Such work is a form of comparative psychophysics in which the behavioral output is continuous, easy to measure completely and precisely, and, for some modes of behavior, little influenced by fluctuations in the motivational state, and in which the relevant input can be readily identified, manipulated, and simulated. More than 25 papers have appeared in recent years. I cannot review here those dealing primarily with the signal value or social significance of EODs and their modulations (Pimentel de Souza 1969, 1971, 1973, Bell et al 1974, Kramer 1974, 1976a–d, 1978, 1979, Kramer & Bauer 1976, Westby 1974a,b, 1975, Moller 1976, Hopkins 1972, 1974a–c, 1977, Hopkins & Bass 1981, Serrier 1979, Toerring & Belbenoit 1979). These and earlier studies indicate that the EOD repetition rate and its modulations are the main if not the only parameters varied in connection with social stimulation and communication in most species. Some species have two or three forms of EOD produced by the same individual, differing in waveform and ampli-tude, and employed in different contexts. Some species have different types of EOD in males and females or in juveniles and adults.

Certain forms of response manifest not only a sensitivity to pulse rate modulation and power spectrum but to precise phase relations and timing both within and between individual EODs, the fish's own and its neighbor's. Feedback experiments show critical periods: several mormyriform species are much more sensitive during a period from 2 ms to 20 ms after the fish's own EOD than at other times (J. H. Meyer, C. C. Bell, and W. Heiligenberg, unpublished); *Gymnotus* abruptly changes from high sensitivity to very low sensitivity about 1 ms after each EOD and then back to high sensitivity about 15 ms later—out of a usual EOD period of 20 ms (Bullock 1969, MacDonald & Larimer 1970, Westby 1975). *Hypopomus artedi* discrimi-nates between electric pulses with identical spectral amplitudes but different spectral phase functions (Heiligenberg & Altes 1978). Species differences in waveform are especially notable among the Mormyridae, which discharge 0.2–6 ms pulses at <1 to >50 per second; the broad power spectrum has a peak that varies between species from 0.15 to 15 kHz. Behavioral evidence in the field and by playback experiments and tank experiments with mixtures of species shows recognition of the individual's own species waveform (Hopkins 1980, Moller & Serrier 1980, Hopkins & Bass 1981; See *Physiology,* Mesencephalon, Mormyriforms, below); the same accomplishment has independently evolved among gymnotiform pulse species (Hopkins & Heiligenberg 1978, Heiligenberg & Bastian 1980b).

The most studied form of behavior in electroreceptive fish is the jamming avoidance response (JAR) of electric fish. This is due to its ready quantifiability, nonhabituating reproducibility, and the attractiveness for neuronal analysis; it is higher in level than a spinal reflex and is really a form of social response that particularly involves the midbrain. Yet we have a head start in knowledge of the few principal receptor classes involved and the single command unit, which is a pacemaker driving the EOD 1:1, hence controlling the single dimension of the behavior: all-or-none EOD temporal sequence. Discovered (Watanabe & Takeda 1963) and most extensively examined in the highly regularly discharging wave species of the genus *Eigenmannia* (Gymnotiformes, Sternopygidae), a remarkably similar JAR has been found in several members of the family Apteronotidae and in the unrelated mormyriform fish, *Gymnarchus* (Bullock et al 1975), which can be regarded as a case of convergent evolution. One wave fish is known to lack any JAR, the very low frequency and low voltage *Sternopygus* of the same family as *Eigenmannia* (see *Physiology,* Medulla, below). Pulse species have developed somewhat different forms of response that likewise accomplish an avoidance of jamming. I deal with these in turn: in this section only with the behavioral analysis; with the physiology, in a section below.

Earlier work has shown that the adequate stimulus for the JAR of *Eigenmannia* is the difference between its EOD rate and that of a neighbor or substituted sine wave current in the water; this difference, $F_{stimulus}$-$F_{fish,}$ is called the ΔF and the JAR is a shift in pacemaker frequency in the direction to increase ΔF, optimally to ca. 10–15 Hz (Bullock et al 1972a,b). These authors described the time course, stimulus intensity function, and JAR amplitude/ΔF amplitude function and found a best ΔF of 3–4 Hz, and a ceiling and floor EOD frequency beyond which the fish would not go. From these quantities a model predicted quite well the form of responses to step, sine wave-FM and sawtooth-FM stimuli of any rate of change. Later Scheich (1977a–c) introduced the technique of curarizing to eliminate the fish's own EOD, substituting an artificial EOD which could be controlled in waveform and driven cycle by cycle from the still ongoing command unit in the medulla, recorded with a microelectrode.

Under his conditions, particularly the transverse field geometry, Scheich found that a necessary and sufficient clue for the fish to recognize both the magnitude and the sign of the ΔF was given by the phase-specific harmonics in the normal EOD, imitated by slant-clipping a substituted sine wave in the right polarity. This makes the envelope of the beating with the simulated neighbor asymmetrical in opposite senses for plus and for minus ΔF. Neurons in a hierarchical order capable of analyzing this temporal asymmetry are mentioned below, under Physiology.

Subsequently Heiligenberg et al (1978a,b), Heiligenberg & Bastian (1979, 1980a), and Heiligenberg (1980a) introduced a new technique and a more natural field geometry. The substitute for the curarized fish's own EOD, called S_1, was delivered between an electrode placed in the alimentary canal and one just behind the tail, thus approximating the real EOD field shape. The simulated neighbor, S_2, was delivered between two other electrodes asymmetrical to the fish, for instance, transversely. Under these conditions slant clipping is not necessary, pure sine waves are adequate for both S_1 and S_2; the fish's pacemaker shows a normal JAR. Evidently the spatial inequality of the two fields provides a clue, in different degrees of "contamination" of the fish's "own" S_1 with S_2 in different parts of the skin, presumably read by the receptors in terms of the direction of changes in phase and in amplitude during each beat cycle. If amplitude is increasing, while phase (for example of zero crossings in a positive direction) is advancing over a portion of the body surface, relative to the rest, it signifies a ΔF of one sign; if amplitude is increasing while phase is retarding, it signifies the opposite sign and triggers the appropriate direction of modulation of the pacemaker. Neurons relevant to this are mentioned below.

The evidence therefore shows that the fish has two distinct methods of extracting the sign of the ΔF from the receptor input. Heiligenberg regards the analysis of spatial differences as more important, but normally both this and the waveform clue are available. It is noteworthy that the brain is not required to identify some input as "own" and some as "other fish's EOD field," or to form an image. The evaluation could adequately occur by a distributed system; the minority and majority subsets need not be anatomically defined.

Further experiments showed that S_1 does not have to be at the pacemaker frequency or harmonically related to it. Within a wide range, any frequency will do, providing that S_2 is 3–4 Hz above or below it. A convenient graphical method of predicting the response from arbitrary and unnatural sequences is a two-dimensional state-plane plot of amplitude against phase at each moment; adequate stimuli are sequences of points, normally but not necessarily continuous, that define a closed figure in a certain direction. On the basis of such graphs, JARs to simultaneous stimulation by two or more simulated neighbors of different ΔF and intensity can be predicted successfully (Partridge & Heiligenberg 1980)—meaning that the brain acts as though the effect of a complex graph (e.g. one looking like flower petals) is the integral of incremental contributions by line segments, with the effect of a single line segment depending on its location and orientation within the state-plane. Perhaps the processing in the brain is more complex than this, but it is significant—and an elegant result of behavioral analyses—that a normal response to a social stimulus in a vertebrate can be so adequately accounted for, quantitatively, by such a simple, physiologically and anatomically undemanding model. Indeed, each of two models is adequate, the above (Heiligenberg's) and the preceding one (Scheich's), although Scheich's is not quantitative and cannot predict the response to arbitrary stimuli. Nevertheless, both mechanisms are present, if unequal in importance in this species.

Pulse species, too, respond to the proximity of other fish with EODs that threaten to jam or coincide with their own. Since the EOD intervals are relatively long (tens to hundreds of ms) and irregular (usually from 10 to several hundred percent of the mean), there is not the continuous beating of sinusoidal EODs, and coincidence is rare. Nevertheless, in species or states with rather steady intervals (coefficient of variation ca. 20%), two fish at not too different repetition rates will fire at progressively changing phases such that each fish scans through advancing or retarding phases relative to the other. The finding is that in such gymnotiforms a transient shift of EOD rate is commanded by the pacemaker, depending on the species, on the direction of the phase change (sign of the ΔF), and its rate—hence, the number of EOD cycles during which a foreign pulse occurs within a specified time window, and the recent history of stimulation or rest. This serves to reduce the probability of coincidence and number of consecutive cycles with a foreign EOD in the critical time window, as well as to speed up the march of phases. For such reasons this behavior has been called a jamming avoidance response and its characteristics have been examined in several species (Bullock 1969, MacDonald & Larimer 1970, Westby 1975, 1979, Scheich et al 1977, Heiligenberg et al 1978a,b, Baker 1980, 1981, Heiligenberg 1980a,b). Experiments with curarized fish demonstrate that the JAR is controlled by electroreceptor input alone and without reference to any internal pacemaker-related signal; this applies only to gymnotiforms. To elicit a JAR (transient in pulse species, in contrast to the tonic JAR of wave species), a train of strong, EOD-like stimulus pulses (S_1) suffices, simulating the normal in spatial and temporal pattern, plus a train of weak pulses (S_2) of slightly different rate and geometry. The behavioral data permit a model of the electroreceptive feedback and its slight perturbation by a weak S_2, in terms of excitatory (E) and inhibitory (I) processes triggered at specific phases of the S_1 cycle by S_2. S_2 pulses that just precede S_1 pulses activate an E process, causing acceleration of the pacemaker; those at near-zero latencies activate an I process, preventing response to the first, and/or causing deceleration; the process activated first dominates the response. Neurons especially relevant are mentioned below.

Mormyrids, all of which are pulse species, are generally more irregular and seldom experience a slow scanning of one fish's train by another's. Therefore the succession of several cycles with S_1 progressively led or followed by S_2 is rare. One reason for this is a special form of behavior that is often but not invariably seen under circumstances close to these: so-called "echoing" (Russell et al 1974). One fish fires its EOD pulse at a fixed, short interval after the other, for example, 8 ms. This prevents jamming since the sensitive window is narrow. As we see below, these fish have, and use, a central signal corollary of the pacemaker command not found in the gymnotiforms.

Some fish of the latter group (*Gymnotus*) employ almost the same trick; two fish avoid coincidence by firing their EODs alternately for thousands of cycles (several minutes), with a fixed time difference. The time of firing is not necessarily in the insensitive part of the cycle. Westby (1975, 1979) suggested that a dominant fish may impose an EOD on the other during a sensitive window!

Although they earlier proposed (Heiligenberg et al 1978a) that a basically common JAR was first evolved in pulse species, implying that wave species are derived, Heiligenberg et al (1978b) later emphasized that the JARs are importantly different. Not only do pulse species have more than one form of jamming avoidance, but wave species may have a relict mechanism less dependent on T units and a superimposed, more influential mechanism crucially dependent on both P and T units (Baker, 1980, 1981, Heiligenberg 1980a,b).

Avoiding coincidence is not the only form of response, though it is by far the most common. Occasionally two wave fish will actively phase lock for periods of minutes (Langner & Scheich 1978). From the point of view of object location, this is not necessarily a serious jamming situation; experimentally it is found that object detection is almost as good with $\Delta F = 0$ as with a large ΔF (>20 Hz) or no simulated neighbor. Jamming is harmful, judging by performance, when the beat frequency is >0.5 and <6; steady or very slowly changing electroreceptive input on top of the fish's own EOD field is compatible with object detection, as is rapid beating whose corresponding input flicker is removed by low pass filters in the brain (see below and Behrend 1977, Scheich 1977a,c). But it seems likely that there is some other significance to the high precision active phase coupling, with jitter of only a few microseconds, seen from time to time in some very high frequency wave species (*Sternarchorhynchus* sp. and *Sternarchorhampus* sp.).

Langner & Scheich (1978) suggested that it might be a form of electrical hiding from a dominant, conspecific fish. It appears to be more complex than the JAR shown by the same fish. The JAR is only manifest when a neighbor or S_2 is within 10–15 Hz. Active phase coupling starts and ends suddenly, not gradually, and can occur when ΔF is 100 Hz so that the coupling is to every *nth* cycle of the stimulus, n being up to 10. For example, the phase was precisely maintained for some minutes by a fish at 1346 Hz to a stimulus at 1224 Hz, a frequency ratio of 11:10, making a standing pattern on the oscilloscope, until the fish reverted to its normal resting frequency of 1334 Hz, which makes a rapidly moving pattern. The minimum pathway is believed to have four synapses and to take several EOD cycles; electrotonic junctions would seem to be required to sustain the temporal precision.

Anatomy

Recent years have seen marked advance in tracing connections of the electrosensory system, especially by axonal transport methods. This typically means sources of afferents and destinations of efferents at the level of nuclei. For certain taxa and brain structures the cell types within a nucleus are also becoming known and, to some degree, the connections of these types within the nucleus. Because of space limitation, the microstructure of the periphery, including the sense organs, must be omitted; the review of Szabo (1974a) and recent papers of Srivastava & Szabo (1974), Srivastava (1975, 1978a,b), Baillet-Derbin (1978), and Denizot (1978) may be consulted for further citations. Table 1 summarizes the current status and terminology of the receptors.

MEDULLA   Four separate somatotopic maps are found (Bell & Russell 1978b) within the electroreceptor region of the posterior lateral line lobe (PLLL) of **mormyriforms**—three in its cortex and one in its nucleus. The nucleus receives from knollenorgans (K units), the ventrolateral zone of the cortex from ampullary receptors (F units), and the two dorsolateral zones from mormyromasts (D units), probably each from a different subtype. Mainly ipsilateral, a few terminals from the contralateral posterior lateral line nerve occur in the cortex. Details—including endings in the intermediate cell and fiber layer and the granule layer, the synapses and the second order neurons—are given by Bell & Russell (1978) and by Maler (1973), Maler et al (1973a), Szabo & Ravaille (1976), Szabo et al (1978), Szabo & Libouban (1979), Bell (1981), and Bell et al (1981). One of the targets of interest is the population of adendritic, large, round cells that receive many small boutons of two types as well as a few large (2–9 μm) club endings with gap junctions. The club endings themselves also receive presynaptic terminals with gap junctions and these, in turn, receive boutons with the characteristics of chemical junctions. The club endings and round cells are believed to be part of a high speed pathway to the nucleus exterolateralis mesencephali, part of the torus semicircularis (TS), from the knollenorgans (K units), bilaterally (Enger et al 1976b, Szabo et al 1979a, Haugedé-Carré 1980 cited by Bell et al 1981).

The **gymnotiform** PLLL is also complex, with seven main laminae and 11 cell types (Maler 1979), which for the most part are not to be homologized with those of mormyriforms. The essential features of both taxa are that they have a true cortex, many more cell types than receptor fiber types, intrinsic circuits, regional differentiation, segregation of terminals and second order cells for the several receptor types, and descending input, particularly from lobus caudalis (Maler et al 1973a,b, 1974, 1981, Maler 1974). C. Carr (in preparation) reports somatotopic maps in PLLL.

1. F unit (ampullary) input appears to end on a class of cells isolated from the rest of the PLLL circuitry, called multipolar, with large receptive fields; however, the PLLL representation of the ampullary input is not as well known here as it is in mormyrids.
2. The T unit input goes to a private set of second order cells, with some convergence and with gap junctions. These junctions, large efferent axons, and other cytological particulars correlate with the high velocity and synchrony of impulses in this pathway. It is a highly specialized system for preserving tightly phase-locked signals 1:1 with the EOD and projecting them to the nucleus magnocellularis mesencephali of the TS, analogous to the K unit system in mormyriforms from the PLLL nucleus to nucleus exterolateralis mesencephali.
3. The P unit input goes to two sets of second order cells, called basal pyramids and nonbasal pyramids, as well as to granule cells in a defined surrounding zone. The distribution of excitatory chemical and electrical synapses from both P axons and granule cell axons and of inhibitory synapses from granule cell axons leads Maler et al (1981) to predict that the basal pyramids are center-surround units excited by good conductors of small size near their receptive field center and inhibited by such objects off center, whereas the nonbasal pyramids are similarly excited in the center and inhibited off center by nonconducting objects. Such units are physiologically known (Scheich 1977c), but the identity with these anatomical types remains to be confirmed. Additional complications and types of responses are expected from cell types whose connections have not been as well analyzed and from descending input. Maler et al (1981) contributed cytological detail and correlations of general neurobiological interest.

The EOD command nucleus is of special interest as a center, with three features: spontaneity, an extreme range of regularity (from a coefficient of variation of $<0.01\%$ at 1000 Hz to $>100\%$ at a mean of ca 1 Hz, in different species), and modulation by specific inputs that alter the output reliably and meaningfully from $<0.1\%$ to $>>100\%$. Smaller, pacemaker cells, from 14 to 400 in different species, are electrically coupled and make gap junctions with larger, relay cells, also electrically coupled, that project to spinal electromotor neurons. Both types of cells receive chemical synapses of several kinds as well as gap junctions (Ellis & Szabo 1980, Akert et al 1980, Tokunaga et al 1980). There are disagreements on essential points between the description for *Apteronotus* by the latter authors and that by Elekes & Szabo (1981). Because of its value for general neurobiology, this nucleus deserves further attention. Inputs to it come from the TS (Ebbesson & Scheich 1980) and a prepacemaker nucleus (Heiligenberg et al 1980).

MESENCEPHALON   Bell et al (1981) and used horseradish peroxidase and tritiated amino acids to trace connections both to and from PLLL in mormyriforms; they found that somatotopically corresponding parts of each of the three cortical maps of PLLL project, bilaterally, to one point in the nucleus lateralis mesencephali (see also Haugedé-Carré 1980). Thus, pathways from ampullary and two subtypes of mormyromast receptors are distinct up to the midbrain but there converge, while the fast knollenorgan pathway to the nucleus exterolateralis mesencephali remains separate. The nucleus lateralis mesencephali in turn has a somatotopic projection to the nucleus praeeminentialis mesencephali. This projects back doubly, both directly, point to point, bilaterally, to the PLLL, and indirectly to the lobus caudalis cerebelli, which has a major output to the PLLL cortex. Several other targets of PLLL efferents are described as well as sources of descending afferents to the PLLL. In at least one of the targets the receptor types from the PLLL cortex and nucleus finally do come together (nucleus medialis ventralis of the TS).

In gymnotiforms the electric (PLLL) and other inputs to the TS and outputs of the TS have been recently studied with HRP by Ebbesson & Scheich (1980, 1981), Scheich & Ebbesson (1981), and Carr et al (1981; see also C. Carr, L. Maler, W. Heiligenberg, and E. Sas on the torus, to be submitted). The accounts overlap but do not agree fully; in this summary the differences cannot be detailed. Electrosensory input is confined to the dorsal torus semicircularis, which has nine major laminae; the ventral torus (laminae 10–15) receives acoustic and mechanoreceptive inputs. A massive, predominantly contralateral input from the PLLL comprises P, T, and ampullary units of the lobe. P units end in laminae 3, 5, 7 and possibly 8b and 8d; T units in lamina 6 which is also marked by extensive axon collaterals throughout the layer. Lamina 5 cells project back bilaterally to the PLLL. Eurydendroid cells of the contralateral lobus caudalis go to 8b; some reciprocal connections are described from 8–11. The descending nucleus of V projects to 8a, 8c, and 9. Laminae 5, 6, 7, 9, and 11 have extensive commissural connections with the opposite TS. The ipsilateral tectum has reciprocal and topographic connections with several laminae. Laminae 7, 8, and 9 project forward bilaterally to a specific part of the thalamus; 8d projects into ipsilateral reticular formation; all except 1, 6, and 8b project topographically into the ipsilateral nucleus praeeminentialis. Additional targets of TS output are said to include the pacemaker nucleus, the anterior octavolateral nucleus, and other loci in the rhombencephalon.

CEREBELLUM   The broad picture that seems to have slowly emerged is that, unlike the vestibulocerebellum, those parts related to electrosense receive from the receptors only indirectly, as least in most species. The mormyrid lobus caudalis receives ampullary and mormyromast information chiefly from the nucleus praeeminentialis and feeds back in two ways. Granule cells of the lobus caudalis send axons to form the bulk of the molecular layer of the PLLL and its efferent cells project back upon the praeeminentialis and the nucleus lateralis mesencephali, hence indirectly back to the PLLL cortex. A defined area of the valvula cerebelli receives ampullary and mormyromast information chiefly from the nucleus lateralis mesencephali, both directly and via the pretectum and the thalamus. It projects back upon many of the midbrain electrosensory structures. Knollenorgan information reaches a different part of the valvula, coming from the nucleus medialis ventralis of the TS, the nucleus exterolateralis, and the nucleus isthmi; reciprocal connections again appear to be rich. These summary statements are condensed from work on mormyriforms (Maler 1974, Szabo et al, 1979b,c, Haugedé-Carré 1979, 1980, Libouban & Szabo 1977, 1980, Bell et al 1981, Finger et al 1981). Work on gymnotiforms has not been published.

TELENCEPHALON   Platt et al (1974) reported evoked potentials in the telencephalon of gymnotiforms. Baker & Carr (1980) found that a distinct area of the telencephalon takes up 2-deoxyglucose when the gymnotiform *Eigenmannia* is stimulated to give the jamming avoidance response, and not in control animals; known electric centers in the brain stem are also labeled: PLLL, lobus caudalis, eminentia granularis, nucleus praeeminentialis, and TS. Nonelectric areas are not labeled. The electric areas of the telencephalon have not yet been identified in terms of the recognized divisions of the cerebrum of teleosts.

Physiology

RECEPTORS   Table 1 summarizes the types and present understanding of the receptors. Since the reviews of Fessard & Szabo (1974), Bennett (1978), and Viancour (1979c), Viancour has reexamined the physiological properties of a large sample of tuberous receptors quantitatively, plotting the dispersion of each of a number of measures and looking for correlations. Under his conditions, especially his stimulus geometry, there is no significant evidence of subclasses, particularly of the P and T types of Table 1 (Viancour 1979b), which have been recognized by others (Bullock & Chichibu 1965, Scheich et al 1973, Scheich 1977b, Scheich & Maler 1976, Hopkins 1976, Heiligenberg & Bastian 1980a, Heiligenberg 1980a,b) and even have apparently quite separate central projections (see Anatomy, above). Viancour concludes that under the conditions and geometry in intact, noncurarized fish with their own EOD, functional subclasses might arise from an otherwise uniform receptor population. Nevertheless, his findings deserve to be followed up, in particular to plot the distribution of several properties that he examined in a small sample and only divided into high and low. If sufficient spread occurs, and there is some clustering of properties, the earlier accounts and the models of Scheich & Bullock (1974) and Heiligenberg & Bastian (1980a) that require. P and T units would be satisfied. Still, the apparent discrepancy between Viancour's low correlations and chiefly unimodal histograms and the older, qualitative reports implying rarity of intermediate units calls for renewed attention.

Baker (1981) subdivided one of the pulse species receptor types, the burst duration coders (B units of Table 1) into four subtypes, with intermediates. These are distinguished by their responses to stimuli simulating the fish's EOD pulse train (S_1) scanned by that of another fish (S_2) at a slightly different EOD rate; each subtype is excited or inhibited by certain S_1/S_2 intervals. They are partly understandable by their sensitivity to inward or outward current, or both. Further analysis led Baker (1981) to conclude that the differentiation represented by these four subtypes is not essential to explain simple, reflex jamming avoidance responses. Their value has not yet been identified, but various more complex social responses to the EODs of neighbors are known and could employ these subtypes.

Discrimination between objects by detecting differences in complex impedance and not only by their differences in ohmic resistance was suggested, with evidence at the receptor level, by Scheich & Bullock (1974) and extended by Feng & Bullock (1977). Some tuberous receptors more than others are able to respond differently to local, purely ohmic shunts in the water and to capacitive shunts of the same impedance at their sinusoidal

EOD frequency. Behavioral assessment of this ability has been accomplished by J. H. Meyer (to be submitted; see above).

Bastian (1981a) has recorded from afferent nerve fibers in the high frequency wave fish, *Apteronotus,* during modulations of the EOD field similar to those caused by objects. He gives the firing rate as a function of amplitude modulation depth and rate. He also quantifies the receptive field of the unit afferent and the affected area of skin for an object of given size and distance (see above). For these tuberous receptors (P units), normally in the presence of an EOD field of several millivolts, it is the increment or amplitude modulation sensitivity that is relevant to their role, not the high absolute threshold—about 10^3 times higher than for ampullary receptors in fresh water siluriforms. Bastian finds that there is a marked best amplitude modulation (AM) rate, equivalent to 64 Hz. At this rate a near threshold firing rate change in an afferent unit requires ca. 20 μV/cm of AM in one of his figures. AM rate is influenced by object distance, size, and velocity. A comparison study of the second order unit responses to objects (Bastian 1981b) is mentioned below (Medulla).

In a study of the tuning curves of tuberous receptors, Viancour (1977, 1979b,c) first confirmed earlier authors that the well-tuned units in each fish have nearly the same best frequency and that among specimens, as among species, this varies in good correlation with the EOD rate of the fish. Then he showed that there is a significant deviation: the best frequency is generally a little lower than the EOD. The interesting suggestion put forward was that the receptors are tuned, not to the peak of the power spectrum of the continuous wave, but to that of the single cycle; it is one of those nonintuitive facts that even for a pure sine wave the latter is lower, by about 15%. If this suggestion is correct, it once more underlines what Scheich & Bullock (1974) and Scheich (1974, 1977a–c) concluded on other grounds, namely, that these receptors operate, not in the frequency domain, but in the time domain, cycle by cycle.

Viancour (1979b,c) pointed out that the sharpness of the tuning curves in many units is as good as that of cat cochlear units with similar best frequencies. Since there is no mechanical event or basilar membrane analysis to which to attribute the tuning, the tuberous receptors support the notion of a "second filter" in the mammalian ear, following the cochlear microphonic. Viancour and also Watson & Bastian (1979) showed that an important factor in tuning is receptor oscillation. In a wave species and a pulse species, although some receptors do not oscillate, others give evidence of damped oscillation of the receptor potential; its frequency is highly correlated with the best frequency of that unit. In wave species this is seen by comparing fish with various EOD rates; in pulse species there is a wide range of best frequencies in each fish (Bastian 1976b) and a trend along the body, correlated with the local EOD shape (Bastian 1977). The properties of the oscillation strongly suggest an active fluctuation of the receptor membrane. This is reminiscent of the long-known rhythmic receptor potentials of various electric fish (Szabo 1962, Bennett 1971, Roth & Szabo 1972 Pimentel de Souza, 1976a,b). The obvious question—How does it come about that receptor tuning and EOD waveform are so matched?—has not been elucidated as yet; plasticity of one or both has not been reported.

CENTRAL NERVOUS SYSTEM   This subject was reviewed by Bell (1979) but neither there nor here is there room for a full synthesis of the available information, which is rich in detail, in behavioral and anatomical corollaries. We must be content with indicative highlights at medullary, midbrain, and cerebellar levels.

Medulla   The physiology of units in the first nucleus, the PLLL of **gymnotiforms,** like that of units in the mammalian cochlear nuclei, continues to reveal new insights into integrative processing of input (Enger & Szabo 1965, Schlegel 1974, Scheich 1974, 1977c, Szabo et al 1975, Bastian & Heiligenberg 1980a,b, Matsubara 1981, Bastian 1981b). The two basic types of tuberous afferents in each species (Table 1) here diverge into at least six types of second order units, including those of opposite sign of excitatory stimuli, those sensitive to direction, and others. Several of the types in wave species retain basic properties of T units, following the EOD $1:1$ with tight phase coupling and little jitter. They, therefore, accurately signal the time of threshold crossing and thus, with highly periodic stimuli, the time of zero crossing. JAR stimuli cause systematic shifts of zero crossing of the combined S_1 and S_2 fields, leading and lagging relative to the EOD (S_1), once each beat cycle. By summing a number of afferents, T units in the PLLL achieve a reliability to a small fraction of a millisecond. Other types of second order units retain basic properties of P afferents, principally encoding the voltage of the field by the probability of firing on each EOD cycle. Subtypes vary widely in the phase of a beat cycle at which they fire maximally, as well as in their sensitivity to harmonics. These differences mean that some units are excited by small objects of higher conductivity than the water, others by low conductivity objects, and still others by neither but by large shunts. *Sternopygus,* which lacks a JAR, has a special variety of P units that ignores amplitude changes affecting a large receptor population, such as those due to beats with neighboring fish of different EOD rate; these units are excited by local increases in amplitude that affect a small part of the fish and by local decreases in amplitude that affect an adjacent area, such as would be caused by objects or edges nearby. The broad categories of P and T in wave species are segregated anatomically. Maler et al (1981) have identified them with defined cell types.

In the high frequency wave fish, *Apteronotus,* where most receptor fibers are of the P type, Bastian (1981b) distinguished tonic and phasic classes of second order units in the PLLL. The phasic cells adapt nearly 100 times as fast as first order afferents. Of these phasic PLLL units, Bastian studied two subclasses, E and I, respectively excited and inhibited by an increased amplitude of EOD. E and I units differ in latency and in best AM rate; E = ca. 64 Hz, I = 2–16 Hz. Some 6–15 receptor units converge on each E or I unit and the latter are about 16 times as sensitive, chiefly due to decreased background firing and variance. This puts their threshold in the range of 2 μV/cm at optimum AM frequency, which is reasonably close to the behavioral threshold of 0.2 μV/cm measured by Knudsen (1974). Bastian's (1981a) measurements of EOD modulation with objects allowed him to specify from this the size, distance, and rate of movement of just detectable objects.

In **mormyriforms** the PLLL second order units likewise preserve the distinction between the two main receptor types, K and D units (Table 1) (Zipser 1971, Schlegel 1974, Zipser & Bennett 1976a,b, Szabo et al 1975, 1979a, Enger et al 1976c) and are sharply segregated. Mormyromast input subdivides into two zones of PLLL cortex, probably from different subtypes of afferents; each zone has a somatotopic map. Intracellular recording in the PLLL dorsolateral cortex has shown two modes of activation of the second order cells: a direct monosynaptic input by a single primary fiber and a disynaptic delayed input via granule cells. Inhibitory influence from surrounding receptors influences only the disynaptic input. Spontaneous activity is modulated, normally by objects that distort the fish's own EOD. Some units are sensitive to direction of motion but, on present knowledge, there is notably less variety among second order units in mormyriforms than in gymnotiforms. Unlike the latter group, mormyriforms show an effect of a central signal corollary to the EOD command; it facilitates the transmission of the mormyromast information, at the time of the EOD. Knollenorgan (K unit) input ends in the PLLL nucleus on cells that show a lower threshold than mormyromast units, little latency change with intensity, and an ability to follow high frequency stimuli. These are the cells with large club endings and electrical synapses, mentioned above. They are sharply and deeply inhibited during the 1 ms beginning 3 ms after an EOD, by the action of the central, EOD-command-related signal peculiar to mormyriforms. Thus the normal stimulus for these units in the PLLL nucleus is the EOD of other fish, at some distance and often feeble.

Mesencephalon   Tuberous receptor input to the midbrain has been studied in **gymnotiforms** (Scheich 1974, 1977c, Scheich & Bullock 1974, Szabo et al 1975, Scheich & Maler 1976, Schlegel 1977, 1979, Bastian & Heiligenberg 1980a,b, Heiligenberg 1980a,b) chiefly in the wave species where the torus semicircularis (TS, homolog of the inferior colliculus) is extremely large and highly differentiated; it has conspicuous lamination, as we have noted above under Anatomy. P and T cells of the third order are still distinct, in properties, and are especially numerous in two distinct laminae in the dorsal part of the torus, the T layer (lamina 6) being superficial to a deep P (lamina 9). The P units show more precise limitation of their firing to a certain part of the beat cycle when a second fish with a different EOD rate comes within range; that is, their amplitude discrimination is increased, relative to the afferents. This also means that the temporal difference in the firing phase of the beat, between the two situations called $+\Delta F$ and $-\Delta F$, is clearer, due to the asymmetry of the beat which results from the form of the EOD, approximating a slant-clipped sine wave. The T units of the TS show greater precision in their firing phase within the (ca. 3 ms) EOD cycle, i.e. the jitter is less than 1/4 of the already very small jitter of afferent units; this puts the constancy of latency in the μs range. The possibility that some of this reduction of variance is due to recording not from single units but from well-synchronized field potentials of small groups of units needs to be assessed.

The layer of T units in *Eigenmannia,* lamina 6, is regarded as homologous to the nucleus mesencephalicus magnocellularis in pulse species of gymnotiforms (Szabo et al 1975). It is the terminus of the fast pathway from one class of electroreceptors, the M or the pulse marker units in pulse species, the T or phase coder units in wave species of gymnotiforms, equivalent to the K or knollenorgan units of mormyriforms. Even with a relay in the medulla, well-synchronized impulses reach the mesencephalic nucleus with a latency of 0.8–1.5 ms after an electric field pulse. This specialized fast electrosensory system (Sotelo et al 1975) must presumably eventually interact with the rest of the system, but the data from gymnotiforms indicate that it is still quite discrete at the TS level, with little sign of other influence.

There are somatotopic maps of the fish's body in the deep P (ninth) and in the T (sixth) layers, mainly contralateral, and they are in register. Optimal field orientations are also congruent for P and T units in adjacent layers, but the distribution of preferred orientations has not been systematically examined. Some units are much more selectively tuned for field orientation than are afferent or second or the usual third other units; others are without any preferred orientation. Presumably both represent convergence. Convergence in laminae of the ventral part of the TS occurs in various combinations.

Among other complex units are those capable of discriminating the sign of the ΔF in a JAR situation, i.e. whether another fish or an S_2 stimulus is above or below the fish's own EOD frequency, or that of an S_1 stimulus. Three types of such cells have been reported:

1. "ΔF_p"units or ΔF decoders of the P system were found by Scheich (1977c) in the TS of *Eigenmannia*; they fire more often while $S_2 = + \Delta F$ and less while $S_2 = -\Delta F$ than they do with S_1 only and no S_2, providing that the ΔF is in the JAR range. They show a marked best ΔF of ca. 3–5 Hz. Scheich could propose a simple basis for such units in the hypothesized convergence of observed kinds of third order P units, which reach their firing maxima at different times in the beat cycle and which differ in degree of sensitivity to the phase of the harmonics in the S_1, such as are present in the normal EOD.

2. Scheich found a second type of ΔF decoder belonging to the T system, the "ΔF_T" units, which differ from ordinary T units in firing at a slightly different phase of the S_1 cycle for $+\Delta F$ and for $-\Delta F$.

3. Bastian & Heiligenberg (1980a,b), using different geometric conditions for S_1 and S_2 that yield stronger JARs, even with pure sine waves, found a class of units in the TS that burst like P units at a certain phase of the beat cycle and shift this phase markedly between the $+\Delta F$ and $-\Delta F$ situations; this is only true if S_1 and S_2 have different geometries so there is a difference between parts of the skin in the mixture and the polarities of currents from S_1 and S_2. Experiment proved that these units indeed encode phase differences between parts of the body in the contamination of S_1 by S_2. Such units cannot by themselves give unequivocal information as to the sign of ΔF but can in conjunction with cells sensitive only to the amplitude modulation of the beat.

The parts of the TS so far studied are quite unconcerned with object detection, but are specialized for the analysis necessary to the jamming avoidance response, and very likely other social behaviors not yet studied. One candidate behavior is the occasional active phase coupling in some high frequency wave species (Langner & Scheich 1978), mentioned above. In respect to performance, it may give perspective to note that whereas humans can hear the difference between one beat combination and another, for example, in a nearly limiting case, $300 + 303$ Hz versus $297 + 300$ Hz, providing that the intensities of the two tones are nearly equal, *Eigenmannia* detects the presence, sign, and magnitude of a sine wave added to its own field when the ΔF is $\ll 1$ Hz and the intensity of the stimulus is 60 dB weaker than its EOD (Bullock et al 1972b).

A few units located deep in the torus semicircularis have been reported that are sensitive to other modalities: acoustic, vibratory, and visual, alone or in addition to high frequency electric fields. A few units sensitive to low frequency electric fields—presumably ampullary input—have been seen.

Tuberous receptor input to the midbrain has been less studied in **mormyriforms** (Bennett & Steinbach 1969, Zipser & Bennett 1976b, Enger et al 1976b,c, Szabo et al 1979a), where the TS is quite different in histology from that of the gymnotiforms. There is also a fast pathway in mormyriforms, coming from the K afferents or knollenorgans and ending in the nucleus exterolateralis mesencephali, equivalent to the nucleus magnocellularis mesencephali, in gymnotiforms. The latency of response to a pulse in the water is 2.5–3 ms; it is bilateral. As expected from the command-related inhibition of its PLLL source, no response is seen to the fish's own EOD in nucleus exterolateralis. However, another part of the TS, the nucleus lateralis, shows long-latency, long-lasting, high-threshold responses to pulses within a limited time window including the fish's own EOD. It preserves the features of mormyromast input. Hopkins & Bass (1981) postulate a central integration of knollenorgan input from different patches of skin to yield a two-spike code for the EOD of females of a mormyrid species. Behaviorally the males can distinguish between EOD-like pulses <1 ms long, with a different shape but an identical power spectrum (shifted phase spectrum). Receptors fire a single spike on the transition from outside negative to positive. A receiving fish normally senses EOD current from its neighbor flowing in one side of its body and out the other; thus one set of receptors fires on the rising and one on the falling phase, 0.4 ms later. Males can measure this interval and confuse normal female EODs with rectangular pulses that cause the same interval between spikes in left and right receptors.

Cerebellum   Electrosensitive units have been found in **gymnotiform** high frequency wave species in a circumscribed part of the caudal lobe, in response to physiological stimulation (Bastian 1974, 1975, 1976a). Mainly Purkinje cells, these units show slowly adapting increase or decrease from a prevailing high level of irregular firing, when adequately stimulated. Usually motion is important and often the response is quite sensitive to the direction of motion of an object that distorts the fish's own EOD field or which is itself a source of current field. Each unit has a receptive field, commonly with an excitatory center and inhibitory flanks. Something of a map of the aquatic space around the fish is represented in the cerebellum.

Some units are bimodal, responding also to tail-bending proprioceptive stimuli (tested in the absence of an electric field), or to visual stimuli in a restricted receptive field. Some units show a constant response amplitude for object movement at different distances, out to about 25 mm, although the electric field is markedly attenuated with distance. Other units show a weaker response with a power function similar to that in the sensory afferents (Bastian 1976a). Some units show a best distance, i.e. fire most to objects at a certain coordinate in three dimensional space. The evidence adds up to a significant role for the cerebellum in electrolocation of objects. This is also suggested by reversible cerebellar local cooling experiments that, without blocking the jamming avoidance response, blocked the opercular breathing rate response to looming objects—a task involving a minimum of motor coordination but some sophisticated input processing (Bombardieri & Feng 1977b). In contrast, social stimuli that are quite effective in causing response elsewhere (midbrain and EOD command center) indicate a minimal role of the cerebellum in electrocommunication. Behrend (1977) shed light on this in a study that revealed a low pass filter property of (if not ahead of) the cerebellum such that jamming by a social partner with an EOD different by less than 10 Hz ($\Delta F < 10$ Hz) allows the low frequency amplitude beats to interfere with Purkinje cell modulation by moving objects, but above 10 Hz the beat signal does not get through the filter to interfere. Hence the normal JAR (jamming avoidance response) mediated by the midbrain keeps the $\Delta F > 10$ Hz to improve electrolocation.

In **mormyriform** fishes, one expects an even more elaborate involvement of the cerebellum in electroreception, because of the extreme development of that organ in both size and neuronal architecture. Russell & Bell in 1978 gave some evidence that this is so. Remarkable is that two separate areas are found. Input from knollenorgans goes to a small, paired ventrolateral area of the valvula; ampullary and mormyromast inputs converge on a large area of medial valvula. This area is also influenced by mechanical stimuli —which is expected from the rather high sensitivity of ampullary receptors in mormyriforms to gentle touch (Szabo 1970).

Also special to mormyriforms, and absent in gymnotiforms, is a dependence of the response to an electric event upon its time in relation to the EOD command. A special corollary potential has been found in the cerebellum as well as in the brain stem (Bennett & Steinbach 1969, Zipser & Bennett 1976a,b) that accompanies the firing of the EOD command center. It has a simple inhibitory effect on responses in the knollenorgan region but a mixture of effects in the mormyromast region (Russell & Bell 1978).

DEVELOPMENT

Young electric fish are rare and it is only with the success of Kirschbaum (1975, 1979) in breeding one species each of gymnotiform and mormyriform fishes that material has become available for studying the ontogeny of the behavior, and anatomy and physiology of the electric system. Space precludes an informative review but attention should be called to the papers on development of electrocytes, the temporary larval and later adult electric organs and their EODs (Kirschbaum & Westby 1975, Kirschbaum 1977a,b, Westby 1977, Westby & Kirschbaum 1977, 1978, Srivastava 1978b, Mellinger et al 1978, Denizot et al 1978, Belbenoit 1979) and their motor neurons (Kirschbaum et al 1979). Much less attention has been given to the development of electroreceptors (Kirschbaum & Denizot 1975). A detailed anatomical study has been done on the development of the gigantic cerebellum of the mormyrid (Haugedé-Carré et al 1977, Haugedé-Carré et al 1979).

SUMMARY

Surprisingly few species are intermediate between those with clear electric receptors and central afferent pathways and those lacking such specializations. Taxa possessing them include agnathans, holocephalans, elasmobranchs, dipneustans, chondrosteans, polypteriforms, and three orders of teleosts: siluriforms, gymnotiforms, and mormyriforms. Holosteans and most teleosts lack electroreception. This distribution bespeaks independent invention of this modality at least three times. So far, no clear evidence of electroreception has been found in other groups (amphibians, invertebrates, etc).

It seems best at present to distinguish clearly between *ampullary receptors* of nonteleosts (Lorenzinian ampullae) and those in the three teleost orders (teleost ampullae). Although both are low frequency (ca. 0.1–50 Hz) band pass filters, there are basic differences, including the polarity of current that excites; even among the three teleost orders it is not clear whether their ampullary organs should be regarded as homologous. Each animal has a spectrum of ampullary receptors on several variables but subclasses have not been defined. Adaptive differences are known between fresh water and marine species. Something is known about the Ca current receptor potential and chemical transmission to the nerve ending. There seem to be no centrifugal (efferent) axons to the receptors.

Tuberous electroreceptors are peculiar to gymnotiforms and mormyriforms (continally discharging weakly electric fish). Although in both they are high frequency (ca. 70–3000 Hz) band pass filters, their independent origin is manifest especially in structural differences. Two subclasses of tuberous receptors coexist in each fish, and varieties of these have been defined in some species. These differ in the way they encode stimulus intensity; several codes are known in addition to a classical mean frequency code. Many tuberous receptors are tuned close to the peak of the power spectrum of the electric organ discharge (EOD) of that fish. Active receptor potential oscillation contributes to this tuning. In some taxa one subclass is specialized for communication, i.e. detecting the EOD of other fish; the other is specialized for electrolocation, i.e. detecting modulations of the fish's own EOD. Some receptors discriminate objects of ohmic from those with complex impedance, by the way in which the EOD is distorted. Some receptors mainly encode stimulus amplitude; others, with high resolution encode the time (phase) of the EOD field or of mixtures of the EOD's of several fish. Efferents are apparently lacking.

Central systems for ampullary input in nonteleosts begin with a dorsal octavolateral nucleus, which receives all and only this input. Something is known of the further connections, especially with the mesencephalic torus semicircularis (TS), as well as projections to cerebellum and telencephalon. More is known about central ampullary units and systems in teleosts, including mapping, field orientation preference, and frequency sensitivity. There is no dorsal nucleus but a special part of the medial nucleus receives the primary afferents. Electroreceptor structures are mainly distinct from acoustic, vestibular, and mechanoreceptive lateral line structures at least to the diencephalon and probably to the telencephalon.

Central systems for tuberous input are known in considerable detail, anatomically and physiologically. There is multiple mapping in the medullary as well as cerebeller and midbrain centers; the separate receptor types form distinct systems. Parallel processing for the electrolocation and electrocommunication functions is the result. The cerebellum continually processes sensory input, including electroreceptive input, even when no movement is under way. In the African order (Mormyriformes) there is a well-developed corollary discharge of the EOD that inhibits the input from the electrocommunication receptors just during the EOD, confining them to reporting on EODs of other fish; the corollary discharge inhibits the electrolocating receptors during the period between EODs, confining them to reporting on the local intensity of the fish's own EOD. The South American order (Gymnotiformes) lacks a corollary discharge but nevertheless shows gating, timed by sensory input. Some gymnotiforms that have quasisinusoidal EODs, usually highly regular, though subject to frequency modulation for social signaling, exhibit a behavioral jamming avoidance response. This is an EOD rate shift caused by a neighbor whose EOD is only a few Hz different. Two distinct neuronal mechanisms have been found in the highly laminated midbrain TS capable of measuring small differences in EOD rate and determining the sign of the difference unequivocally within a fraction of a beat cycle. Although most evidence does not require it, the midbrain and cerebellum, possibly higher centers too, probably generate a kind of image or representation of the electrical world near the fish, in three dimensions, continually updated and sensitive to movement.

Literature Cited

Akert, K., Tokunaga, A., Sandri, C., Bennett, M. V. L. 1980. Synaptic organization of electromotor command pathway in the gymnotid, *Sternarchus albifrons*. *Proc. Intl. Union Physiol. Sci.* 14:293 (Abstr.)

Akoev, G. N., Alekseev, N. P., Il'inskii, O. B. 1971. Modern theories of electroreception. *Prog. Physiol. Sci.* 2:335–51 (In Russian)

Akoev, G. N., Andrianov, G. N., Volpe, N. O. 1980a. Effects of divalent ions and drugs on electric and thermal sensitivity of the ampullae of Lorenzini. *Proc. Intl. Union Physiol. Sci.* 14:294 (Abstr.)

Akoev, G. N., Ilyinsky, O. B. 1973. Some functional characteristics of the electroreceptors (the ampullae of Lorenzini) of elasmobranchs. *Experientia* 29:293–94

Akoev, G. N., Ilyinsky, O. B., Zadan, P. M. 1976a. Physiological properties of electroreceptors of marine skates. *Comp. Biochem. Physiol. A* 53:201–9

Akoev, G. N., Ilyinsky, O. B., Zadan, P. M. 1976b. Responses of electroreceptors (ampullae of Lorenzini) of skates to electric and magnetic fields. *J. Comp. Physiol.* 106:127–36

Akoev, G. N., Volpe, N. O., Zhadan, G. G. 1980b. Analysis of effects of chemical and thermal stimuli on the ampullae of Lorenzini of the skates. *Comp. Biochem. Physiol. A* 65:193–201

Akutsu, Y., Obara, S. 1974. Calcium dependent receptor potential of the electroreceptor of marine catfish. *Proc. Jpn. Acad.* 50:247–51

Andrianov, G. N., Brown, H. R., Ilyinsky, O. B. 1974. Responses of central neurons to electrical and magnetic stimuli of the ampullae of Lorenzini in the Black Sea Skate. *J. Comp. Physiol.* 93:287–99

Andrianov, G. N., Enin, L. D. 1977. Acoustico-lateralis system. Morphological and functional properties of cerebral structures of the lateral line system in

fish. In *Sensory Systems,* 2:125–46. Leningrad: Nauka (In Russian)

Andrianov, G. N., Ilyinsky, O. B. 1973. Some functional properties of central neurons connected with the lateral-line organs of the catfish (*Ictalurus nebulosus*). *J. Comp. Physiol.* 86:365–76

Andrianov, Yu. N., Volkova, N. K. 1976. Some morphological and functional properties of the lateral line system of the dwarf catfish. *Neurophysiology* 7:160–64

Andrianov, Yu. N., Broun, G. R. 1976. Perception of the magnetic field by the electroreceptor system in fishes. *Neurophysiology* 7:338–39

Baillet-Derbin, C. 1978. Cytodifferentiation of the regenerating electrocyte in an electric fish. *Biol. Cell* 33:15–24

Baker, C. L. Jr. 1980. Jamming avoidance behavior in gymnotoid electric fish with pulse-type discharges: Sensory encoding for a temporal pattern discrimination. *J. Comp. Physiol.* 136:165–81

Baker, C. L. Jr. 1981. Sensory control of pacemaker acceleration and deceleration in gymnotiform electric fish with pulse-type discharges. *J. Comp. Physiol.* 161:197–206

Baker, C. L. Jr., Carr, C. E. 1980. 2-deoxyglucose identification of electroreceptive pathways in the South American weakly electric fish, *Eigenmannia*. *Neurosci. Abstr.* 6:605 (Abstr.)

Bastian, J. 1974. Electrosensory input to the corpus cerebelli of the high frequency electric fish *Eigenmannia virescens*. *J. Comp. Physiol.* 90:1–24

Bastian, J. 1975. Receptive fields of cerebellar cells receiving exteroceptive input in a gymnotid fish. *J. Neurophysiol.* 38:285–300

Bastian, J. 1976a. The range of electrolocation: A comparison of electroreceptor responses and the responses of cerebellar neurons in a gymnotid fish. *J. Comp. Physiol.* 108:193–210

Bastian, J. 1976b. Frequency response characteristics of electroreceptors in weakly electric fish (Gymnotoidei) with a pulse discharge. *J. Comp. Physiol.* 112: 165–80

Bastian, J. 1977. Variations in frequency response of electroreceptors dependent on receptor location in weakly electric fish (Gymnotoidei) with a pulse discharge. *J. Comp. Physiol.* 121:53–64

Bastian, J. 1981a. Electrolocation I: An analysis of the effects of moving objects and other electrical stimuli on the electroreceptor activity of *Apteronotus albifrons. J. Comp. Physiol.* In press

Bastian, J. 1981b. Electrolocation II: The effects of moving objects and other electrical stimuli on the activities of two categories of posterior lateral line lobe cells in *Apteronotus albifrons. J. Comp. Physiol.* In press

Bastian, J., Heiligenberg, W. 1980a. Neural correlates of the jamming avoidance response of *Eigenmannia. J. Comp. Physiol.* 136:135–52

Bastian, J., Heiligenberg, W. 1980b. Phase sensitive midbrain neurons in *Eigenmannia:* Neural correlates of the jamming avoidance response. *Science.* 209:828–31

Bauer, R. 1979. Electric organ discharge (EOD) and prey capture behavior in the electric eel, *Electrophorus electricus. Behav. Ecol. Sociobiol.* 4:311–19

Bauswein, E. 1977. Effect of calcium on the differentiating operation of the ampullar electroreceptor in *Ictalurus nebulosus. J. Comp. Physiol.* 121:381–94

Behrend, K. 1977. Processing information carried in a high frequency wave: Properties of cerebellar units in the high frequency electric fish. *J. Comp. Physiol.* 118:357–71

Belbenoit, P. 1979. Electric organ discharge of *Torpedo* (Pisces) basic pattern and ontogenetic changes. *J. Physiol. Paris* 75:435–41

Belbenoit, P., Moller, P., Serrier, J., Push, S. 1979. Ethological observations on the electrical organ discharge behaviour of the electric catfish, *Malapterurus electricus* (Pisces). *Behav. Ecol. Sociobiol.* 4:321–30

Bell, C. C. 1979. Central nervous system physiology of electroreception, a review. *J. Physiol. Paris* 75:361–79

Bell, C. C. 1981. Central distribution of octavolateral afferents and efferents in a teleost (Mormyridae). *J. Comp. Neurol.* 95:391–414

Bell, C. C., Finger, T. E., Russell, C. J. 1981. Central connections of the posterior lateral line lobe in mormyrid fish. *Exp. Brain Res.* 42:9–22

Bell, C. C., Myers, J. P., Russell, C. J. 1974. Electric organ discharge patterns during dominance related behavioral displays in *Gnathonemus petersii* (Mormyridae). *J. Comp. Physiol.* 92:201–28

Bell, C. C., Russell, C. J. 1976. Termination of tonic electroreceptors in lateral line lobe of mormyrids. *Neurosci. Abstr.* 2:177 (Abstr.)

Bell, C. C., Russell, C. J. 1978a. Effect of electric organ discharge on ampullary receptors in a mormyrid. *Brain Res.* 145:85–96

Bell, C. C., Russell, C. J. 1978b. Termination of electroreceptor and mechanical lateral line afferents in the mormyrid acousticolateral area. *J. Comp. Neurol.* 182:367–82

Bennett, M. V. L. 1965. Electroreceptors in mormyrids. *Cold Spring Harbor Symp. Quant. Biol.* 30:245–62

Bennett, M. V. L. 1967. Mechanisms of electroreception. In *Lateral Line Detectors,* ed. P. Cahn, 20:313–93. Bloomington: Indiana Univ. Press, 496 pp.

Bennett, M. V. L. 1970. Comparative physiology: Electric organs. *Ann. Rev. Physiol.* 32:471–528

Bennett, M. V. L. 1971. Electroreception. In *Fish Physiology,* ed. W. S. Hoar, D. S. Randall, 11:493–574. New York: Academic. 600 pp.

Bennett, M. V. L. 1978. Mechanism of afferent discharge from electroreceptors: Implications for acoustic reception. In *Evoked Electrical Activity in the Auditory Nervous System,* ed. R. F. Naunton, C. Fernandez, pp. 83–89. New York: Academic. 588 pp.

Bennett, M. V. L., Clusin, W. T. 1978. Physiology of the ampulla of Lorenzini, the electroreceptor of elasmobranchs. In *Sensory Biology of Sharks, Skates and Rays,* ed. E. S. Hodgson, R. F. Mathewson, 5:483–505. Arlington, Va.: Off. Naval Res. 666 pp.

Bennett, M. V. L., Clusin, W. T. 1979. Transduction at electroreceptors: Origins of sensitivity. In *Membrane Transduction Mechanisms,* ed. R. A. Cone, J. E. Dowling, pp. 91–116. New York: Raven. 236 pp.

Bennett, M. V. L., Steinbach, A. B. 1969. Influence of electric organ control system on electrosensory afferent pathways in mormyrids. In *Neurobiology of Cerebellar Evolution and Development,* ed. R. Llinas, pp. 207–14. Chicago, Ill: Am. Med. Assoc. 931 pp.

Berge, J. A. 1979. The perception of weak electric A. C. currents by the European eel, *Anguilla anguilla. Comp. Biochem. Physiol. A* 62:915–19

Blakemore, R. 1975. Magnetotactic bacteria. *Science* 190:377–79

Blakemore, R. P., Frankel, R. B., Kalmijn, A. J. 1980. South-seeking magnetotactic bacteria in the southern hemisphere. *Nature* 286:384–85

Bodznick, D. A., Northcutt, R. G. 1980. Segregation of electro- and mechanoreceptive inputs to the elasmobranch medulla. *Brain Res.* 195:313–22

Bodznick, D. A., Northcutt, R. G. 1981. Electroreception in lampreys: Evidence that the earliest vertebrates were electroreceptive. *Science* 212:465–67

Bombardieri, R. A., Feng, A. S. 1977a. Behavioral analysis of range of object detection (electrolocation) in the weakly electric fish, *Apteronotus albifrons. Neurosci. Abstr.* 3:363 (Abstr.)

Bombardieri, R. A., Feng, A. S. 1977b. Deficit in object detection (electrolocation) following interruption of cerebellar function to the weakly electric fish *Apteronotus albifrons. Brain Res.* 130: 343–47

Boord, R. L., Campbell, C. B. G. 1977. Structural and functional organization of the lateral line system of sharks. *Am. Zool.* 17:431–41

Boord, R. L., Northcutt, R. G. 1979. Ascending projections of anterior and posterior lateral line lobes of the clearnose skate, *Raja eglanteria. Anat. Rec.* 193:487–88

Branover, G. G., Vasil'yev, A. S., Gleizer, S. I., Tsinober, A. B. 1971. A study of the behavior of eels in natural and artifical magnetic fields and an analysis of its reception mechanism. *J. Ichthyol.* 11:608–14

Bretschneider, F., Peters, R. C., Peele, P. H., Dorresteijn, A. 1980. Functioning of catfish electroreceptors: Statistical distribution of sensitivity and fluctuations of spontaneous activity. *J. Comp. Physiol.* 137:273–79

Bromm, B., Hensel, H., Tagmat, A. T. 1976. The electrosensitivity of the isolated ampulla of Lorenzini in the dogfish. *J. Comp. Physiol.* 111:127–36

Broun, G. R., Govardovskii, V. I., Zhadan, G. G. 1980. Investigation of the mechanism of temperature sensitivity of the electroreceptors of ampullae of Lorenzini. *Neurophysiology* 12:54–60

Broun, G. R., Ilyinsky, O. B., Volkova, N. K. 1972. The study of some properties of electroreceptor structures of the lateral line in skates. *Fiziol. Zh. SSSR im. I.M. Sechenova* 58:1499–1505

Brown, F. A. 1971. Some orientational influences of nonvisual terrestrial electromagnetic fields. *Ann. NY Acad. Sci.* 188:224–41

Brown, H. R., Govardovsky, V. I., Ilyinsky, O. B. 1980. Thermal sensitivity of Lorenzinian ampullae. *Proc. Intl. Union Physiol. Sci.* 14:339 (Abstr.)

Brown, H. R., Ilyinsky, O. B. 1978. The ampullae of Lorenzini in the magnetic field. *J. Comp. Physiol.* 126:333–41

Bullock, T. H. 1969. Species differences in effect of electroreceptor input on electric organ pacemakers and other aspects of behavior in electric fish. *Brain Behav. Evol.* 2:85–118

Bullock, T. H. 1973. Seeing the world through a new sense: Electroreception in fish. *Am. Sci.* 61:316–25

Bullock, T. H. 1974. An essay on the discovery of sensory receptors and the assignment of their functions together with an introduction to electroreceptors. In *Handbook of Sensory Physiology III/3,* ed. A. Fessard, pp. 1–12. New York: Springer-Verlag. 333 pp.

Bullock, T. H. 1979a. What is the interest for general neurophysiology of electroreception in certain fishes? *J. Physiol. Paris* 75:315–17

Bullock, T. H. 1979b. Processing of ampullary input in the brain: Comparisons of sensitivity and evoked responses among siluroids and elasmobranchs. *J. Physiol. Paris* 75:397–407

Bullock, T. H. 1981. Physiology of the tectum mesencephali in elasmobranchs. In *The Vertebrate Tectum,* ed. H. Vanegas. New York: Plenum. In press

Bullock, T. H., Behrend, K., Heiligenberg, W. 1975. Comparison of the jamming avoidance responses in gymnotoid and gymnarchid electric fish: A case of convergent evolution of behavior and its sensory basis. *J. Comp. Physiol.* 103:97–121

Bullock, T. H., Chichibu, S. 1965. Further analysis of sensory coding in electroreceptors of electric fish. *Proc. Natl. Acad. Sci.* 54:422–29

Bullock, T. H., Corwin, J. T. 1979. Acoustic evoked activity in the brain of sharks. *J. Comp. Physiol.* 129:223–34

Bullock, T. H., Fernandes-Souza, N., Graf, W., Heiligenberg, W., Langner, G., Meyer, D. L., Pimentel-Souza, F., Scheich, H., Viancour, T. A. 1979. Aspects of the use of electric organ discharge and electroreception in Amazonian Gymnotoidei and other fishes. *Acta Amazonica* 9:549–72 (In Portuguese)

Bullock, T. H., Hagiwara, S., Kusano, K., Negishi, K. 1961. Evidence for a category of electroreceptors in the lateral line of gymnotid fishes. *Science* 134:1426–27

Bullock, T. H., Hamstra, R. H. Jr., Scheich, H. 1972a. The jamming avoidance response of high frequency electric fish. I. General features. *J. Comp. Physiol.* 77:1–22

Bullock, T. H., Hamstra, R. H. Jr., Scheich, H. 1972b. The jamming avoidance response of high frequency electric fish. II. Quantitative aspects. *J. Comp. Physiol.* 77:23–48

Butsuk, S. V., Bessonov, B. I. 1981. Direct current electric field in some teleost species: Effect of medium salinity. *J. Comp. Physiol.* 141:277–82

Carr, C., Maler, L., Heiligenberg, W., Sas, E. 1981. Connections of the torus semicircularis in two genera of weakly electric fish. *Anat. Rec.* In press

Carré, F. see Haugedé-Carré, F.

Chichibu, S. 1981. In *Adv. Physiol. Sci.* 30:165–78

Clusin, W. T., Bennett, M. V. L. 1977a. Calcium-activated conductance in skate electroreceptors. Current clamp experiments. *J. Gen. Physiol.* 69:121–43

Clusin, W. T., Bennett, M. V. L. 1977b. Calcium-activated conductance in skate electroreceptors. Voltage clamp experiments. *J. Gen. Physiol.* 69:145–82

Clusin, W. T., Bennett, M. V. L. 1979. The oscillatory responses of skate electroreceptors to small voltage stimuli. *J. Gen. Physiol.* 73:685–702

Denizot, J. P. 1978. Enzyme activity during the metabolism of glycogen. II. Cytochemical study of glycogen synthetase in the sensory cells of the tuberous organ of *Gnathonemus petersii* (Mormyridae). *Histochemistry* 55:117–28

Denizot, J. P., Kirschbaum, F., Westby, G. W. M., Tsuji, S. 1978. The larval electric organ of the weakly electric fish *Pollimyrus* (*Marcusenius*) *isidori* (Mormyridae, Teleostei). *J. Neurocytol.* 7:165–81

Derbin, C. 1974. Ultrastructure of the ampullary receptor organs in a mormyrid fish *Gnathonemus petersii*, part 3. *J. Ultrastruct. Res.* 46:254–67

Dijkgraaf, S., Kalmijn, A. J. 1962. Verhaltungsversuche zur Funktion der Lorenzinischen Ampullen. *Naturwissenschaften* 49:400

Dijkgraaf, S., Kalmijn, A. J. 1963. Untersuchungen über die Funktion der Loren-

zinischen Ampullen an Haifisch. *Z. Vgl. Physiol.* 47:438–56

Dijkgraaf, S., Kalmijn, A. J. 1966. Versuche zur biologischen Bedeutung der Lorenzinischen Ampullen bei den Elasmobranchiern. *Z. Vgl. Physiol.* 53:187–94

Dunning, B. B. 1973. *A quantitative and comparative analysis of the tonic electroreceptors of Gnathonemus, Gymnotus, and Kryptopterus*. PhD thesis, Univ. Minn.

Ebbesson, S. O. E., Scheich, H. 1980. Afferent and efferent connections of the torus semicircularis in the electric fish, *Eigenmannia virescens*. *Neurosci. Abstr.* 6:332 (Abstr.)

Ebbesson, S. O. E., Scheich, H. 1981. Connections of the torus semicircularis in the weakly electric fish *Eigenmannia virescens* I-V. *Cell Tissue Res.* In press

Elekes, K., Szabo, T. 1981. Synaptology of the command (pacemaker) nucleus in the brain of the weakly electric fish *Sternarchus* (*Apterontus*) *albifrons*. *Neuroscience* 6:443–60

Ellis, D. B., Szabo, T. 1980. Identification of different cell types in the command (pacemaker) nucleus of several gymnotiform species by retrograde transport of horseradish peroxidase. *Neuroscience* 5:1917–29

Enger, P. S., Kristensen, L., Sand, O. 1976a. The perception of weak electric D.C. currents by the european eel (*Anguilla anguilla*). *Comp. Biochem. Physiol. A* 54:101–3

Enger, P. S., Libouban, S., Szabo, T. 1976b. Fast conducting electrosensory pathway in the mormyrid fish *Gnathonemus petersii*. *Neurosci. Lett.* 2:133–36

Enger, P. S., Libouban, S., Szabo, T. 1976c. Rhombo-mesencephalic connections in the fast conducting electrosensory system of the mormyrid fish, *Gnathonemus petersii*. An HRP study. *Neurosci. Lett.* 3:239–44

Enger, P. S., Szabo, T. 1965. Activity of central neurons involved in electroreception in some weakly electric fish (Gymnotidae). *J. Neurophysiol.* 28:800–18

Feng, A. S. 1977. The role of the electrosensory system in postural control in the weakly electric fish *Eigenmannia virescens*. *J. Neurobiol.* 8:429–37

Feng, A. S., Bullock, T. H. 1977. Neuronal mechanisms for object discrimination in the weakly electric fish *Eigenmannia virescens*. *J. Exp. Biol.* 66:141–58

Fessard, A. 1974. Physiology of electroreceptors. I. Introduction. See Bullock 1974, pp. 60–64

162   BULLOCK

Fessard, A., Szabo, T. 1961. Mise en evidence d'un recepteur sensible a l'electricité dans la peau d'un mormyre. *CR Acad. Sci. Paris* 253:1859–60

Fessard, A., Szabo, T. 1974. Physiology of Receptors. II. Peripheral mechanisms of electroreceptors in teleosts. See Bullock 1974, pp. 64–95

Fields, R. D., Lange, G. D. 1980. Electroreception in the ratfish (*Hydrolagus colliei*). *Science* 207:547–48

Finger, T. E. 1980. Nonolfactory sensory pathway to the telencephalon in a teleost fish. *Science* 210:671–73

Finger, T. E., Bell, C. C., Russell, C. J. 1981. Electrosensory pathways to the valvula cerebelli in mormyrid fish. *Exp. Brain Res.* 42:23–33

Frankel, R. B., Blakemore, R. P., Kalmijn, A. J., Denham, C. R. 1979. Navigational compass in magnetotactic bacteria. *Bioelectromagnet. Symp.* 79:485

Gelinek, E. 1978. On the ampullary organs of the South American paddle-fish *Sorubim lima* (Siluroidea, Pimelodidae). *Cell Tissue Res.* 190:357–69

Gilson, M. K., Kalmijn, A. J. 1980. Statistical mechanics of geomagnetic orientation in sediment bacteria. *Biol. Bull.* 159:459–60

Gleizer, S. I., Khodorkovsky, V. A. 1971. An experimental determination of geomagnetic receptors in the European eel. *Dokl. Acad. Sci. USSR* 201:964–67 (In Russian)

Gottlieb, N. D., Caldwell, W. E. 1967. Magnetic field effects on the compass mechanism and activity level of the snail *Helisoma duryi endiscus. J. Genet. Psychol.* 111:85–102

Hagiwara, S., Kusano, K., Negishi, K. 1962. Physiological properties of electroreceptors of some gymnotids. *J. Neurophysiol.* 25:430–49

Hagiwara, S., Morita, H. 1963. Coding mechanisms of electroreceptor fibers in some electric fish. *J. Neurophysiol.* 26:551–67

Haugedé-Carré, F. 1979. The mesencephalic exterolateral posterior nucleus of the mormyrid fish *Bryenomyrus niger:* Efferent connections studies by the HRP method. *Brain Res.* 178:179–84

Haugedé-Carré, F. 1980. *Contribution a l'etude des connexions du torus semicircularis et du cervelet chez certains mormyrides.* PhD thesis. L'Univ. Pierre et Marie Currie, Paris

Haugedé-Carré, F., Kirschbaum, F., Szabo, T. 1977. On the development of the gigantocerebellum in the mormyrid fish *Pollimyrus (Marcusenius) isidori. Neurosci. Lett.* 6:209–13

Haugedé-Carré, F., Szabo, T., Kirschbaum, F. 1979. Development of the gigantocerebellum of the weakly electric fish *Pollimyrus isidori. J. Physiol. Paris* 75:381–95

Heiligenberg, W. 1975a. Electrolocation and jamming avoidance in the electric fish *Gymnarchus niloticus* (Gymnarchidae, Mormyriformes). *J. Comp. Physiol.* 103:55–67

Heiligenberg, W. 1975b. Theoretical and experimental approaches to spatial aspects of electrolocation. *J. Comp. Physiol.* 103:247–72

Heiligenberg, W. 1976. Electrolocation and jamming avoidance in the mormyrid fish *Brienomyrus. J. Comp. Physiol.* 109:357–72

Heiligenberg, W. 1977. Principles of electrolocation and jamming avoidance in electric fish. A neuroethological approach. In *Studies of Brain Function,* ed. V. Braitenberg, 1:1–85. New York: Springer-Verlag. 85 pp.

Heiligenberg, W. 1980a. The evaluation of the electroreceptive feedback in a gymnotoid fish with pulse-type electric organ discharges. *J. Comp. Physiol.* 138:173–85

Heiligenberg, W. 1980b. The jamming avoidance response in the weakly electric fish *Eigenmannia.* A behavior controlled by distributed evaluation of electroreceptive afferences. *Naturwissenschaften* 67:499–507

Heiligenberg, W., Altes, R. S. 1978. Phase sensitivity in electroreception. *Science* 199:1001–4

Heiligenberg, W., Baker, C., Bastian, J. 1978a. The jamming avoidance response in gymnotoid pulse-species: A mechanism to minimize the probability of pulse-train coincidence. *J. Comp. Physiol.* 124:211–24

Heiligenberg, W., Baker, C., Matsubara, J. 1978b. The jamming avoidance response in *Eigenmannia* revisited: The structure of a neuronal democracy. *J. Comp. Physiol.* 127:267–86

Heiligenberg, W., Bastian, J. 1979. Streaking in a state-plane: The analysis of stimulus parameters controlling the jamming avoidance response in *Eigenmannia. Neurosci. Abstr.* 5:469 (Abstr.)

Heiligenberg, W., Bastian, J. 1980a. The control of *Eigenmannia's* pacemaker by distributed evaluation of electroreceptive afferences. *J. Comp. Physiol.* 136:113–33

Heiligenberg, W., Bastian, J. 1980b. Species specificity of electric organ discharges in sympatric gymnotoid fish of the Rio Negro. *Acta Biol. Venez.* 10:187–203

Heiligenberg, W., Finger, T., Matsubara, J., Carr, C. 1980. Missing links in the neuronal hardware of the jamming avoidance response (JAR) in the electric fish, *Eigenmannia. Neurosci. Abstr.* 6:604 (Abstr.)

Hetherington, T. E., Bemis, W. E. 1979. Morphological evidence of an electroreceptive function of the rostral organ of *Latimeria chalumnae. Am. Zool.* 19:986 (Abstr.)

Hetherington, T. E., Wake, M. H. 1979. The lateral line system in larval *Ichthyophis* (Amphibia: Gymnophiona). *Zoomorphology* 93:209–25

Higuchi, T., Nagai, T., Umekita, S.-I., Obara, S. 1980. The afferent neurotransmitter in the ampullary electroreceptors: L-glutamate mimics the natural transmitter. *Neurosci. Lett.* Suppl. 4, S 7

Hopkins, C. D. 1972. Sex differences in electric signaling in an electric fish. *Science* 176:1035–37

Hopkins, C. D. 1974a. Electric communication in fish. *Am. Sci.* 62:426–37

Hopkins, C. D. 1974b. Electric communication in the reproductive behavior of *Sternopygus macrurus* (Gymnotoidei). *Z. Tierpsychol.* 35:518–35

Hopkins, C. D. 1974c. Electric communication: Functions in the social behavior of *Eigenmannia virescens. Behaviour* 50: 270–306

Hopkins, C. D. 1976. Stimulus filtering and electroreceptors in three species of gymnotoid fish. *J. Comp. Physiol.* 111:171–207

Hopkins, C. D. 1977. Electric communication. In *How Animals Communicate,* ed. T. Sebeok, 13:263–89. Bloomington: Indiana Univ. Press. 1128 pp.

Hopkins, C. D. 1980. Evolution of electric communication channels of mormyrids. *Behav. Ecol. Sociobiol.* 7:1–13

Hopkins, C. D. 1981a. On the diversity of electric signals in a community of mormyrid electric fish in West Africa. *Am. Zool.* 21:211–22

Hopkins, C. D. 1981b. The neuroethology of electric communication. *Trends Neurosci.* 4:4–6

Hopkins, C. D., Bass, A. H. 1981. Temporal coding of species recognition signals in an electric fish. *Science* 212:85–87

Hopkins, C. D., Heiligenberg, W. F. 1978. Evolutionary designs for electric signals and electroreceptors in gymnotoid fishes of Surinam. *Behav. Ecol. Sociobiol.* 3:113–34

Hoshimiya, N., Shogen, K., Matsuo, T., Chichibu, S. 1980. The *Apteronotus* EOD field: Waveform and EOD field simulation. *J. Comp. Physiol.* 135:283–90

Il'inskii, O. B., Enin, L. D., Volkova, N. K. 1972. Evoked potentials of the medulla oblongata of the skate in response to stimulation of lateral line nerves. *Neurophysiology* 3:213–18

Jørgensen, J. M. 1980. The morphology of the Lorenzinian ampullae of the sturgeon *Acipenser ruthenus* (Pisces: Chondrostei). *Acta Zool. Stockholm* 61: 87–92

Jørgensen, J. M., Flock, Å., Wersall, J. 1972. The Lorenzinian ampullae of *Polyodon spathula. Z. Zellforsch.* 130:362–77

Kalmijn, A. J. 1966. Electro-perception in sharks and rays. *Nature* 212:1232–33

Kalmijn, A. J. 1971. The electric sense of sharks and rays. *J. Exp. Biol.* 55:371–83

Kalmijn, A. J. 1974. The detection of electric fields from inanimate and animate sources other than electric organs. See Bullock 1974, pp. 147–200

Kalmijn, A. J. 1978a. Experimental evidence of geomagnetic orientation in elasmobranch fishes. In *Animal Migration, Navigation and Homing,* ed. K. Schmidt-Koenig, W. T. Keeton, pp. 347–53. New York: Springer-Verlag. 462 pp.

Kalmijn, A. J. 1978b. Electromagnetic guidance systems in fishes. In *Proc. Biomagnetic Effects Workshop, Lawrence Berkeley Lab.,* ed. T. S. Tenforde, pp. 8–10

Kalmijn, A. J. 1978c. Electric and magnetic sensory world of sharks, skates, and rays. See Bennett & Clusin 1978, pp. 507–28

Kalmijn, A. J. 1979. Biological sensors for the detection of electric and magnetic fields. *Intl. Union Radio Sci. Symp.* 79:301

Kalmijn, A. J. 1981. Biophysics of geomagnetic field detection. *IEEE Trans. Magnet.* In press

Kalmijn, A. J., Blakemore, R. P. 1978. The magnetic behavior of mud bacteria. See Kalmijn 1978a, pp. 354–55

Kalmijn, A. J., Kolba, C. A., Kalmijn, V. 1976. Orientation of catfish (*Ictalurus nebulosus*) in strictly uniform electric fields: I. Sensitivity of response. *Biol. Bull.* 151:415 (Abstr.)

Kalmijn, V., Kolba, C. A., Kalmijn, A. J. 1976. Orientation of catfish (*Ictalurus nebulosus*) in strictly uniform electric fields. II. Spatial discrimination. *Biol. Bull.* 151:415–16 (Abstr.)

Kastoun, E. 1971. Elektrische Felder als Kommunikationsmittel beim Zitterwels. *Naturwissenschaften* 58:459

Keeton, W. T., Larkin, T. S., Windsor, D. M. 1974. Normal fluctuations in the earth's magnetic field influence pigeon orientation. *J. Comp. Physiol.* 95:95–103

Kirschbaum, F. 1975. Environmental factors control the periodical reproduction of tropical electric fish. *Experientia* 31:1159

Kirschbaum, F. 1977a. Electric organ ontogeny: Distinct larval organ precedes the adult organ in weakly electric fish. *Naturwissenschaften* 64:387–88

Kirschbaum, F. 1977b. Ontogeny of electric organs in weakly electric fish *Eigenmannia* and *Pollimyrus* (*Marcusenius*). *Proc. Intl. Union Physiol. Sci.* 13:387 (Abstr.)

Kirschbaum, F. 1979. Reproduction of the weakly electric fish *Eigenmannia virescens* (Rhamphichthyidae, Teleostei) in captivity. I. Control of gonadal recrudescence and regression by environmental factors. *Behav. Ecol. Sociobiol.* 4:331–55

Kirschbaum, F., Denizot, J. P. 1975. Sur la differenciation des électrorecepteurs chez *Marcusenius* sp. (Mormyrides) et *Eigenmannia virescens* (Gymnotides), poissons électriques à faible décharge. *CR Acad. Sci. Ser. D* 281:419–21

Kirschbaum, F., Denizot, J. P., Tsuji, S. 1979. On the electromotor neurons of both electric organs of *Pollimyrus isidori* (Mormyridae, Teleostei). *J. Physiol. Paris* 75:429–33

Kirschbaum, F., Westby, G. W. M. 1975. Development of the electric discharge in mormyrid and gymnotid fish (*Marcusenius* sp. and *Eigenmannia virescens*) *Experientia* 31:1290

Kleerekoper, H., Sibakin, K. 1956. An investigation of the electrical "spike" potentials produced by the sea lamprey *Petromyzon marinus* in the water surrounding the head region. *J. Fish Res. Board Canada* 13:375–83

Knudsen, E. I. 1974. Behavioral thresholds to electric signals in high frequency electric fish. *J. Comp. Physiol.* 91: 333–53

Knudsen, E. I. 1976a. Midbrain responses to electroreceptive input in catfish. *J. Comp. Physiol.* 106:51–67

Knudsen, E. I. 1976b. Midbrain units in catfish: Response properties to electroreceptive input. *J. Comp. Physiol.* 109:315–35

Knudsen, E. I. 1977. Distinct auditory and lateral line nuclei in the midbrain of catfishes. *J. Comp. Neurol.* 173:417–32

Knudsen, E. I. 1978. Functional organization in electroreceptive midbrain of the catfish. *J. Neurophysiol.* 41:350–64

Koester, D. M., Boord, R. L. 1978. The central projections of first order anterior lateral line neurons of the clearnose skate, *Raja eglanteria. Am. Zool.* 18:587

Kramer, B. 1974. Electric organ discharge interaction during interspecific agonistic behaviour in freely swimming mormyrid fish. A method to evaluate two (or more) simultaneous time series of events with a digital analyzer. *J. Comp. Physiol.* 93:203–35

Kramer, B. 1976a. Kommunikation bei schwachelektrischen Fischen (*Gnathonemus petersii*, Mormyridae). Bewirken elektrische Signale Verhaltensänderungen? *Verh. Dtsch. Zool. Ges.* 1976:261 (Abstr.)

Kramer, B. 1976b. The attack frequency of *Gnathonemus petersii* towards electrically silent (denervated) and intact conspecifics, and towards another mormyrid (*Brienomyrus niger*). *Behav. Ecol. Sociobiol.* 1:425–46

Kramer, B. 1976c. Flight-associated discharge pattern in a weakly electric fish, *Gnathonemus petersii* (Mormyridae, Teleostei). *Behaviour* 59:88–95

Kramer, B. 1976d. Electric signalling during aggressive behaviour in *Mormyrus rume* (Mormyridae, Teleostei). *Naturwissenschaften* 63:48

Kramer, B. 1978. Spontaneous discharge rhythms and social signalling in the weakly electric fish *Pollimyrus isidori* (Cuvier et Valenciennes) (Mormyridae, Teleostei). *Behav. Ecol. Sociobiol.* 4:61–74

Kramer, B. 1979. Electric and motor responses of the weakly electric fish, *Gnathonemus petersii* (Mormyridae), to play-back of social signals. *Behav. Ecol. Sociobiol.* 6:67–79

Kramer, B., Bauer, R. 1976. Agonistic behavior and electric signalling in a mormyrid fish, *Gnathonemus petersii. Behav. Ecol. Sociobiol.* 1:45–61

Langner, G., Scheich, H. 1978. Active phase coupling in electric fish: Behavioral control with microsecond precision. *J. Comp. Physiol.* 128:235–40

Larsell, O. 1967. *The comparative anatomy and history of the cerebellum from myxinoids through birds.* Minneapolis: Univ. Minn. Press

Libouban, S., Szabo, T. 1977. An integration center of the mormyrid fish brain: The auricula cerebelli. An HRP study. *Neurosci. Lett.* 6:115–19

Libouban, S., Szabo, T. 1980. Valvular afferents and efferents in the cerebellum of mormyrid fish, *Gnathonemus petersii. Proc. Intl. Union Physiol. Sci.* 14:548 (Abstr.)

Lindauer, M., Martin, H. 1972. Magnetic effect on dancing bees. In *Animal Orientation and Navigation,* ed. S. R. Galler, K. Schmidt-Koenig, G. J. Jacobs, R. E. Belleville, pp. 559–78. Washington DC: NASA. 606 pp.

Lissmann, H. W. 1958. On the function and evolution of electric organs in fish. *J. Exp. Biol.* 35:156–91

Lissmann, H. W. 1963. Electric location by fishes. *Sci. Am.* 1963:50–59

Lissmann, H. W., Machin, K. E. 1958. The mechanism of object location in *Gymnarchus niloticus* and similar fish. *J. Exp. Biol.* 35:451–86

Lissmann, H. W., Mullinger, A. M. 1968. Organization of ampullary electric receptors in Gymnotidae (Pisces). *Proc. R. Soc. London Ser. B* 169:345–78

MacDonald, J. A., Larimer, J. L. 1970. Phase-sensitivity of *Gymnotus carapo* to low-amplitude electrical stimuli. *Z. Vgl. Physiol.* 70:322–34

Machin, K. E., Lissmann, H. W. 1960. The mode of operation of the electric receptors in *Gymnarchus niloticus. J. Exp. Biol.* 37:801–11

Mago-Leccia, F. M. 1978. Los peces de la familia Sternopygidae de Venezuela. *Acta Cient. Venez.* 29: Suppl. 1, pp. 5–89

Mago-Leccia, F., Zaret, T. 1978. The taxonomic status of *Rhabdolichops troscheli* (Kaup, 1856), and speculations on gymnotiform evolution. *Environ. Biol. Fish* 3:379–84

Maler, L. 1973. The posterior lateral line lobe of a mormyrid fish—a Golgi study. *J. Comp. Neurol.* 152:281–99

Maler, L. 1974. The acousticolateral area of bony fishes and its cerebellar relations. *Brain Behav. Evol.* 10:130–45

Maler, L. 1979. The posterior lateral line lobe of certain gymnotoid fish: Quantitative light microscopy. *J. Comp. Neurol.* 183:323–64

Maler, L., Finger, T., Karten, H. J. 1974. Differential projections of ordinary lateral line receptors and electroreceptors in the gymnotid fish, *Apteronotus (Sternarchus) albifrons. J. Comp. Neurol.* 158:363–82

Maler, L., Karten, H. J., Bennett, M. V. L. 1973a. The central connections of the posterior lateral line nerve of *Gna-*

thonemus petersii. J. Comp. Neurol. 151:57–66

Maler, L., Karten, H. J., Bennett, M. V. L. 1973b. The central connections of the anterior lateral line nerve of *Gnathonemus petersii. J. Comp. Neurol.* 151:67–84

Maler, L., Sas, E. K. B., Rogers, J. 1981. The cytology of the posterior lateral line lobe of high frequency weakly electric fish (Gymnotidae): Dendritic differentiation and synaptic specificity in a simple cortex. *J. Comp. Neurol.* 195:87–140

Matsubara, J. A. 1981. Neural correlates of a non-jammable electrolocation system. *Science* 211:722–25

Matsubara, J., Heiligenberg, W. 1978. How well do electric fish electrolocate under jamming? *J. Comp. Physiol.* 125:285–90

McCleave, J. D., Rommel, S. A., Cathcart, C. L. 1971. Weak electric and magnetic fields in fish orientation. *Ann. NY Acad. Sci.* 188:270–82

McCormick, C. A. 1978. *Central projections of the lateralis and eighth nerves in the bowfin, Amia calva.* PhD thesis. Univ. Michigan, Ann Arbor

McCormick, C. A. 1981a. Central projections of the lateral line and eighth nerves in the bowfin, *Amia calva. J. Comp. Neurol.* 197:1–15

McCormick, C. A. 1981b. The organization of the octavolateralis area in actinopterygian fishes: A new hypothesis. *J. Morphol.* In press

McCready, P. J., Boord, R. L. 1976. The topography of the superficial roots and ganglia of the anterior lateral line nerve of the smooth dogfish *Mustelus canis. J. Morphol.* 150:527–38

McCreery, D. B. 1977a. Two types of electroreceptive lateral lemniscal neurons of the lateral line lobe of the catfish *Ictalurus nebulosus;* connections from the lateral line nerve and steady-state frequency response characteristics. *J. Comp. Physiol.* 113:317–39

McCreery, D. B. 1977b. Spatial organization of receptive fields of lateral lemniscus neurons of the lateral line lobe of catfish *Ictalurus nebulosus. J. Comp. Physiol.* 113:341–53

Mellinger, J., Belbenoit, P., Ravaille, M., Szabo, T. 1978. Electric organ discharge development in *Torpedo marmorata,* Chondrichthyes. *Dev. Biol.* 67:167–88

Meyer, D. L., Heiligenberg, W., Bullock, T. H. 1976. The ventral substrate response. A new postural control mechanism in fishes. *J. Comp. Physiol.* 109:59–68

Mikhailenko, N. A. 1971. Biological significance and dynamics of electrical discharges in weak electrical fishes of the Black Sea. *Zool. Zh.* 50:1347–52 (In Russian)

Mikhailenko, N. A. 1973. Organ of sound formation and electrogeneration in the Black Sea stargazer (Uranoscopidae). *Zool. Zh.* 52:1353–59 (In Russian)

Moller, P. 1976. Electric signals and schooling behavior in a weakly electric fish, *Marcusenius cyprinoides* L. (Mormyriformes). *Science* 193:697–99

Moller, P., Serrier, J. 1980. Specificity of electric organ discharge in social spacing of mormyrid fish. *Proc. Intl. Union Physiol. Sci.* 14:591 (Abstr.)

Moller, P., Serrier, J., Belbenoit, P. 1976. Electric organ discharge of the weakly electric fish *Gymnarchus niloticus* (Mormyriformes) in its natural habitat. *Experientia* 32:1007–8

Murray, R. W. 1960. Electrical sensitivity of the ampullae of Lorenzini. *Nature* 187:957

Murray, R. W. 1962. The response of the ampullae of Lorenzini of elasmobranchs to electrical stimulation. *J. Exp. Biol.* 39:119–28

Murray, R. W. 1965. Electroreceptor mechanisms: The relation of impulse frequency to stimulus strength and responses to pulsed stimuli in the ampullae of Lorenzini of elasmobranchs. *J. Physiol.* 180:592–606

Murray, R. W. 1974. The ampullae of Lorenzini. See Bullock 1974, pp. 125–46

Northcutt, R. G. 1980. Anatomical evidence of electroreception in the coelacanth (*Latimeria chalumnae*). *Zentralbl. Veterinaer med. Reihe C* 9:289–95

Northcutt, R. G. 1981a. Evolution of the telencephalon in nonmammals. *Ann. Rev. Neurosci.* 4:301–50

Northcutt, R. G. 1981b. Audition and the central nervous system of fishes. In *Hearing and Sound Communication in Fishes,* ed. R. R. Fay, A. N. Popper, W. N. Tavolga. New York: Springer-Verlag. 16:331–55

Northcutt, R. G., Bodznick, D. A., Bullock, T. H. 1980. Most non-teleost fishes have electroreception. *Proc. Intl. Union Physiol. Sci.* 14:614 (Abstr.)

Obara, S. 1974. Receptor cell activity at 'rest' with respect to the tonic operation of a specialized lateralis receptor. *Proc. Jpn. Acad.* 50:386–91

Obara, S. 1976. Mechanism of electroreception in ampullae of Lorenzini of the ma-

rine catfish *Plotosus.* In *Electrobiology of Nerve, Synapse, and Muscle,* ed. J. P. Reuben, D. P. Purpura, M. V. L. Bennett, and E. R. Kandel, pp. 129–46. New York: Raven. 390 pp.

Obara, S., Bennett, M. V. L. 1972. Mode of operation of ampullae of Lorenzini of the skate, *Raja. J. Gen. Physiol.* 60: 534–57

Obara, S., Higuchi, T., Umekita, S., Matsumoto, Y. 1980. Glutamate as an agonist to the afferent neurotransmitter in the ampullary electroreceptor of the marine catfish, *Plotosus. Proc. Intl. Union Physiol. Sci.* 14:617 (Abstr.)

Obara, S., Oomura, Y. 1972. Receptor mechanism of a specialized lateral line organ in the sea catfish, *Plotosus anguillaris. J. Physiol. Soc. Jpn.* 34:298 (Abstr.)

Obara, S., Oomura, Y. 1973. Disfacilitation as the basis for the sensory suppression in a specialized lateralis receptor of the marine catfish. *Proc. Jpn. Acad.* 49:213–17

Obara, S., Sugawara, Y. 1979. Contribution of Ca to the electroreceptor mechanism in *Plotosus* ampullae. *J. Physiol. Paris* 75:335–40

Okitsu, S., Umekita, S.-I., Obara, S. 1978. Ionic compositions of the media across the sensory epithelium in the ampullae of Lorenzini of the marine catfish, *Plotosus. J. Comp. Physiol.* 126:115–21

Partridge, B. L., Heiligenberg, W. 1980. Three's a crowd? Predicting *Eigenmannia's* responses to multiple jamming. *J. Comp. Physiol.* 136:153–64

Paul, D. H., Roberts, B. L. 1977a. Studies on a primitive cerebellar cortex. I. The anatomy of the lateral-line lobes of the dogfish, *Scyliorhinus canicula. Proc. R. Soc. London Ser. B* 195:453–66

Paul, D. H., Roberts, B. L. 1977b. Studies on a primitive cerebellar cortex. II. The projection of the posterior lateral-line nerve to the lateral-line lobes of the dogfish brain. *Proc. R. Soc. London Ser. B* 195:467–78

Paul, D. H., Roberts, B. L. 1977c. Studies on a primitive cerebellar cortex. III. The projection of the anterior lateral-line nerve to the lateral-line lobes of the dogfish brain. *Proc. R. Soc. London Ser. B* 195:479–96

Paul, D. H., Roberts, B. L., Ryan, K. P. 1977. Comparisons between the lateral-line lobes of the dogfish and the cerebellum: An ultrastructural study. *J. Hirnforsch.* 18:335–43

Pearson, A. A. 1936. The acoustico-lateral centers, and the cerebellum, with fiber connections, of fishes. *J. Comp. Neurol.* 65:201–94

Peters, R. C., Bretschneider, F. 1972. Electric phenomena in the habitat of the catfish *Ictalurus nebulosus* LeS. *J. Comp. Physiol.* 81:345–62

Peters, R. C., Bretschneider, F., Schreuder, J.-J.A. 1975. Influence of direct current stimulation on the ion-induced sensitivity changes of the electroreceptors (small pit organs) of the brown bullhead, *Ictalurus nebulosus* LeS. *Netherlands J. Zool.* 25:389–97

Peters, R. C., Buwalda, R. J. A. 1972. Frequency response of the electro-receptors ("small pit organs") of the catfish, *Ictalurus nebulosus* LeS. *J. Comp. Physiol.* 79:29–38

Peters, R. C., Loos, W. J. G., Gerritsen, A. 1974. Distribution of electroreceptors, bioelectric field patterns, and skin resistance in the catfish, *Ictalurus nebulosus* LeS. *J. Comp. Physiol.* 92:11–22

Peters, R. C., van Wijland, F. 1974. Electro-orientation in the passive electric catfish, *Ictalurus nebulosus* LeS. *J. Comp. Physiol.* 92:273–80

Pfeiffer, W. 1968. Die Fahrenholzschen Organe der Dipnoi und Brachiopterygii. *Z. Zellforsch. Mikrosk. Anat.* 90:127–47

Phillips, J. B. 1977. Use of the earth's magnetic field by orienting cave salamanders (*Eurycea lucifuga*). *J. Comp. Physiol.* 121:273–88

Pickens, D. E., McFarland, W. N. 1964. Electric discharge and associated behavior in the stargazer. *Anim. Behav.* 12:283–88

Pimentel de Souza, F. 1969. Analyse des autoactivités des électrorecepteurs d'un mormyre et leur stabilité. *J. Physiol. Paris* 61: Suppl. 2, p. 266

Pimentel de Souza, F. 1971. Regulation de l'autoactivité des électrorecepteurs du mormyre *Gnathonemus petersii. CR Acad. Sci. Paris* 272:2465–67

Pimentel de Souza, F. 1973. Ação controladora das descargas efetuadoras do peixe-elétrico, *Gnathonemus petersii,* sobre o nível de emissão de seus próprios receptores elétricos—um metodo de aplicação estatística e de computadores especificos na pesquisa fundamental em neurofisiologia. *Rev. Bras. Tecnol.* 4:171–76 (In Portuguese)

Pimentel de Souza, F. 1976a. Origins of the spontaneous receptor potential in the tuberous organ in a mormyrid fish. *Gen. Pharmacol.* 7:141–43

Pimentel de Souza, F. 1976b. Regulation of the electroreceptor potential frequency by the electric discharge of *Gnathonemus petersii. J. Comp. Physiol.* 111:115–25

Platt, C. J., Bullock, T. H., Czéh, G., Kovačvić, N., Konjević, Dj., Gojković, M. 1974. Comparison of electroreceptor, mechanoreceptor, and optic evoked potentials in the brain of some rays and sharks. *J. Comp. Physiol.* 95:323–55

Richardson, N. E., McCleave, J. D., Albert E. H. 1976. Effect of extremely low frequency electric and magnetic fields on locomotor activity rhythms of Atlantic salmon (*Salmo salar*) and American eels (*Anguilla rostrata*). *Environ. Pollut.* 10:65–76

Rommel, S. A. Jr., McCleave, J. D. 1972. Oceanic electric fields: Perception by American eels? *Science* 176:1233–35

Roth, A. 1968. Electroreception in the catfish, *Amiurus nebulosus. Z. Vgl. Physiol.* 61:196–202

Roth, A. 1969. Elektrische Sinnesorgane beim Zwergwels *Ictalurus nebulosus* (*Amiurus nebulosus*). *Z. Vgl. Physiol.* 65:368–88

Roth, A. 1971. Zur Funktionsweise der Elektrorezeptoren in der Haut von Welsen (*Ictalurus*): Der Einfluss der Ionen im Susswasser. *Z. Vgl. Physiol.* 75:303–22

Roth, A. 1972. Wozu dienen die Elektrorezeptoren der Welse? *J. Comp. Physiol.* 79:113–35

Roth, A. 1973a. Ampullary electroreceptors in catfish: Afferent fiber activity before and after removal of the sensory cells. *J. Comp. Physiol.* 87:259–75

Roth, A. 1973b. Electroreceptors in Brachioterygii and Dipnoi. *Naturwissenschaften* 60:106

Roth, A. 1975. Central neurons involved in the electroreception of the catfish *Kryptopterus. J. Comp. Physiol.* 100:135–46

Roth, A. 1978. Further indications of a chemical synapse in the electroreceptors of the catfish. *J. Comp. Physiol.* 126:147–50

Roth, A., Szabo, T. 1972. The receptor potential and its functional relationship to the nerve impulse analyzed in a sense organ by means of thermal and electrical stimuli. *J. Comp. Physiol.* 80:285–308

Russell, C. J., Bell, C. C. 1978. Neuronal responses to electrosensory input in mormyrid valvula cerebelli. *J. Neurophysiol.* 51:1495–1510

Russell, C. J., Myers, J. P., Bell, C. C. 1974. The echo response in *Gnathonemus petersii* (Mormyridae). *J. Comp. Physiol.* 92:181–200

Russell, I. J., Sellick, P. M. 1976. Measurement of potassium and chloride ion concentrations in the cupulae of the lateral lines of *Xenopus laevis. J. Physiol.* 257:245–55

Sand, A. 1938. The function of the ampullae of Lorenzini, with some observations on the effect of temperature of sensory rhythms. *Proc. R. Soc. London Ser. B* 125:524–53

Scheich, H. 1974. Neuronal analysis of wave form in the time domain: Midbrain units in electric fish during social behavior. *Science* 185:365–67

Scheich, H. 1977a. Neural basis of communication in the high frequency electric fish, *Eigenmannia virescens* (jamming avoidance response). I. Open loop experiments and the time domain concept of signal analysis. *J. Comp. Physiol.* 113:181–206

Scheich, H. 1977b. Neural basis of communication in the high frequency electric fish *Eigenmannia virescens* (jamming avoidance response). II. Jammed electroreceptor neurons in the lateral line nerve. *J. Comp. Physiol.* 113:207–27

Scheich, H. 1977c. Neural basis of communication in the high frequency electric fish *Eigenmannia virescens* (jamming avoidance response). III. Central integration in the sensory pathway and control of the pacemaker. *J. Comp. Physiol.* 113:229–55

Scheich, H., Bullock, T. H. 1974. The detection of electric fields from electric organs. See Bullock 1974, pp. 201–56

Scheich, H., Bullock, T. H., Hamstra, R. H. Jr. 1973. Coding properties of two classes of afferent nerve fibers: High-frequency electroreceptors in the electric fish, *Eigenmannia. J. Neurophysiol.* 36:39–60

Scheich, H., Ebbesson, S.O.E. 1981. Inputs to the torus semicircularis in the electric fish *Eigenmannia virescens. Cell Tissue Res.* 215:531–36

Scheich, H., Gottschalk, B., Nickel, B. 1977. The jamming avoidance response in *Rhamphichthys rostratus:* An alternative principle of time domain analysis in electric fish. *Exp. Brain Res.* 28:229–33

Scheich, H., Maler, L. 1976. Laminar organization of the torus semicircularis related to the input from two types of electroreceptors. In *Afferent and Intrinsic Organization of Laminated Structures in the Brain,* Suppl. 1/*Exp. Brain Res.*, ed. O. Creutzfeldt, pp. 566–79. New York: Springer-Verlag. 579 pp.

Schlegel, P. A. 1974. Activities of rhom-

bencephalic units in mormyrid fish. *Exp. Brain Res.* 19:300–14

Schlegel, P. 1975. Elektroortung bei schwach elektrischen Fischen: Verzerrungen des elektrischen Feldes von *Gymnotus carapo* und *Gnathonemus petersii* durch Gegenstande und ihre Wirkungen auf Afferente. *Biol. Cybernet.* 20:197–212

Schlegel, P. A. 1977. Electroreceptive single units in the mesencephalic magnocellular nucleus of the weakly electric fish *Gymnotus carapo. Exp. Brain Res.* 29:201–18

Schlegel, P. 1979. Single unit activities in the mesencephalon of *Sternarchus. J. Physiol. Paris* 75:421–28

Schouten, V. J. A., Bretschneider, F. 1980. Functioning of catfish electroreceptors: Influence of calcium and sodium concentration on the skin potential. *Comp. Biochem. Physiol. A* 66:291–95

Sellick, P. M., Johnstone, B. M. 1975. Production and role of inner ear fluid. *Prog. Neurobiol.* 5:337–62

Serrier, J. 1979. Electric organ discharge (EOD) specificity and social recognition in mormyrids (Pisces). *Neurosci. Lett. Suppl.* 3:S62 (Abstr.)

Sotelo, C., Rethelyi, M., Szabo, T. 1975. Morphological correlates for electrotonic transmission in the magnocellular mesencephalic nucleus of the weakly electric fish *Gymnotus carapo. J. Neurocytol.* 4:587–607

Srivastava, C. B. L. 1975. Fine structure of sensory cells of the tuberous organ (electroreceptor) of *Sternarchus albifrons* (Gymnotidae), with special reference to the membrane at the receptive and synaptic sites. *Proc. Natl. Acad. Sci. India Sect. B* 45:203–12

Srivastava, C. B. L. 1978a. Evidence for receptor nerve endings in tendons and related tissues of an electric teleost, *Gnathonemus petersii. Arch. Anat. Microsc. Morphol. Exp.* 66:253–61

Srivastava, C. B. L. 1978b. Differentiation of electric organ from muscle precursor in the regenerating tail of a weakly electric teleost: A morphogenetic approach. *Indian J. Exp. Biol.* 16:762–67

Srivastava, C. B. L., Szabo, T. 1974. Auxiliary structures of tuberous organ (electroreceptor) of *Sternarchus albifrons* (Gymnotidae). *J. Ultrastr. Res.* 48: 69–91

Suga, N. 1967. Electrosensitivity of specialized and ordinary lateral line organs of the electric fish, *Gymnotus carapo,* in *Lateral Line Detectors,* ed. P. Cahn, pp.

395–409. Bloomington, Ind.: Indiana Univ. Press. 496 pp.

Szabo, T. 1962. The activity of cutaneous sensory organs in *Gymnarchus niloticus. Life Sci.* 7:285–86

Szabo, T. 1963. Elektrorezeptoren und Tatigkeit des elektrischen organs der Mormyriden. *Naturwissenschaften* 50:447

Szabo, T. 1970. Über eine bisher unbekannte Funktion der sog. ampullaren Organe bei *Gnathonemus petersii. Z. Vgl. Physiol.* 66:164–75

Szabo, T. 1974a. Anatomy of the specialized lateral line organs of electroreception. In *Handbook of Sensory Physiology,* III/3, ed. A. Fessard, pp. 13–58. New York: Springer-Verlag. 333 pp.

Szabo, T. 1974b. Central processing of messages from tuberous electroreceptors in teleosts. See Bullock 1974, pp. 95–124

Szabo, T., Enger, P. S., Libouban, S. 1979a. Electrosensory systems in the mormyrid fish, *Gnathonemus petersii:* Special emphasis on the fast conducting pathway. *J. Physiol. Paris* 75:409–20

Szabo, T., Fessard, A. 1965. Le fonctionnement des electrorecepteurs etudies chez les Mormyres. *J. Physiol. Paris* 57: 343–60

Szabo, T., Haugedé-Carré, F., Libouban, S. 1979b. Cerebellar afferents in weakly electric mormyrid fish. *Neurosci. Lett. Suppl.* 3:S144 (Abstr.)

Szabo, T., Libouban, S. 1979. On the course and origin of cranial nerves in the teleost fish *Gnathonemus* determined by ortho- and retrograde horseradish peroxidase axonal transport. *Neurosci. Lett.* 11:265–70

Szabo, T., Libouban, S., Haugedé-Carré, F. 1979c. Convergence of common and specific sensory afferents to the cerebellar auricle (auricula cerebelli) in the teleost fish *Gnathonemus* demonstrated by HRP method. *Brain Res.* 168: 619–22

Szabo, T., Ravaille, M. 1976. Synaptic structure of the lateral line lobe nucleus in mormyrid fish. *Neurosci. Lett.* 2:121–27

Szabo, T., Ravaille, M., Libouban, S. 1978. Club endings of primary afferent fibres identified by anterograde horseradish peroxidase labelling. An EM study. *Neurosci. Lett.* 9:7–15

Szabo, T., Sakata, H., Ravaille, M. 1975. An electrotonically coupled pathway in the central neuron system of some teleost fish, *Gymnotidae and Mormyridae. Brain Res.* 95:459–74

Szamier, R. B., Bennett, M. V. L. 1980. Ampullary electroreceptors in the fresh water ray, *Potamotrygon. J. Comp. Physiol.* 138:225–30

Taylor, E. H. 1970. The lateral-line sensory system in the caecilian family, Ichthyophiidae (Amphibia: Gymnophiona). *Univ. Kansas Sci. Bull.* 48:861–68

Teague, B. D., Gilson, M., Kalmijn, A. J. 1979. Migration rate of mud bacteria as a function of magnetic field strength. *Biol. Bull.* 157:399

Teeter, J. H., Bennett, M. V. L. 1981. Synaptic transmission in the ampullary electroreceptor of the transparent catfish, *Kryptopterus. J. Comp. Physiol.* 142: 371–77

Teeter, J. H., Szamier, R. B., Bennett, M. V. L. 1980. Ampullary electroreceptors in the sturgeon *Scaphirhynchus platorynchus* (Rafinesque). *J. Comp. Physiol.* 138:213–23

Toerring, M. J., Belbenoit, P. 1979. Motor programmes and electroreception in mormyrid fish. *Behav. Ecol. Sociobiol.* 4:369–79

Tokunaga, A., Akert, K., Sandri, C., Bennett, M. V. L. 1980. Cell types and synaptic organization of the medullary electromotor nucleus in a constant frequency weakly electric fish, *Sternarchus albifrons. J. Comp. Neurol.* 192:407–26

Umekita, S.-I., Matsumoto, Y., Abe, T., Obara, S. 1980. The afferent neurotransmitter in the ampullary electroreceptors: Stimulus-dependent release experiments refute the transmitter role of L-glutamate. *3rd Annu. Mtg. Jpn. Neurosci. Soc.* No. 164 (Abstr.)

Viancour, T. A. 1977. Review of electroreceptor and peripheral electrosensory system physiology. *Proc. Intl. Union Physiol. Sci.* 12:656 (Abstr.)

Viancour, T. A. 1979a. Electroreceptors of a weakly electric fish. I. Characterization of tuberous receptor organ tuning. *J. Comp. Physiol.* 133:317–21

Viancour, T. A. 1979b. Electroreceptors of a weakly electric fish. II. Individually tuned receptor oscillations. *J. Comp. Physiol.* 133:328–39

Viancour, T. A. 1979c. Peripheral electrosense physiology: A review of recent findings. *J. Physiol. Paris* 75:321–33

Watanabe, A., Takeda, K. 1963. The change of discharge frequency by A.C. stimulus in a weak electric fish. *J. Exp. Biol.* 40:57–66

Watson, D., Bastian, J. 1979. Frequency response characteristics of electroreceptors in the weakly electric fish, *Gymnotus carapo. J. Comp. Physiol.* 134: 191–202

Westby, G. W. M. 1974a. Further analysis of the individual discharge characteristics predicting social dominance in the electric fish, *Gymnotus carapo. Anim. Behav.* 23:249–60

Westby, G. W. M. 1974b. Assessment of the signal value of certain discharge patterns in the electric fish, *Gymnotus carapo,* by means of playback. *J. Comp. Physiol.* 92:327–41

Westby, G. W. M. 1975. Has the latency dependent response of *Gymnotus carapo* to discharge-triggered stimuli a bearing on electric fish communication? *J. Comp. Physiol.* 96:307–41

Westby, G. W. M. 1977. Ontogeny of electric organ discharge in the mormyrid fish *Pollimyrus. Proc. Intl. Union Physiol. Sci.* 13:809 (Abstr.)

Westby, G. W. M. 1979. Electrical communication and jamming avoidance between resting *Gymnotus carapo. Behav. Ecol. Sociobiol.* 4:381–93

Westby, G. W. M., Kirschbaum, F. 1977. Emergence and development of the electric organ discharge in the mormyrid fish *Pollimyrus isidori.* I. The larval discharge. *J. Comp. Physiol.* 122: 251–71

Westby, G. W. M., Kirschbaum, F. 1978. Emergence and development of the electric organ discharge in the mormy-

rid fish, *Pollimyrus isidori.* II. Replacement of the larval by the adult discharge. *J. Comp. Physiol.* 127:45–59

Wiltschko, W. 1968. Über den Einfluss statischer Magnetfelder auf die Zugorientierung der Rotkehlchen (*Erithacus rubecula*). *Z. Tierpsychol.* 25:537–59

Wiltschko, W., Wiltschko, R. 1976. Interrelation of magnetic compass and star orientation in night-migrating birds. *J. Comp. Physiol.* 109:91–99

Zhadan, G. G., Zhadan, P. M. 1976. Effect of sodium, potassium, and calcium ions on electroreceptor function in catfish. *Neurophysiology* 7:312–18

Zhadan, P. M., Zhadan, G. G., Il'inskii, O. B. 1976. Functional features of electroreceptors (small pit organs) in the catfish. *Neurophysiology* 7:225–31

Zipser, B. 1971. *The electrosensory system of mormyrids.* PhD thesis. Yeshiva Univ., New York

Zipser, B., Bennett, M. V. L. 1976a. Responses of cells of posterior lateral line lobe to activation of electroreceptors in a mormyrid fish. *J. Neurophysiol.* 39:693–712

Zipser, B., Bennett, M. V. L. 1976b. Interaction of electrosensory and electromotor signals in lateral line lobe of a mormyrid fish. *J. Neurophysiol.* 39:713–21

References added in proof:

Finger, T. E., Bullock, T. H. 1982. Evoked potential studies on a thalamic center for the lateral line system in the catfish *Ictalurus nebulosus. J. Neurobiol.* In Press

Tong, S.-L., Bullock, T. H. 1981. Cerebellar responses to electroreceptor input in catfish. *Neurosci. Abstr.* 7:79 (Abstr.)

Ann. Rev. Neurosci. 1982. 5:171-87

SIGNALING OF KINESTHETIC INFORMATION BY PERIPHERAL SENSORY RECEPTORS

P. R. Burgess and Jen Yu Wei

Department of Physiology, College of Medicine, University of Utah, Salt Lake City, Utah 84108

F. J. Clark

Department of Physiology and Biophysics, University of Nebraska Medical Center, Omaha, Nebraska 68105

J. Simon

Department of Physiology, Faculty of Medicine, Universidad Complutense, Ciudad Universitaria, Madrid-3, Spain

KINESTHESIA[1]: SENSATIONS ASSOCIATED WITH JOINT POSITION AND MOVEMENT

The relative position of our skeletal body parts is determined by the angles of our joints. In the absence of visual cues, joint angle information becomes conscious through a mental image of our body (body image). Introspective attention to the body image indicates that we have the following sensory information concerning a particular joint: knowledge of 1. the angle both when the joint is moving and stationary, and an awareness of 2. the direction and 3. the speed of angular changes. These three types of information can vary independently (e.g. a given position can be reached by movements

---

[1]In the strict sense, "kinesthesia" refers only to sensations associated with joint movement. There seems to be no general term for sensations experienced both during joint movement and when the joint is stationary. Therefore, in this review we use the term kinesthesia to include sensations that occur under both static and dynamic conditions.

0147-006X/82/0301-0171$02.00

in two or more directions that can occur at different speeds) and so each is considered a different attribute of kinesthetic sensibility. We propose to refer to each of these different types of conscious information by a separate term: we will call type 1 *joint position information;* type 2, *joint direction information;* and type 3, *joint speed information.*

There is also an intensive aspect to kinesthetic experience. We consider the sensory intensity of a stimulus to be proportional to its ability to attract a person's attention and, in the context of kinesthesia, the term refers to stimuli that affect the body image. Kinesthetic intensity is less when a joint is stationary than during joint movement and the intensity increases as the rate of movement increases. If the rate of movement is kept constant there is little change in perceived intensity as the joint is moved passively over most of its range. Kinesthetic intensity differs from conscious information about joint position, speed, and direction since it is present in the absence of joint movement and is enhanced in proportion to the speed of rotation without being much influenced by its direction.

The body image, as we perceive it, is invested with skin and flesh, and when stimuli unrelated to joint angle are applied to the skin and deeper tissues, the body image is enhanced. For example, if the tips of two fingers touch a surface, there is an immediate increase in the clarity with which the relative position of the two fingers is perceived. Since the cutaneous input from the fingertips contains no information about finger position, this input must facilitate a signal from other receptor populations capable of measuring joint angle. Such a facilitatory interaction is reasonable; part of the information available about a cutaneous or subcutaneous stimulus is its location, and this is assessed in terms of the body image.

Non-angle-related inputs that refresh the body image but do not shift it may be said to have a "secondary" or "facilitatory" role in kinesthesia. Receptors whose discharge changes as a function of joint angle have a potential "primary" or "specific" role since they might produce an appropriate shift in body image during a change in joint angle. If these receptors are tonic, they might also provide ongoing signals that specify the position of the joint in the absence of movement.

Before considering how the different types of conscious information are elaborated by the central nervous system, we review the role of various peripheral receptors in the measurement of joint angle.

SENSORY RECEPTORS THAT MEASURE JOINT ANGLE

Articular Receptors

Articular mechanoreceptors have been considered important for many years in supplying joint position information (see Matthews 1982, this volume), but recent evidence suggests that they are less well designed for this function than was originally believed. The knee joint of the cat has been studied most extensively; and it has been found that articular receptors signal mainly whether the knee is at or near the end of its range (Skoglund 1956, Burgess & Clark 1969, McCall et al 1974, Clark & Burgess 1975, Grigg 1975, McIntyre et al 1978, Carli et al 1979, Ferrell 1980). Monkey knee joint articular receptors behave similarly (Grigg & Greenspan 1977). This suggests that articular receptor activity might produce the deep pressure sensations that are felt in the vicinity of the joint as the end of the range is approached. One way to assess the sensory function of these receptors is to inject local anesthetic into the knee joint cavity. This silences almost all articular receptors in the cat (Clark et al 1979, Ferrell 1980) and monkey (Clark et al 1979) knee joint except some of those that respond only to external pressure on the articular tissues (Clark 1975). When human subjects are similarly injected, simple motor tasks such as walking are unimpaired and the most obvious sensory defect that can be identified by introspection is a reduction in the deep pressure sensations at the end of the range. Careful tests have shown no deterioration in subjects' ability to detect a slow passive change in knee joint angle of 5 deg (Clark et al 1979), which ordinarily is a difficult task. Therefore, the evidence is good that knee articular receptors do not have an important primary or facilitatory role in signaling joint position but do contribute to deep pressure sensations.

The next question is whether the knee joint is typical. Elbow (Millar 1975) and wrist (Tracey 1979) articular receptors resemble those of the knee in signaling primarily near the end of the range. However, those in the hip (Carli et al 1979) respond over a large fraction of this joint's working range. Nevertheless, removal of almost all hip articular receptors during joint replacement surgery produces little kinesthetic impairment (Grigg et al 1973). No electrophysiological recordings have been published of activity from articular receptors in the finger joints, but eliminating the metacarpalphalangeal articular receptors by joint replacement surgery (Cross & McCloskey 1973) or injecting the cavity with local anesthetic (F. J. Clark, K. W. Horch, and P. R. Burgess, unpublished observations) does not impair kinesthetic position sensations. However, joint pressure and pain sensations are noticeably dulled by intracapsular anesthetic injections.

In summary, there is no evidence at present that articular receptors in any joint are important for the conscious awareness of joint position. Other kinesthetic attributes have yet to be studied quantitatively but casual observations suggest that information about the speed and direction of joint movement is little impaired by the loss of articular receptor activity. Instead, articular mechanoreceptors appear to contribute to the deep pressure sensations that occur toward the end of a joint's range. Although some articular receptors respond at both ends of a joint's range and so do not provide an obvious angle-related signal, others respond only at one end. To further test whether the latter can shift the body image, it would be desirable to investigate their action in the absence of muscle and cutaneous receptor activity. So far, an experiment of this type has not been successfully carried out.

Cutaneous Receptors

Cutaneous receptors provide angle-related signals when the skin covering one side of a joint is stretched (Hulliger et al 1979) or when the position of a joint brings skin surfaces into contact. Anesthetizing the skin around the knee joint, either alone or in combination with an injection into the joint cavity, produces no alteration in appreciation of joint position or change in body image that can be detected introspectively. Performance in a difficult task involving discrimination of slowly produced 5 deg angular changes was not diminished under these conditions (Clark et al 1979). However, when a finger was anesthetized, which eliminated both cutaneous and deep sensibility, there was a profound sense of loss and the finger largely disappeared from the body image. It is not surprising that kinesthetic sensibility was impaired under these conditions (Browne et al 1954, Provins 1958, Goodwin et al 1972), but it is difficult to know how much of this was due to the loss of non-angle-related inputs necessary for the elaboration of the body image and how much was due to the loss of specific kinesthetic signals. It is clear that signals from skin and/or deeper connective tissue receptors around the interphalangeal joints can shift the body image, although comparatively rapid movements of the joint are required (Gandevia & McCloskey 1976) and the actual position of the joint is largely unknown.

In summary, there is no evidence at present that cutaneous receptors around the knee make an important contribution to conscious information about the position of this joint. This presumably applies also to other joints proximal to the hand and foot. However, it would be desirable to test further whether angle-related signals from these cutaneous receptors can shift the perceived angle of a proximal joint by investigating their action in the absence of input from joint and muscle receptors. After a digit is anesthetized, its body image representation is less clear and what is left of the digital image fails to shift appropriately with changes in joint angle. This indicates that receptors in the finger supply important primary or facilitatory kinesthetic inputs. Neither input appears to arise from articular receptors since selective anesthesia or removal of these receptors produces little impairment.

Muscle Receptors

If neither joint receptors nor cutaneous receptors (except perhaps for the digits) provide an important primary input for our awareness of joint position, muscle receptors become the most likely candidates. Moreover, there is direct evidence that muscle receptors are involved in kinesthesia. Vibrating the tendon of a muscle produces an illusion, after a delay of a second or so, that the joint is being displaced in a direction so as to stretch the vibrating muscle (Eklund 1972, Goodwin et al 1972, Matthews 1982). The illusion incorporates both a sense of joint movement and a sense of altered joint position. The effective vibrations are at amplitudes that would excite mainly the primary spindle endings, and perhaps some secondaries, but not the tendon organs to any extent. This does not rule out a role for tendon organs in kinesthesia (Rymer & D'Almeida 1980), but in the discussion below the emphasis is on the receptors in muscle spindles.

Table 1 lists the muscles that cross the cat ankle joint. A number of these also cross either the knee or the toe joints, and plantaris and extensor digitorum longus cross all three. Of the five muscles that cross only the ankle joint, there are two—soleus and tibialis anterior—that signal primarily on the flexion-extension axis (Table 2, Figure 1). Figures 2 and 3 show average input-output functions for the primary (Figures 2A and B) and the secondary (Figure 3A) muscle spindle endings in these two muscles. The input-output functions were constructed from measurements made during a staircase stimulus sequence that started at that end of the range where the muscle was unstretched and proceeded to the opposite end of the range and back again in steps of 6–8 deg. Each position was held for 16–18 sec and the rate of movement from one position to the next was 40 deg/sec. The upper and lower curves in Figures 2A and 3A show the discharge during movement (dynamic input-output functions). In constructing the dynamic input-output function for an individual receptor, each value was plotted at the position where the movement terminated. A number of these individual dynamic functions have been averaged to generate the average dynamic functions in Figures 2A and 3A. Enclosed within these dynamic curves are static input-output functions constructed from measurements made at each position 2 sec and 15 sec after the movements were over. The 15 sec curves lie within the 2 sec functions. These data were collected from cats sufficiently deeply anesthetized that they had little motor tone. Present evidence suggests that relaxed human subjects also have little alpha or gamma tone (Vallbo 1974a, Burke et al 1976, 1978) and the input-output functions in Figures 2 and 3 would presumably resemble those of primate spindles in relaxed muscles when the ankle joint is moved passively (Hagbarth et al 1973, Cheney & Preston 1976). We assume that this is the case in the discussion below.

HOW MUCH OF A JOINT'S RANGE IS SIGNALED BY MUSCLE RECEPTORS UNDER PASSIVE AND ACTIVE CONDITIONS?   Under passive conditions, muscle receptors signal over the joint's entire range, with ago-

Table 1   Muscles influencing ankle joint[a]

| | Joints crossed between origin and insertion | | |
Muscle	Knee	Ankle	Toe
Medial gastrocnemius	+	+	0
Lateral gastrocnemius	+	+	0
Plantaris	+	+	+
Soleus	0	+	0
Flexor digitorum longus	0	+	+
Flexor hallucis longus	0	+	+
Tibialis posterior	0	+	0
Peroneus longus	0	+	0
Peroneus brevis	0	+	0
Peroneus tertius	0	.+	+
Tibialis anterior	0	+	0
Extensor digitorum longus	+	+	+

[a] The muscles influencing the ankle are listed together with any other joints crossed. Plantaris does not cross the toe joints directly but inserts into the flexor digitorum brevis.

nist and antagonist muscles dividing the range about equally between them (Figures 2 and 3). When the joint is moved into the noncoded region for a particular muscle, the muscle's tendon goes slack, as can easily be verified with one's own Achilles tendon. This means that in order to read spindle discharge in terms of joint angle, it must be known whether the extrafusal muscle fibers are contracted since extrafusal contraction, by taking up the slack in the tendon, makes it possible for the spindles in a muscle to signal over the entire range. However, extrafusal contraction alone tends to unload

Table 2   Responses of primary spindle receptors in certain muscles to different positions of the ankle joint[a]

	Soleus	Tibialis anterior	Peroneus longus	Peroneus brevis	Tibialis posterior
Flexion	++	0	0	+	0
Flexion + add & cw twist[b]	+++	0	++	+++	0
Flexion + abd & ccw twist	+++	0	0	0	+++
Extension	0	++	+	0	0
Extension + add & cw twist	0	++	+++	++	0
Extension + abd & ccw twist	0	+++	0	0	++

[a] The number of plus signs indicates the relative strength of the response, +++ designating the maximal response to the most effective stimulus. In each case, the stimulus was strong, i.e. near the end of the range. All values are based on measurements from at least three animals.

[b] add = adduction; abd = abduction; cw = clockwise; ccw = counterclockwise.

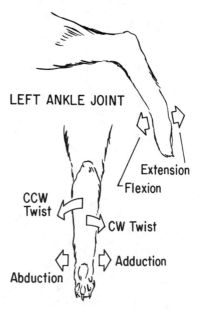

Figure 1 A drawing of the left ankle joint of a cat illustrates the axes of joint rotation tested. Adduction was often combined with clockwise twist and abduction with counterclockwise twist.

the spindles and it is not surprising, therefore, that the gamma motor system is recruited together with the alpha motor system during most motor acts (Vallbo 1974b, Hagbarth et al 1975). The level of gamma activity must also be known to interpret spindle discharge in terms of joint angle since increased fusimotor drive can alter spindle discharge in the absence of a change in joint angle. The afferent response to a given level of fusimotor drive changes with the length of the muscle, due to the length-tension relationship of the intrafusal fibers (Lewis & Proske 1972), and this also has to be adjusted for. A further complication would arise if the intrafusal fibers fatigued, since the relationship between the gamma output and the afferent signal would then be altered at all muscle lengths. Fortunately, the fusimotor system appears to be relatively fatigue resistant (Emonet-Dénand & Laporte 1978).

"PLACE" AND "FREQUENCY" CODES FOR SIGNALING JOINT ANGLE
A fundamental question in kinesthesia is the nature of the code used by the peripheral receptors to signal joint position. A *place code* is one in which individual receptors are spatially tuned so that each signals over only a limited portion of the range. The joint angle is then specified by just which receptors are active. Figures 2 and 3 indicate that although some place

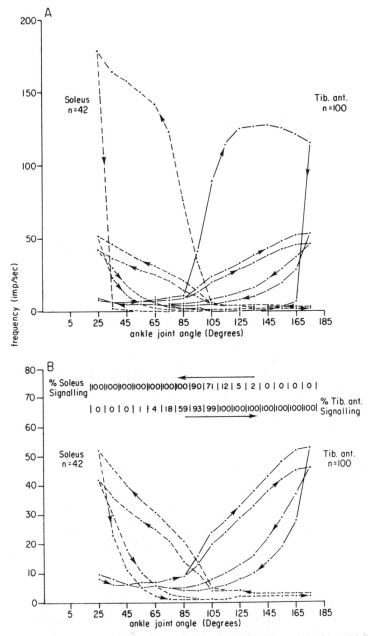

Figure 2   The responses of 100 primary (phasic) muscle spindle receptors from tibialis anterior and 42 primary endings from soleus have been averaged and are graphed as a function
of ankle joint angle. Dynamic input-output (I-O) functions are shown in *A* as well as 2 sec
and 15 sec static functions (see text). The I-O functions of individual receptors were averaged

coding is possible as a muscle comes under tension, over much of a muscle's signaling range the position of the joint during muscle lengthening is coded by an increase in the discharge of fibers that are already active. During a strong isometric contraction, the possibility of recruiting spindle receptors at different positions during muscle lengthening would be even more limited because most of the spindles would have been set into activity by fusimotor activity (Vallbo 1974b, Burke et al 1978). Thus, joint position appears to be signaled to an important degree by the level of activity in two populations of afferent fibers, each from an antagonistic muscle or muscle group. This might be called an *opponent frequency code.*

EVIDENCE FOR CENTRAL PROCESSING OF THE PERIPHERAL FREQUENCY CODE: RATE SENSITIVITY, ADAPTATION, AND LINEAR DIRECTIONALITY   Proceeding on the evidence that the level of spindle discharge is involved in specifying joint angle, Figures 2 and 3 indicate that if the frequency of discharge is read directly, the angle cannot be known with any precision. During movement, spindle discharge exhibits rate sensitivity. This property is well developed among primary endings (Figure 2A) and their activity is determined much more by the speed of joint movement than by joint angle (Cooper 1961, Harvey & Matthews 1961). The secondary endings are also rate-sensitive; the difference in discharge between the dynamic input-output functions and the 2 sec static input-output functions during muscle lengthening in Figure 3A would result in an angular error of about 21 deg for tibialis anterior secondary receptors and 20 deg for soleus secondaries. (How this calculation was made is shown in Figure 3B.)

Adaptation of the discharge after the limb becomes stationary ("static adaptation") also changes the relationship between spindle discharge frequency and joint angle. In the case of the secondary endings, this would cause a position error of 4 deg for tibialis anterior and 5 deg for soleus

---

by dividing the flexion-extension axis into 10 deg bins and summing the discharge frequencies of all the measurements that fell within a particular bin and dividing by the number of receptors contributing. If an individual receptor contributed more than one measurement to a bin, these were averaged before being added to the other measurements. In *B*, the 2 sec and 15 sec curves in *A* have been expanded and the percentage of primary endings signaling at each angle is indicted for both tibialis anterior and soleus. An ending was considered to be signaling if its discharge had increased by an amount equal to 5% of its total change in frequency over the range coded. The percentages given apply to the 2 sec static responses during muscle lengthening. The 5% level was exceeded a little sooner during movement and a little later for the 15 sec static case. Thirty-eight percent of the soleus primary endings and 54% of the tibialis anterior endings had resting activity at angles where their discharge was not influenced by changes in joint position.

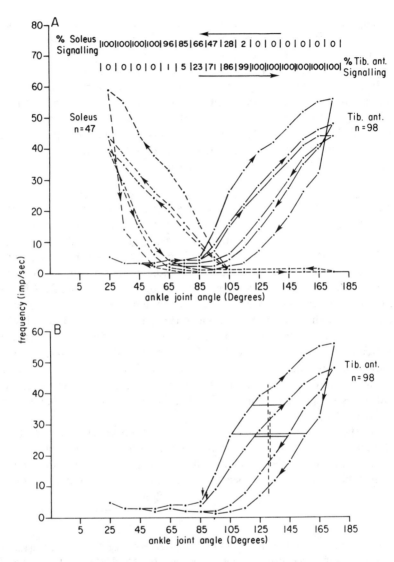

Figure 3 *A* shows the responses of 98 secondary (tonic) muscle spindle receptors in tibialis anterior and 47 secondary endings in soleus, averaged and graphed as in Figure 2, and under the same stimulus conditions. The percentage of secondary endings signaling at each angle is indicated for both muscles. Eight percent of the soleus secondary endings and 20% of the tibialis anterior secondaries had resting activity at angles where changes in joint position were not coded. *B* illustrates how rate and linear directionality errors were calculated. The responses during movement and 2 sec after movement was over are shown for secondary endings in tibialis anterior. The threshold, defined as an increase equal to 5% of the frequency change over the coded region, occurred at 87 deg for the average movement response and at 89 deg for the 2 sec static curve during muscle lengthening (vertical arrows). Linear directionality

receptors during the 13 sec that elapsed between the 2 sec and 15 sec readings (Figure 3A). The error would be greater in the case of the primary endings, amounting to 10 deg for tibialis anterior and 16 deg for soleus receptors during this same 13 sec period (Figure 2B). These error calculations were made as described in Figure 3B and refer to muscle lengthening.

Still another disparity between frequency of discharge and joint position results from the "linear directionality" of the endings. Linear directionality refers to a difference in neural discharge at a given joint position when that position is reached by moving in different directions along the same path. Linear directionality is evident both during movement and after the joint becomes stationary. Figure 3B shows how we estimated the angular error introduced by linear directionality.

Linear directionality is exaggerated during movement because of the rate sensitivity of the endings. The primary endings (Figure 2A) were actually silenced during muscle shortening and for this reason primary endings are unable to signal joint position or joint speed during shortening at rates higher than about 10 deg/sec. Linear directionality is greater 2 sec after movement ceases than after 15 sec because of the static adaptation of the receptors. The primary endings in soleus had 50 deg of linear directionality error at 2 sec and 34 deg at 15 sec. The corresponding values for tibialis anterior primaries were 56 deg and 34 deg.

Secondary endings have less linear directionality than primary endings but the errors are nevertheless appreciable. During movements at 40 deg/-sec, the errors were 52 deg for secondaries in soleus and 50 deg for those in tibialis anterior. Under static conditions, the secondary endings in soleus had an error of 33 deg at 2 sec and 26 deg at 15 sec. The corresponding values for tibialis anterior secondaries were 22 deg and 14 deg.

If left uncorrected, errors of this magnitude would seriously degrade an

---

errors were estimated by measuring the angular (horizontal) distance between the lengthening and shortening I-O functions where they were equidistant from a vertical line placed half way through the coded region. The dotted vertical line at 131 deg is at the half-way point for the dynamic response and the dotted line at 132 deg is half way for the 2 sec static function. Accordingly, the length of the middle horizontal line (50 deg) is the linear directionality error of the dynamic functions and the length of the lowest horizontal line (22 deg) gives this error for the 2 sec static functions. Rate errors during muscle stretch were calculated in a similar fashion as the angular (horizontal) distance between the dynamic I-O function and the 2 sec static I-O function where these functions were equidistant from the vertical line half way through the dynamic response (uppermost horizontal line, 21 deg). Adaptation errors were considered to be equal to the angular distance between the 2 sec and 15 sec static I-O functions where they were equidistant from the vertical line half way through the 2 sec static response. The 15 sec static functions have been omitted from *B* for simplicity but are shown in *A*. Rate, adaptation, and linear directionality errors calculated in this fashion may have no particular biological relevance and are meant simply to express the frequency differences produced by rate sensitivity, adaptation, and linear directionality in such a way that they can be related to joint angle.

animal's information about limb position. Human subjects experience little rate or adaptation error: if a relaxed limb is moved briskly to a particular angle and then stopped and held stationary, the perceived position of the joint remains fixed in spite of the decline in receptor discharge. During a step-wise stimulus sequence like that used in the animal studies, relaxed human subjects experience linear directionality errors of less than 5 deg. Present evidence suggests that primates and cats have spindles with similar properties, and that relaxed muscles receive little fusimotor support. Therefore, it is likely that the frequency of discharge of the receptors is not used directly to specify joint angle but that the central nervous system "processes" this input in some way.

COULD SOLEUS AND TIBIALIS ANTERIOR SPINDLES SPECIFY ANKLE POSITION ON THE FLEXION-EXTENSION AXIS WITHOUT THE PARTICIPATION OF RECEPTORS IN OTHER MUSCLES?   Soleus and tibialis anterior are the two muscles of those specific to the ankle joint that would appear to be best situated to signal flexion and extension. However, tibialis anterior spindles are excited by ankle abduction and by counterclockwise twist of the foot and soleus receptors are excited by abduction, adduction, clockwise, and counterclockwise twist (Table 2, Figure 1). If tibialis anterior spindle activity were to be inhibited within the central nervous system by activity arising from spindles in tibialis posterior (Table 2), and if soleus spindle activity were to be inhibited by tibialis posterior and peroneal spindle discharge, the tibialis anterior and soleus signals would become more specific for ankle extension and flexion.

ROLE OF MULTIJOINT MUSCLES IN SIGNALING JOINT ANGLE   Of the 12 muscles that cross the ankle joint, only 5 are confined to the ankle joint alone (Table 1). It is not yet obvious how the spindles in multijoint muscles can signal joint angle since their discharge at a given ankle joint position can be greatly altered by changes in the position of the knee or toe joints. For example, the spindles in extensor digitorum longus are well excited by ankle extension but their discharge at any position over the extension half of the range can be greatly enhanced by toe flexion. One way in which multijoint muscle receptors might contribute to joint position signaling is for their discharge to be channeled into different central circuits depending on changes in activity from joint specific muscles. Thus, spindles in extensor digitorum longus would contribute to the signaling of ankle extension if tibialis anterior were being stretched but the short toe extensors were not, and would contribute to the signaling of toe flexion if the short toe extensors were being lengthened but tibialis anterior was not. Some support for this idea comes from tendon vibration studies. Vibrating the tendon of tibialis anterior causes an illusion of ankle extension when the joint is stationary. Vibrating the tendon of extensor digitorum longus does not produce any very clear illusions unless the ankle is passively extended, in which case the ankle extension is felt to be greater than it actually is, or unless the toes are passively flexed, which produces an illusory exaggeration of toe flexion. It would appear that the illusions from a multijoint muscle are referred primarily to the joint being moved. However, these results must be interpreted with caution because tendon vibration produces patterns of input that are unlikely to occur in nature and the nervous system may respond in ways that are not indicative of its normal behavior.

SOME POSSIBLE CENTRAL NEURAL CIRCUITS FOR KINESTHETIC SENSIBILITY

In this section we speculate on how the various kinds of conscious kinesthetic information described at the beginning of this review are elaborated by the central nervous system. Only the passive case is considered; i.e. the joint is rotated by an external force with the subject relaxed. Present evidence suggests a good correspondence between the behavior of anesthetized cat and relaxed human ankle joint muscle spindles under these conditions (Hagbarth et al 1973) and therefore the detailed information available for the cat (Figures 2 and 3) is used to discuss how the central circuits might function.

Joint Position Information

CORRECTING FOR RECEPTOR RATE SENSITIVITY, ADAPTATION, AND LINEAR DIRECTIONALITY   During muscle stretch, the problem faced by the kinesthetic system is similar to that confronted by the tactile system during skin indentation. The rate sensitivity of cutaneous mechanoreceptors causes a largely rate dependent discharge to occur while the skin is being indented, and this discharge then declines after the movement stops because of receptor adaptation. There is evidence for an integrator in the circuit signaling skin indentation depth which helps convert the rate sensitive receptor signal into a largely rate independent awareness of altered skin position (D. A. Poulos, K. W. Horch, R. P. Tuckett, J. Mei, and P. R. Burgess, unpublished observations). A similar mechanism could serve for joint position information. What is required is that the excitation set up by a nerve impulse persist in the integrator so that it can add to the excitation produced by the next impulse. This persistent excitation would be independent of any nerve impulses produced by the integrator itself. The rate at which the integrator charges and discharges would have to be matched to the dynamic properties of the receptors, if the system were to function properly. However, this model, which works well for the skin, encounters a serious problem when applied to spindle receptor behavior during muscle shortening. The receptor discharge after a shortening movement gradually builds up rather than declining and an integrator would exaggerate this build up, thereby enhancing the error.

The situation in the light of present understanding can be summarized as follows. The central nervous system must be equipped to extract a reliable position signal from spindle discharge that is both rate and directionally sensitive. Integration of the input (in the mathematical sense) can help compensate for rate sensitivity during muscle stretch but there are problems in producing a properly controlled reduction in the level of the integrator during muscle shortening. However, since humans, and presumably other animals, have relatively small rate and linear directionality errors under passive conditions, the sensory input must have been "corrected" in some way.

IS THE FREQUENCY CODE IN THE PERIPHERY CONVERTED INTO A PLACE CODE WITHIN THE CENTRAL NERVOUS SYSTEM? Another basic question about conscious joint position information is whether the frequency code in the periphery is converted into a place code within the central nervous system. It is generally thought that "spatial" sensations use place codes centrally; i.e. when the perceived location of something that is being sensed changes, this occurs because there is a change in the locus of neural activity in the brain. If we apply this idea to kinesthesia, a perceived change in the location of a body part due to an alteration in joint angle should be associated with a wave or front of activity that moves through the brain. Moreover, the subjective experience is one of the whole anatomical entity, skin and flesh alike, being relocated. The most literal interpretation of this is that muscle spindle receptors propel the central somatotopic representations of the skin and deeper tissues through brain space.

Joint Direction Information

Perhaps the easiest way to detect the fact that joint movement is occurring as well as the direction of that movement is to compare the discharge of the primary and secondary endings. When the muscle is lengthening, the discharge of the primary endings is enhanced much more than that of the secondaries. During muscle shortening, the discharge of the primary endings is greatly suppressed, that of the secondaries less so. Measuring the discharge of the primary endings alone would not, for receptors lacking resting activity, distinguish muscle shortening from the absence of movement in a noncoded portion of the range. An alternative method for determining the direction of movement—by assessing whether the discharge of the secondary endings is increasing or decreasing—would not be so prompt as the comparison of primary and secondary discharge because of the great sensitivity of the primary endings to changes in the direction of movement.

Joint Speed Information

Although primary muscle spindle receptors provide a signal largely proportional to the speed of joint movement during muscle lengthening, they are silent during passive muscle shortening at velocities above about 10 deg/sec, in the absence of fusimotor support. If the same method is to be used by the central nervous system for measuring speed of joint rotation during both passive lengthening and shortening, the secondary endings will have to provide the signal (Figures 2A and 3A). One possibility is to measure the rate of change of the secondary discharge.

Kinesthetic Intensity Sensations

Kinesthesia is like most other spatial sensory experiences in that moving from one location to another is not associated with any obvious change in perceived intensity; however, the discharge of the muscle spindle receptors increases progressively as the joint is moved from an intermediate position to the end of the range (Figures 2 and 3). This is the only instance known to us where progressively increasing afferent input is not reflected in an increasingly intense sensation. The kinesthetic intensity circuit may monitor the size of the wave of activity that we have postulated results from conversion of the peripheral frequency code into a central place code. A mechanism of this sort would also account for the fact that intensity sensations are about the same whether a muscle is lengthening or shortening, even though the overall level of spindle discharge is much higher during lengthening. The fact that the intensity is felt to increase with the rapidity of joint rotation would be explained if the intensity circuit receives additional inputs from nonspindle muscle and connective tissue receptors that are rapidly adapting.

The speculative character of this discussion on the properties of central neural circuits for kinesthetic sensibility indicates how little is actually known about information-processing in this sensory modality. How the circuits might have to change to accommodate the effects of efferent activity is equally unknown and adds additional complexity to the problem. More research will be required to determine whether the circuitry we have postulated is actually present.

ACKNOWLEDGMENTS

The authors wish to thank Gary Frederickson, John Fisher, Barry Evans, Carol Reeves, and Ken Horch for their contributions to this work. Our research was supported by grant BNS76-18764 from the National Science Foundation and by grants NSO8769, NSO7938, and TWO2029 from the National Institutes of Health.

186   BURGESS, CLARK, SIMON & WEI

Literature Cited

Browne, K., Lee, J., Ring, P. A. 1954. The Sensation of passive movement at the metatarso-phalangeal joint of the great toe in man. *J. Physiol.* 126:448–58

Burgess, P. R., Clark, F. J. 1969. Characteristics of knee joint receptors in the cat. *J. Physiol.* 203:317–35

Burke, D., Hagbarth, K.-E., Löfstedt, L. 1978a. Muscle spindle responses in man to changes in load during accurate position maintenance. *J. Physiol.* 276:159–64

Burke, D., Hagbarth, K.-E., Löfstedt, L., Wallin, B. G. 1976. The responses of human muscle spindle endings to vibration during isometric contraction. *J. Physiol.* 261:695–711

Burke, D., Hagbarth, K.-E., Skuse, N. F. 1978b. Recruitment order of human spindle endings in isometric voluntary contractions. *J. Physiol.* 285:101–12

Carli, G., Farabollini, F., Fontani, G., Meucci, M. 1979. Slowly adapting receptors in cat hip joint. *J. Neurophysiol.* 42:767–78

Cheney, P. D., Preston, J. B. 1976. Classification and response characteristics of muscle spindle afferents in the primate. *J. Neurophysiol.* 39:1–8

Clark, F. J. 1975. Information signaled by sensory fibers in medial articular nerve. *J. Neurophysiol.* 38:1464–72

Clark, F. J., Burgess, P. R. 1975. Slowly adapting receptors in cat knee joint: Can they signal joint angle? *J. Neurophysiol.* 38:1448–63

Clark, F. J., Horch, K. W., Bach, S. M., Larson, G. F. 1979. Contributions of cutaneous and joint receptors to static knee-position sense in man. *J. Neurophysiol.* 42: 877–88

Cooper, S. 1961. The responses of the primary and secondary endings of muscle spindles with intact motor innervation during applied stretch. *Q. J. Exp. Physiol.* 46:389–98

Cross, M. J., McCloskey, D. I. 1973. Position sense following surgical removal of joints in man. *Brain. Res.* 55:443–45

Eklund, G. 1972. Position sense and state of contraction; the effects of vibration. *J. Neurol. Neurosurg. Psychiatr.* 35:606–11

Emonet-Dénand, F., Laporte, Y. 1978. Effects of prolonged stimulation at high frequency of static and dynmic axons on spindle primary endings. *Brain Res.* 151:593–98

Ferrell, W. R. 1980. The adequacy of stretch receptors in the cat knee joint for signalling joint angle throughout a full range of movement. *J. Physiol.* 299: 85–99

Gandevia, S. C., McCloskey, D. I. 1976. Joint sense, muscle sense, and their combination as position sense, measured at the distal interphalangeal joint of the middle finger. *J. Physiol.* 260:387–407

Goodwin, G. M., McCloskey, D. I., Matthews, P. B. C. 1972. The contribution of muscle afferents to kinaesthesia shown by vibration induced illusions of movements and by the effects of paralysing joint afferents. *Brain* 95:705–48

Grigg, P. 1975. Mechanical factors influencing response of joint afferent neurons from cat knee. *J. Neurophysiol.* 38:1473–84

Grigg, P., Finerman, G. A., Riley, L. H. 1973. Joint-position sense after total hip replacement. *J. Bone Joint Surg.* 55A:1016–25

Grigg, P., Greenspan, B. J. 1977. Response of primate joint afferent neurons to mechanical stimulation of knee joint. *J. Neurophysiol.* 40:1–8

Hagbarth, K.-E., Wallin, G., Burke, D., Löfstedt, L. 1975. Effects of the Jendrassik manoeurve on muscle spindle activity in man. *J. Neurol. Neurosurg. Psychiatr.* 38:1143–53

Hagbarth, K.-E., Wallin, G., Löfstedt, L. 1973. Muscle spindle responses to stretch in normal and spastic subjects. *Scand. J. Rehab. Med.* 5:156–59

Harvey, R. J., Matthews, P. B. C. 1961. The response of de-efferented muscle spindle endings in the cat's soleus to slow extension of the muscle. *J. Physiol.* 157:370–92

Hulliger, M., Nordh, E., Thelin, A.-E., Vallbo, A. B. 1979. The responses of afferent fibres from the glabrous skin of the hand during voluntary finger movements in man. *J. Physiol.* 291:233–49

Lewis, D. M., Proske, U. 1972. The effect of muscle length and rate of fusimotor stimulation on the frequency of discharge in primary endings from muscle spindles in the cat. *J. Physiol.* 222:511–35

Matthews, P. C. B. 1982. Where does Sherrington's "muscular sense" originate? Muscles, joints, corollary discharges? *Ann. Rev. Neurosci.* 5:In press

McCall, W. D. Jr., Farias, M. C., Williams, W. J., Bement, S. L. 1974. Static and dynamic responses of slowly adapting joint receptors. *Brain Res.* 70:221–43

McIntyre, A. K., Proske, U., Tracey, D. J. 1978. Afferent fibres from muscle receptors in the posterior nerve of the cat's knee joint. *Exp. Brain Res.* 33:415–24

Millar, J. 1975. Flexion-extension sensitivty of elbow joint afferents in cat. *Exp. Brain Res.* 24:209–14

Provins, K. A. 1958. The effect of peripheral nerve block on the appreciation and execution of finger movements. *J. Physiol.* 143:55–67

Rymer, W. Z., D'Almeida, A. 1980. Joint position sense: The effects of muscle contraction. *Brain* 103:1–22

Skoglund, S. 1956. Anatomical and physiological studies of knee joint innervation in the cat. *Acta Physiol. Scand.* 36:Suppl. 124, pp. 1–101.

Tracey, D. J. 1979. Characteristics of wrist joint receptors in the cat. *Exp. Brain Res.* 34:165–76

Vallbo, A. B. 1974a. Afferent discharge from human muscle spindles in non-contracting muscles. Steady state impulse frequency as a function of joint angle. *Acta Physiol. Scand.* 90:303–18

Vallbo, A. B. 1974b. Human muscle spindle discharge during isometric voluntary contractions. Amplitude relations between spindle frequency and torque. *Acta Physiol. Scand.* 90:319–36

Ann. Rev. Neurosci. 1982. 5:189–218
Copyright © 1982 by Annual Reviews Inc. All rights reserved

WHERE DOES SHERRINGTON'S "MUSCULAR SENSE" ORIGINATE? MUSCLES, JOINTS, COROLLARY DISCHARGES?[1]

P. B. C. Matthews

University Laboratory of Physiology, Parks Road, Oxford OX1 3PT, United Kingdom

Are our muscles sentient? In other words, do we receive sensory signals from voluntary muscles that reach consciousness and tell us what is going on in the muscles in the same sort of way that sensory signals from skin tell us what is happening there? Of course, we have pain signals from muscle as in muscular cramp, but do we have more routine signals that help to tell us where our limbs are in space or what muscular effort we are exerting? This might seem a trivial question, one which should have been sorted out at the very beginning of physiological enquiry. But it has not proved an easy question to answer and there are still many unresolved problems of detail, and even matters of principle remain clouded. Moreover, there has been a remarkable ebb and flow of opinion over the fundamentals. The present review starts by outlining the position as it was at the beginning of the century, since in many respects understanding has shown little advance. It then concentrates on recent work, since reviews on the situation up to about 1977 are readily available (Goodwin et al 1972, Matthews 1977, McCloskey 1978, 1981), but enough of the intervening work has been included to make the present account complete in itself. The conclusion summarizes the

[1]This review is based on a lecture given to the tenth annual meeting of the Society for Neuroscience at Cincinnati in November 1980. It has been modified both to suit the printed idiom and also to complement, without major reduplication, the accompanying article by Burgess et al (1982).

189

0147-006X/82/0301-0189$02.00

current position, with the sense organs in muscle allocated the crucial role in kinesthesia, but with the sensation referred to the relevant joint and "body image" rather than to the muscles themselves. The recent reversals of opinion on the subject are illustrated in Table 1.

Sherrington's Views

In 1900 Sherrington wrote a remarkable article for Schäfer's *Textbook of Physiology* in which he reviewed the "muscular sense," which had already been a topic of physiological interest for at least 75 years. His views seem to have been widely followed for the next half century, but then for a variety of perfectly respectable reasons they came to be rejected almost *in toto*. However, in the last ten years the situation has reversed yet again and most of his views are once more acceptable, as outlined in Table 1. Sherrington provided a rather full explanation of what he meant by muscle sense and it is still hard to better. It runs as follows: "The perceptions of muscular sense may be grouped into (*i*) those of posture, (*ii*) of passive movement, (*iii*) of active movement, and (*iv*) those of resistance to movement. Changes in consciousness accompany all movements of the body which occur at speeds above a certain liminal [amount], and over distances of beyond a liminal extent." The implicit assumption is that these different aspects of muscle sense are based on broadly similar mechanisms and blend into each other, though clearly differing in detailed mechanism.

Sherrington's view of the origin of this bundle of sensory experiences was crisply simple, namely that it arises from "specific sense organs in muscles, tendons, and joints." Muscle receptors held pride of place in his thinking, even when there was no ongoing muscular contraction, although he had nothing very much to go on except his dramatic experimental demonstration of a few years earlier of the wealth of afferent innervation of muscle, and that the muscle spindles as well as the tendon organs were afferent end-organs. At that time he seemed quite unaware of how much else they might have to do which could be better carried out quite unconsciously.

Table 1 Suggested sources of "muscular sense"

	Sherrington 1900	1960 to 1970	Now
Muscle & tendon	Yes	No	Yes
Joints	Yes	Uniquely so	Possibly in some cases
Skin	Perhaps	No ?	Perhaps
Corollary discharges	No	?	Effort & weight—Yes Kinesthesia—No, but needed for decoding

Joint receptors were already widely accepted as having an important sensory role in muscle sense and this Sherrington accepted for both active and passive movement. For cutaneous receptors he emphasized that "it is obvious that any extensive action of skeletal musculature must affect . . ., by stretching and flexing the skin, sense organs within it," but he nonetheless concluded that these signals do "not much assist muscular sense."

But what really troubled Sherrington was the idea that muscular sense depends not on afferent input but on motor output; or, following the terminology much more recently introduced by Sperry (1950), that sensation might depend upon corollary discharges transmitted from a motor center to a sensory center at the same time as the dispatch of a motor command. Citing Helmholtz and others Sherrington noted, "The view which dispenses with peripheral organs and afferent nerves for the muscular sense has had powerful adherents . . . It supposes that during . . . a willed movement the outgoing current of impulses from brain to muscle is accompanied by a 'sensation for innervation'." He attacked this idea from every side and concluded that it "remains unproven"; but he does not seem to have appreciated, as it was in the forefront of more recent analyses (Sperry 1950, von Holst & Mittelstaedt 1950, von Holst 1954), that corollary discharges and afferent signals may be complementary rather than exclusive mechanisms.

REJECTION OF SHERRINGTON'S VIEWS   The justification for laboring Sherrington's views, which might be felt to be merely an exercise in history, is that in some respects we are no further advanced now than we were then, and we have had an intervening period in which Sherrington's attribution of muscle sense to muscle receptors was widely regarded as wrong-headed, as well as simply being thought to be incorrect. Instead, from at least 1960 to 1970, joint afferents were seen by most physiologists as being uniquely responsible for muscle sense, or rather for kinesthesia (or kinesthesis), as it was then, and is now, felt more appropriate to term the same range of sensations; indeed, as Burgess et al's article emphasizes, we still lack suitable agreed terms to describe the details of these sensations. This rejection arose partly from continued interest in the classical problems of the oculomotor system; here corollary discharges have long seemed to provide the most likely source of the information by means of which the brain can recognize the constancy of the visual scene, in the face of the displacement of the retinal image of the world that is induced by voluntary movement of the eyes. But probably more important in influencing opinion was the red herring that for a brief period while technique was developing it was thought that there was no group I muscle afferent projection to the cerebral cortex. However, by the time that this had been found in the early 1960s, the rejection of a sensory role for muscle afferents had found a wealth of support from quite reasonable experiments, which do not merit present description, but all of which can be argued to have been flawed (McCloskey 1978, Matthews 1977).

In retrospect, the force with which Sherrington's views were rejected appears remarkable. Thus, in 1959 Rose & Mountcastle concluded their important article in the American *Handbook of Physiology* with the statement, "The sense of position and of movements of the joints depends solely on the appropriate receptors in the joints themselves. There is no need to invoke a mysterious 'muscle' sense to explain kinaesthetic sensations, and to do so runs contrary to all the known facts concerning the muscle stretch receptors." Thirteen years later E. G. Jones (1972), in a scholarly article on the nineteenth century neurologist Bastian who introduced the term kinesthesis, still stated that this had "finally been shown to be mediated by joint mechanisms" and that "all the evidence militates against the specialised receptors of muscle playing any part in conscious sensation."

REINSTATEMENT OF MUSCLE AFFERENTS

The return to the older sanity was initiated by two entirely independent lines of inquiry. The first push was given by Burgess & Clark in 1969 when they performed a particularly detailed single unit electrophysiological study of joint receptors. This threw everything into the melting pot by suggesting that joint receptors did not, in fact, provide the right sort of signals to mediate position sense as it is known psychophysically. Three years later Goodwin, McCloskey & Matthews (1972) described a range of kinesthetic illusions produced by applying vibration to a human tendon which were attributed to excitation of muscle spindle afferents. On the basis of these and other experiments they argued that muscle afferents did indeed contribute to kinesthesia, thus filling the logical gap left by Burgess & Clark's (1969) work. The present state of these lines of attack are outlined below.

Do Joint Receptors Contribute to Kinesthesia?

Whether joint receptors contribute to kinesthesia, rather than the existence of a contribution from the muscle receptors, has become the new uncertainty upon which opinion continues to ebb and flow. The initial part to the question is: Do the joint receptors provide a signal that is suitable related to joint angle to provide the requisite basis for kinesthesia? For the knee joint of the cat and the primate, a large amount of direct recording of single afferents (see references in Burgess et al 1982) has raised two major difficulties for the classical view that they do provide an adequate signal. First, many units can fire tonically in both flexion and extension, suggesting that their signaling about knee position would be ambiguous. Some of this "double-ended" firing might, however, be physiologically irrelevant, since the excitation at one end of the range might perhaps result from using unduly large movements, beyond those normally occurring physiologically. Grigg (1975) found that for 80% of flexion-extension units, "simple flexion" was insufficient to excite them, whereas extension readily did so. Second, and more crucially, most of the afferents in an articular nerve are silent when the knee-joint is in its midposition, and some or all of the few that are then discharging are spindle afferents from the nearby popliteus muscle, rather than from receptors in the joint itself. If there is no signal, then there can be no information about what is happening to the joint; yet, kinesthesia demonstrably does not vanish when a joint is in the middle of its working range.

Ferrell (1980), however, has recently urged that the midrange joint signal from the knee is still significant, though much less than that at the extremes. His sample of joint nerve afferents contained 18% that continued to provide a signal over, or near to, the center of the range while being maximally excited at one of the extremes; Burgess & Clark's (1969) figure was 5% at the most, as they also found later (Clark 1975, Clark & Burgess 1975). Moreover, Ferrell suggested that few of his own samples were spindle afferents, since their number was not significantly altered by ablating the popliteus muscle. But some at least of the midrange receptors were recently conclusively demonstrated to be spindle afferents by observing their excitation on stimulating isolated gamma motor fibers (McIntyre et al 1978). The specificity of the earlier test for spindle afferents, namely their excitation by succinylcholine, had been disputed by Ferrell. But the experimental issue may be less clear cut than it might appear, since Grigg & Greenspan (1977) found that "repeated intense movements into extension" could shift the responsiveness of an extension unit yet further into extension, possibly due to a change in the mechanical state of the underlying tissues. This suggests that what is found may depend upon the mechanical history of the joint.

It would thus be valuable to obtain systematic information of the range of responsiveness of joint receptors in the behaving animal to see what is actually being signaled physiologically, as already begun by Loeb et al (1977). Such recording might also help illuminate the effect of the contraction of nearby muscles in tensing the joint capsule. This certainly can extend the range of responsiveness of a joint unit, but on the evidence from the acute preparation the effect is not large enough to be able to fill the midrange gap (Grigg 1975, 1976). It should also be noted that most electrophysiological studies have been chiefly concerned with whether or not the unit in question will fire tonically at a given position of the knee. Short phasic bursts of firing during movement may occur over a slightly larger angular range than tonic discharges, and may be relevant to the detection of the occurrence of movement and the signaling of its speed and direction, which is what is sometimes tested for psychophysically rather than a sense of absolute position.

Not all joints, however, are poorly supplied with receptors that fail to fire over the middle of the physiological range. Some 80% of the receptors in the hip joint of the cat have recently been found to fire throughout the range of movement, and to vary their firing monotonically with joint angle whichever of its various directions of movement was tested (Carli et al 1979). The rather sparse costovertebral joint receptors of the rabbit have also long been known, almost all (90%), to fire throughout the full range of movement, some being excited maximally by displacement in one direction and some in the other (Godwin-Austen 1969). But the wrist and elbow joints of the cat seem to be like the knee and have a paucity of midrange firing (Millar 1975, Tracey 1979). The same may well be true for the human interphalangeal joints; when single units were sought in the median nerve with a tungsten electrode, and with the fingers held at rest in an "intermediate position," only 2 of 111 units isolated proved to supply joint receptors and neither of these fired tonically except near one of the extremes. If, as the authors noted, midrange firing were common, a larger number of joint receptors should have been detected (Hulliger et al 1979). It seems that for joints as a whole there is wide variation as to how effectively joint receptors signal throughout the physiological range. But, since they do not always do so, joint receptors as a class are thereby excluded from being uniquely responsible for position sense. The question thus inevitably arises whether they contribute at all, even when they happen to be discharging.

EFFECTS OF PARTIAL SENSORY DEPRIVATION   To decide whether a given afferent message actually contributes to a given central outcome—in this case to perceptions of muscle sense—requires a quite different kind of experimentation. The two most direct lines of attack to decide whether the joint discharges have a role in kinesthesia are to test the effects in man; first, of eliminating the joint afferents while leaving all others intact and, second and conversely, of preserving the joint input while eliminating the others. Unfortunately, both sorts of experiment are beset with practical difficulties and only fragmentary results are available. The elimination of joint input seems to have been satisfactorily achieved for the knee by the intracapsular injection of local anesthetic (Clark et al 1979a). There was then no demonstrable effect on the sense of absolute position, as tested by displacing the knee so slowly that there was no subjective awareness of the movement per se. This supports the idea that joint afferents make no contribution to kinesthesia. It should be noted, however, as a matter of logic, that were a given modality to be signaled by several distinct channels operating in parallel, then the elimination of just one of them does not necessarily permit its preexisting contribution to be judged from the deficit. What happens will depend upon just how the various inputs are integrated together; at one extreme, redundancy might be used simply for checking up on what had happened rather than in improving the accuracy of measurement.

Another method of eliminating joint afferents was pursued systematically but unknowingly by surgeons in the course of joint replacement surgery, notably of the hip. If, during the development of these operations in the 1960s, the surgeons had sought the advice of physiologists they would surely have been warned of the danger of destroying position sense and thereby producing an impairment of locomotor function analogous to that seen in tabes dorsalis. But they do not seem to have hesitated and must soon have established that some kinesthesia persists post-operatively, leaving it for much later physiological testing to show the slightness of the deficit. For the hip, slow movements of about 1° can still be detected post-operatively compared with about 0.5° normally (Grigg et al 1973). For the hand or big toe, movements of 10° can still be reliably detected (Cross & McCloskey 1973); the precise threshold was not then determined. Two questions arise. First, is any deficit from the normal specific and due to the inactivation of the joint receptors, or is it nonspecific, resulting from other aspects of the surgery? To this there is no answer. Second, is the persisting kinesthesia still attributable to joint receptors, whether residual or reinnervated ones in the operated joint, or those in more distant joints? All these seem unlikely, leaving the afferent input responsible to originate from skin or muscle. The hip can be expected to be similar to the knee for which cutaneous afferents seem unlikely to contribute, since a band of cutaneous anaesthesia around the knee is without relevant kinesthetic effect (Clark et al 1979a), while for the hand a cutaneous contribution remains a possibility. But the muscle afferents seem likely always to be implicated, as discussed shortly.

The restriction of the afferent input to joint afferents on moving a joint has yet to be achieved, though recently attempted by Gandevia & McCloskey (1976a). They exploited the long-known fact that appropriate positioning of the hand can uncouple the muscles acting on the distal interphalangeal joint of the middle finger (i.e. middle finger fully flexed, others extended). Voluntary movement at this joint is then impossible, showing that the muscles are "disengaged" by virtue of the terminal parts of their tendons being slack, thereby leaving it to joint and cutaneous receptors to signal kinesthesia; in confirmation of this view, local anesthesia of the finger then abolishes its kinesthetic sense, showing that there is no appreciable muscular contribution (from the unaffected forearm muscles) while the finger is so held (Goodwin et al 1972). Gandevia & McCloskey's detailed charting of kinesthesia in this posture, without anesthesia, shows that the subject does possess a reasonable measure of kinesthetic sensitivity, as shown for example by his ability to detect fairly regularly, and recognize the direction of movement, of ramp displacements of 10° at velocities of 5°/sec and upwards. The sensitivity was, however, markedly worse than with the hand in a more normal position and the muscles coupled to the joint so that they were affected by its movement; the same displacement could then be detected when applied more slowly (1-2°/sec) than in the abnormal posture, thus strongly arguing that there was then a muscular contribution to kinesthesia. Once again the evidence excludes the possibility that the joint afferents can be allocated an exclusive role in kinesthesia. Moreover, it has still to be established that the joint receptors were contributing at all. Since the existence and extent of any contribution from the cutaneous receptors could not be decided upon, it inevitably remains an open question as to whether these latter could have been entirely responsible for the awareness of movement when the muscles were disengaged. Attempts to eliminate the cutaneous contribution by iontophoretically applied local anesthetic failed.

Do Cutaneous Afferents Contribute to Kinesthesia?

Recent recording of single unit activity in man has now amply validated Sherrington's view that cutaneous receptors are inevitably excited by movement of a nearby joint, at any rate for the hand. Hulliger et al (1979) studied 103 cutaneous mechanoreceptors in the glabrous skin of the hand during voluntary finger movements and found that the majority were then excited (77%), even though there was no direct touch stimulus to their receptive fields; the study included both slowly and rapidly adapting receptors and over half of each of the four subgroups into which the receptors were divided responded. When the hand was held in a constant position most receptors were silent, but those that were not commonly discharged at slightly different rates for different static positions of a nearby joint (these were mostly of the subtype named SA II which are sensitive to stretching of the skin). Earlier, Knibestöl (1975) illustrated the response of a slowly adapting cutaneous receptor (SA II) in the nail region whose firing was linearly related to the angle at the adjacent interphalangeal joint when it was moved passively.

Thus it must be concluded that cutaneous receptors commonly transmit a signal that is related to what is happening at a nearby joint, especially in relation to its movement, and so the question becomes how far the CNS puts this information to use in the elaboration of kinesthetic sensations. Parenthetically, it may be noted that the conscious experience elicited by an external object touching the hand is entirely different from the experience on finger movement, whether active or passive, even though many of the same receptors must be activated in both cases. This indicates a considerable processing of the afferent information before it is allowed access to consciousness, at any rate for that initiated during movement. It remains quite unknown whether the processing depends crucially upon corollary discharges (possibly sometimes acting to raise the threshold for transmission at relay nuclei) or whether it also depends upon the details, both spatial and temporal, of the afferent impulse pattern, which will usually differ in the two cases (see below for an analogous problem with the muscle afferents).

As with the joint receptors, it has not proved feasible to restrict the afferent input on movement solely to that from the cutaneous receptors while eliminating all the rest. Selective removal of the relevant cutaneous input by local anesthesia has been achieved, but so far only for the knee where it was without detectable effect on position sense (Clark et al 1979a). Local anesthesia of the hand or fingers does produce a considerable impairment both of the acuity and the subjective clarity of position sense (Goodwin et al 1972, Gandevia & McCloskey 1976a). But reversing the previous argument, it is now uncertain how much of this should be attributed to the removal of cutaneous inputs, and how much to the elimination of the joint afferents. Moreover, there is the further uncertainty as to how far any reduction of kinesthetic sensitivity and intensity by cutaneous anesthesia is a specific effect, dependent upon the removal of an afferent input that is quantitatively graded with the stimulus. The effect could equally be nonspecific and due to the removal of a general facilitation, at spinal or higher levels, that aids the transmission of other specific graded inputs. A nonspecific effect might well be especially important for the fingers with their role in grasping and would explain why the movement sensitivity of a finger may be reduced by anesthetizing its neighbor (Gandevia & McCloskey 1976a). Thus, all in all, there has been little advance in the present century. We are now sure that a cutaneous signal must often be there, but we still do not know if it is ever used to appreciable effect in kinesthesia.

Do Muscle Afferents Contribute to Kinesthesia?

In a complete reversal of recent opinion there can now be few who would deny that muscle afferents do contribute to kinesthesia, and indeed play the crucial role. This view is based on three different lines of evidence.

MUSCLE PULLING The most direct and conceptually simple experiment is to pull upon an exposed tendon in a conscious subject so as to alter muscle length but not joint position and then determine, as objectively as possible, the nature of the subject's sensory experience—if any. The pulling must inevitably increase the afferent discharge from the muscle, especially from its muscle spindles. Gelfan & Carter (1967) performed this experiment on a number of patients undergoing acute reparative surgery and concluded that the sensory effects were either localized to the site of incision, over the tendon, or to a "pulling" over the muscle. Their subjects reported nothing which could be called "muscle sense," referred either to the muscle or the joint. Moberg (1972) independently reported a similar result. However, more recently Matthews & Simmonds (1974) pulled upon the flexor tendons at the wrist after they had been exposed in a planned operation under local anesthetic to section the carpal ligament to relieve pressure on the median nerve. All five of their subjects then reported that they were experiencing a movement of the appropriate finger; moreover, the perceived movement was larger than that which could have escaped detection if some real movement had occurred through inadequate immobilization of the distal end of the tendon. A similar result was reported by a colleague to J. Houk (personal communication 1972); pulling upon the exposed biceps tendon produced a sensation of elbow movement which was directionally correct, graded with the stimulus, and eliminated when the belly of biceps was infiltrated with local anesthetic. Spurred by the disagreement on such an important matter, D. I. McCloskey very recently (1981, personal communication) had the extensor tendon of his own big toe exposed acutely and sectioned. When, under carefully controlled conditions the tendon was pulled upon without his knowledge, either sinusoidally or with a ramp displacement, he experienced movement of the toe; the threshold for rapid movements was about 0.5 mm.

The case may now be suggested to be closed since the positive evidence for the occurrence of a kinesthetic effect on muscle pulling seems more cogent than the negative evidence of the failure to observe it; in such peculiar circumstances a subject may easily fail to concentrate sufficiently to become aware of a sensory experience that is admittedly of low "intensity." Those who feel the matter deserves further testing to decide the matter in black and white, but qualitative, terms are invited to follow McCloskey's example; they could then also usefully occupy themselves with various quantitative matters which are ripe for study.

MUSCLE VIBRATION   Cat experiments originally demonstrated the exquisite sensitivity of the primary ending of the muscle spindle to high frequency vibration. For example, in favorable circumstances a peak-to-peak movement of $5 \mu m$ at 200 Hz applied longitudinally to the tendon will suffice to drive the majority of the soleus primary endings in $1:1$ synchrony (Brown et al 1967); the Ia afferents then discharge at 200 impulses/sec, a rate that could otherwise only be produced by a large fast stretch. The secondary ending of the muscle spindle and the tendon organ are both much less sensitive, though the latter become appreciably more sensitive when they are already being excited by muscle contraction. Thus vibration has passed into common use as a way of injecting a massive Ia afferent input into the spinal cord so as to study its central effects. In man, where the vibration is applied percutaneously and so transversely to the tendon, the selectivity of action is less good; both secondary endings and tendon organs are regularly excited to some extent, as well as some cutaneous receptors and Pacinian corpuscles (Burke et al 1976). Thus, while there is no doubt that vibration can be used to create a very high level of Ia activity in man, it must be assumed to be accompanied by a variety of other afferent activity; the effects of vibration should not, therefore, be uncritically ascribed solely to Ia action.

Goodwin, McCloskey & Matthews (1972) made a systematic study of the effect of tendon vibration (100 Hz, about 0.5 mm total movement) on position sense at the elbow. They found it produced the consistent illusion that, even though it was still, the elbow was moving in the direction that it would have been if the vibrated muscle was being stretched; thus, on vibrating biceps tendon, the arm was felt to be extending, while triceps vibration caused an illusion of flexion. As with the classical visual illusions, like the Mach bands, this called for an explanation in terms of normal physiological mechanisms. The most obvious one was that the excess Ia discharge was treated as if it were due to a gross muscle stretch, rather than to the miniscule stretch of vibration which has no counterpart in nature, and led to kinesthetic sensations. These were referred to the joint, instead of the muscle itself, and used in the elaboration of the body image. Since no other explanation was, or still is, readily forthcoming the illusion provides evidence that muscle afferent discharges are in fact utilized in kinesthesia.

The illusion is readily demonstrated objectively by asking the subject to use his unvibrated arm to track what he feels to be happening to his vibrated arm. Although this highlights the effect as a mismatch between the perceived and the real positions of the vibrated arm, the illusion is in fact mainly one of movement; the subject uses his tracking arm to follow a continuous apparent movement of the vibrated arm without bothering very much about its absolute position. This is entirely appropriate if the vibration is acting principally via the spindle primary endings, since these are normally much more powerfully excited by movement (stretching) than they are by maintained stretch. The suggestion that the illusion is solely one of the direction of movement and lacks a continuous gradation of apparent velocity with graded Ia input (Juta et al 1979) has proved to be unfounded; gradation is essential if the illusion is to be taken as a paradigm of normal kinesthesia. The velocity of the illusory movement may be systematically and appropriately altered by varying the amplitude of vibration (Clark et al 1979b) and more significantly by varying its frequency (J. P. Roll, J. P. Vedel, J. C. Gilhodes, and M. F. Tardy, personal communication 1980), both of which may be expected to alter the total amount of Ia activity. A reduction of the illusory velocity with increasing steady contraction of the vibrated muscle had already been described (Goodwin et al 1972, McCloskey 1973) and attributed to the initial spindle discharge progressively approaching the vibration frequency, thereby leaving little scope for an augmentation of Ia firing with vibration; but the interpretation is of limited cogency since the change in muscle contraction must be associated with changes in corollary discharges, thus greatly complicating the argument.

The simple occurrence of the illusion does not depend upon the existence of any particular contractile condition of the muscle and thus of the state of the higher motor centers. It is present when the muscle is contracting isometrically, whether under voluntary drive so as to maintain a constant tension, or reflexly in response to the vibration itself (Goodwin et al 1972, Roll et al 1980). If the reflex is allowed to produce an isotonic shortening of the muscle, the illusion shows itself as reduction of the perceived velocity of shortening below the real velocity. The illusion is also present if the vibrated muscle is lying passive, irrespective of whether its antagonist is also lying passive or is contracting under voluntary drive to produce a steady force. The intensity of the illusory sensation appears to differ for different subjects, and for a given subject may wax and wane with time during a continuous period of vibration. It is always subservient to vision, and a subject does not experience it if he looks at his arm, but otherwise it does not matter if his eyes are opened or closed.

The sensation of joint movement is much the same, though reversed in direction, irrespective of whether the flexor or the extensor is being vibrated. In other words, for kinesthetic purposes the sensorium is quite unconcerned with the precise site of origin of the muscle afferent discharges and simply uses them to allow it to deduce what is happening to the joint and elaborate to the appropriate perception. Likewise, during the normal course of events the CNS may be assumed to compound the information provided by the afferent discharges from both flexors and extensors, but the way in which this is done has yet to be studied in detail. During rapid movements the body may have to rely far more upon the signal from the antagonist, whose spindles will be powerfully excited by the stretch, rather than upon the agonist, whose spindles may tend toward silence—as judged by animal recordings (Goodwin & Luschei 1975, Prochazka et al 1977). Capaday & Cook (1981) have shown that vibration of the antagonist in a rapid elbow movement causes the target to be undershot, presumably partly because of the sensory rather than the reflex effects of the vibration, whereas vibration of the shortening agonist was without effect. But while this certainly shows that the CNS is "monitoring" the activity of the antagonist, it does not show that it is ignoring any information that may be provided from the prime mover; the human unitary recordings suggest that vibration would have little or no effect on the Ia discharge from a rapidly shortening muscle (Burke et al 1976).

The illusion produced by vibration seems to contain elements of a false position as well as the more prominent false velocity. This was shown by Goodwin et al (1972) on the basis of finger-pointing tests and by Eklund (1972) on the basis of matching the position of the leg on vibrating the patellar tendon. It has been dramatically emphasized by Craske (1977), who vibrated the wrist flexors and showed that the subject might then feel that his hand was cocked back into an anatomically impossible position, which he could indicate by pointing. It is interesting that the CNS is prepared to extrapolate beyond the range for which the afferent system involved has ever been calibrated. Craske's experiments also provide further evidence that the illusion should indeed be attributed to muscle receptors, since the vibration was applied directly over the flexor muscle bellies which are close to the elbow, whereas the illusion was experienced at the wrist; the wrist joint receptors should have been largely uninfluenced by the vibration. It is also apparent from this varied work that the illusion is in no way restricted to the elbow, though this does appear to be a favorable site for its demonstration.

Vibration has thus come to be accepted as providing cogent evidence that muscle receptors do contribute to muscle sense and it also provides a tool to help throw light on the mechanisms involved. The precise relative contributions of each of the three main mechanoreceptors during vibration is still somewhat problematical. The spindle primary may confidently be ascribed the role of generating the velocity illusion, both because of its powerful excitation by the vibration and because this does seem appropriate for an ending which is so much more sensitive to dynamic than to static muscle stretch. But the spindle primary also has a static response, so it cannot be decided whether it also contributes to judgments of position. The spindle secondary, however, with much less dynamic sensitivity, seems more appropriately employed in static judgments and it also is excited to some degree by vibration. Low frequencies of vibration applied directly to the muscle are relatively more effective at producing a positional illusion than are high frequencies of vibration applied to the tendon, possibly because of a relatively stronger action on secondary endings (McCloskey 1973). The relation between the relative levels of firing of primary and secondary endings might well be important in the genesis of the sensory experience. The question of whether the tendon organs can contribute to sensation has so far not been furthered by the use of vibration, even though this must excite them to some degree when the muscle is contracting. By symmetry, however, their discharges would be expected to be able to influence consciousness, once it is accepted that the spindle afferent discharges do so; as is described below, there is now positive evidence that they can do so.

ELIMINATION OF JOINT AND CUTANEOUS SIGNALS   The elimination of joint signals by surgery, or by anesthetic injection into the knee, leaves position sense remarkably well preserved. This argues that there must be another source of kinesthetic input and the muscle afferents are virtually the only candidates available. For the hand, the effect of acute elimination of joint and cutaneous signals combined has been widely studied on producing a local sensory loss by injecting local anesthetic into the hand or finger, or by appropriately inflating a pressure cuff around the wrist to produce a progressive anoxia; the long flexor and extensor muscles lying in the forearm remain unaffected, and, depending upon the conditions of the experiment, so may some of the intrinsic muscles of the hand. There have now been repeated observations that on thereby restricting the afferent response on joint movement to the muscle afferents, a significant degree of kinesthesia persists, as initially emphasized by Goodwin et al (1972).

The residual sensations, however, may be of low "intensity," which is perhaps why not all workers have observed them when the anesthesia has been produced in the course of experiments designed to study other matters, and their very existence has sometimes been denied. It is thus important that several quantitative studies have now been made of the residual kinesthetic acuity, since these put the matter beyond doubt. Gandevia & McCloskey (1976a) found that 10° displacements of the anesthetized distal interphalangeal joint were readily detectible for a wide range of velocities of application, with performance tending to be better with the muscles contracted rather than relaxed. The discrimination was particularly good in their recent work (unpublished, personal communication) in which the finger was held extended at all joints, thus permitting both flexor and extensor muscles to be coupled to the movement; in their earlier work with the finger flexed, only the flexor muscles were engaged. Performance with the anesthetized finger was degraded from the normal, but it is noteworthy that it was appreciably better than that obtained with the finger unanesthetized and the muscle afferent signals were eliminated by appropriate positioning of the hand (as already described) and so with the input restricted to joint and cutaneous afferents. Rymer & D'Almeida (1980) found that a movement passively imposed upon the proximal interphalangeal joint of an anesthetized finger could be accurately reproduced by the other hand. Roland & Ladegaard-Pedersen (1977) showed that the extent of compression of a spring held between the thumb and finger of one hand could be accurately reproduced on compressing a similar spring in the other hand when both hands were anesthetized; as discussed below, this seems likely to be attributable to muscle afferent signals rather than to any estimate of the force involved on the basis of corollary discharges.

Goodwin et al (1972) provided an important control on the efficacy of the anesthesia by demonstrating that in an anoxic hand lateral movement of the metacarpophalangeal joint could not be detected, showing that the local skin and joint afferents had been eliminated (and also the intrinsic muscles). Yet the subject could still perceive flexion-extension movements passively imposed at the same joint; these are transmitted via their tendons to the unaffected long muscles in the forearm. Since in all such experiments externally imposed movements can be recognized when the muscles involved are not contracting, the kinesthetic effects can be attributed to the spindle afferents, whether primary or secondary, rather than the tendon organs since in passive muscles these latter were likely to have been significantly excited.

Tendon organs   The supposition that tendon organ discharges can reach consciousness also has recently been confirmed experimentally by Roland & Ladegaard-Pedersen (1977). They asked subjects to compress a spring between the finger and thumb of one hand for a distance determined by the experimenter and then to reproduce the final force by appropriately compressing another spring, of different strength, in the other hand. Sometimes a pair of isometric transducers were substituted for the spring. The information from the relevant skin and joint receptors did not appear to be important for such estimates of force, since performance was only minimally impaired by injecting local anesthetic into one or both hands to eliminate such local receptor activity.

The possibility that the estimate of force depended upon corollary discharges linked to the motor command, and so to the strength of contraction, was excluded by disturbing the normal relation between command and achieved force by weakening the relevant muscles through partial local curarization of one arm; this was achieved by retrogradely injecting a neuromuscular blocking agent into a vein in the arm while its circulation was occluded. The relation between tendon organ firing and muscle tension should have remained the same throughout. The subjects continued to be able to match the forces reasonably accurately, and in particular they did not systematically overestimate the force developed on the curarized side, as would have been expected if they were to have been relying upon corollary discharges; both hands were also anesthetized. If, however, in the isometric situation, a subject was asked to reproduce the "effort" that he had to exert on the weakened side to produce the demand reference force, then the force that he developed on the normal side increased in direct proportion to the degree of weakness. This argues strongly that the experience of "effort" but not that of "force" is attributable to corollary discharges —at any rate in this situation. Accordingly, the awareness of "force" must be attributed to muscle receptors and most probably to the tendon organs, since spindle information seems to elicit sensations of movement or of position (see above).

The curarization experiment of itself, however, does not exclude a spindle contribution. When muscle length is constant, spindle firing is normally in rough correspondence to the level of motor activity (and thus force), because of coactivation of alpha and gamma motor fibers. Since intrafusal as well as extrafusal motor junctions are affected by curarization, the relation between spindle firing and voluntary force, like that for tendon organs, might be little affected by curarization (when for a given motor command both variables will be decreased), and much less so than the relation between corollary discharges and force (when only force will be decreased). A further experiment, however, made it very unlikely that spindles could have been used to provide tension information because their signals seemed to be being used to provide movement information at the very same time as the estimate of force was required; it is improbable that they could signal both mechanical parameters simultaneously. In this case, the subject was required to compare the strengths of two springs, one held in either hand; this is a formalization of our every day ability to judge the ripeness of a cheese or a fruit by squeezing it. Success depends upon the ability to estimate, and to put into relation to each other, both the force exerted and the distance moved. Achievement was little affected by bilateral local anesthesia either alone, or in combination with unilateral local curarization, arguing that both kinds of information must have been derived simultaneously from the muscle receptors. Taken in conjunction with all the other evidence about spindles it seems virtually inevitable that the force information was derived from tendon organs signaling the tension in the muscle. This is in entire conformity with other, less crucial, findings (McCloskey et al 1974, Rymer & D'Almeida 1980).

Thus a wide sweep of experiments from muscle pulling onwards provide mutual support to argue that muscle afferents provide a crucial contribution to muscle sense. Moreover, a separate role can reasonably be attributed to each of the three mechanoreceptors, with the spindle primaries concerned mainly with signaling movement, the spindle secondaries with position, and the tendon organs with force. But, as Burgess et al (1982) discuss in detail, muscle sense is remarkably precise in the light of the imperfections of the receptors in maintaining their absolute calibration under different conditions.

COROLLARY DISCHARGES

Necessity of Corollary Discharges for Spindle Decoding

A far-reaching consequential problem is that once it is accepted that muscle spindles contribute in a quantitatively useful manner to muscle sense, it has to be faced that a given spindle afferent discharge provides information that it is quite ambiguous about what is happening to the muscle concerned. An increase in spindle firing can depend equally upon a muscle stretch or upon an increase in fusimotor activity, and commonly both variables will be changing at the same time. This difficulty of interpretation of the spindle messages has now been well-documented for man by single unit recording, as well as flowing inevitably from what is known about the working of the muscle spindle. Thus, on making a voluntary contraction under isometric conditions the spindle afferents have repeatedly been seen to accelerate their discharge (Vallbo et al 1979), but in this situation there is no experience of an illusory movement. Conversely, on following a visual target to make a slow voluntary change of position of the finger (20° flexion and extension at 2.5°/sec of metacarpophalangeal joint), systematic analysis showed no increase in the firing of the spindles in the loaded tonically active extensor muscle when the muscle was long, in comparison with that found when it was short (Vallbo et al 1981); there was, of course, such an increase when the same positional change was externally imposed upon a subject with the muscle passive. Interestingly, the spindle discharge tended to be slightly greater when the loaded muscle was short than when it was long, suggesting that its fusimotor bias was then the greater. Thus in the steady state there was no overt spindle signal in either primary or secondary afferents to convey positional information, although the subject would normally be presumed to be aware of the change of position.

The desire to avoid becoming entangled in such complexities was a continuing strand in the historically recent rejection of a sensory role for the muscle receptors. But there is now no escape, and the situation would not be improved by attempting to resuscitate a crucial role for the joint receptors, since the discharge of these is now well recognized to depend upon the level of contraction of nearby muscles as well as upon the angle of the joint (Grigg 1975, 1976). The problem is not, however, unique to muscle sense and kinesthetic sensations and is merely a restatement in the particular of how, in von Holst & Mittelstaedt's thought-clarifying terminology (1950), the body manages to distinguish between exafference and reafference—that is between whether a given sensory input depends upon an external stimulus, or is a consequence of the animal's own motor activity; the physiological significance of a given afferent signal is quite different in the two cases. Debate has usually centered upon the problem of the perceptual stability of the visual world in the face of self-induced eye movements;

this will not now be pursued since excellent discussions abound (for example, Evarts 1971, McCloskey 1981, McKay 1973, Teuber 1960). The two standard suggested solutions to the general conundrum are: first, and most classically, that the sensory centers distinguish between internally and externally generated afferent signals on the basis of information conveyed by corollary discharges from the motor centers to the sensory centers; and, second, that the fine details of the afferent messages are in fact different in the two cases and that this can be made use of by the sensorium to distinguish between them, and then to permit the elaboration of the appropriate sensory experience.

To postulate corollary discharges is the conceptually simpler solution for the proprioceptive system as it is for the visual system, and there is no shortage of potential feedback pathways to convey the information at all levels of the CNS. The precise requirement is that the sensory centers should be kept informed of the amount of fusimotor activity, distinguishing its static and dynamic components, both so as to allow for the direct excitatory action of the fusimotor discharge on the afferent firing and so as to permit the calibration of the spindle, since the sensitivity of both primary and secondary afferents is regulated by fusimotor action; normally, voluntary movement will also be occurring.

Such complex use of corollary discharges would present a formidable task to the analyzing centers, but no more so than if they were to have to deduce what is happening from analysis of the interrelationships of the firing patterns of the three main types of muscle afferent (see Matthews 1977). Moreover, it has yet to be shown that the simultaneous equations involved in the latter case have sufficiently few degrees of freedom to be soluble, except in the simplest situations. The pattern of afferent firing seen with voluntary contraction under isometric conditions might well be readily distinguishable from that produced by stretch (i.e. during dynamic stretch the primary afferent firing would increase more than that of the secondary afferents and tendon organ firing would be little affected, whereas on contraction the primary and secondary afferents would more nearly have the same increase in firing and the tendon organs should fire strongly). But more complex situations abound; in the extreme, it would seem impossible, in the absence of corollary discharges, to use afferent signals to perceive a change of position for which the level of fusimotor activity was reset so as to maintain the spindle firing at an approximately constant level, as described by Vallbo et al (1981). In these latter experiments, however, the occurrence of small changes in receptor firing during the dynamic phase of stretch was not excluded. As the authors recognized, any such change, whether for agonist or antagonist, might be used to recognize the occurrence of the movement and also determine its extent, as by extracting a velocity signal and then integrating it, and this could be remembered at the final position. It would not then matter that the final afferent signal was unchanged; it has yet to be shown, however, that there was no steady-state signal from the antagonist as well as from the prime mover.

Attribution of Sensation of Heaviness to Corollary Discharges

A role for corollary discharges in decoding the spindle messages has been made appreciably easier to accept with the recent work of McCloskey and his colleagues (see especially McCloskey 1978, 1981) arguing that corollary discharges regularly provide the basis for the subjective estimate of the heaviness of a lifted object; indeed, they are used in preference to the apparently equally suitable signals from the tendon organs. The evidence that this is so is as follows.

First, as long appreciated, muscular fatigue increases the apparent heaviness of a given weight, as shown by the fact that a weight lifted on the nonfatigued side has to be larger than that on the fatigued side for the sensations to be matched (McCloskey et al 1974). It seems unlikely that the tendon organs are appreciably fatigued, though this has yet to be put to the test. But there can be no doubt that the alpha motoneurons of a fatigued muscle must be activated more powerfully than normal to produce a given contractile force, and that there is thus an enhancement of the motor drive, with associated corollary discharges, from the higher motor centers.

Second, when a tonic vibration reflex is induced in a muscle that is being used to support a weight, there is a subjective lessening of the heaviness of the weight. Conversely, vibration of an antagonist of the muscle that is supporting the weight increases its apparent heaviness (McCloskey et al 1974). Provided that the reflexes operate on the alpha motoneuron at least partly at the spinal level, then the descending commands, and with them their corollary discharges, would vary inversely with the size of the reflex-driven component of the fixed contraction required to support the weight, thus explaining the observed sensory effects. Any excitation of the tendon organs by the vibration should have the opposite effect. However, the reflex is sufficiently complex, and is conceivably associated with supraspinal actions on sensorimotor interactions, for the argument on its own to be of limited force. But the lessening of heaviness with vibration of the agonist strongly suggests that the effects of fatigue cannot depend upon an increase in Ia firing from the fatigued muscle, due to a coactivation of the gamma motoneurons along with the alpha motoneurons, since the effect of fatigue would then be the other way around.

Third, local curarization increases the heaviness of objects lifted by the weakened muscles, as would again be expected if the inevitably augmented motor commands were being signaled to the sensory centers by corollary discharges (Gandevia & McCloskey 1977a,b). There are unfortunately still some difficulties in the quantitative interpretation of such findings, since the percentage increase in the apparent heaviness was usually much less than might be expected on simple proportionality and the degree of weakness (a reduction of muscle strength to 10% could be associated with an increase of heaviness of only 40%, although larger increases in heaviness could be found when the thumb being used was anesthetized). The authors' discussion of the problem does not appear to have entirely resolved the matter; for example, because weights are being matched, rather than the "voluntary input to the motoneurons" being measured (by the sensory experience), it would not seem to matter if the descending command acts in part by controlling the gain of various reflexes. In contrast, when Roland & Ladegaard-Pedersen (1977) asked their subjects to match the "effort" involved in producing a given isometric force with an anesthetized hand controlled by partially curarized muscles, they found that the matching force on the normal size increased in direct proportion to the weakness. This suggests a simpler dependence of the "sense of effort" than of the "sense of heaviness" upon corollary discharges and that the two need not always be identical. But, yet another set of experiments briefly reported that the relation "between force and perceived effort" was not altered by local curarization that produced up to 50% paralysis (Campbell et al 1976), suggesting a peripheral origin for the sensation. Possibly the precise conditions of the experiment are crucial for determining whether a subject pays attention to corollary discharges or to signals from the tendon organs.

Finally, it is noteworthy that the sense of heaviness of an object may be increased, in the absence of other sensory loss, for patients with lesions of various higher parts of the motor system, notably the cerebellum. This is readily demonstrated objectively by the matching of weights on the two sides of the body when the lesion is unilateral (Head & Holmes 1911, Holmes 1917, Gandevia & McCloskey 1977a). An increase in the apparent heaviness of all objects and of the subjective effort to perform any motor task is, of course, a standard symptom of a normal stroke with a mixture of sensory and motor deficits (see also Brodal 1973). This all suggests, as one would expect, that the corollary discharges are read out at a high level and not simply from the spinal motoneurons (as by their current collaterals); however, the corpus collosum is not involved (Gandevia 1978). The irrelevance of the motoneuron firing is equally shown by the action of vibration, which alters apparent heaviness without changing the motor firing required to lift a given weight, as also can various cutaneous inputs (Gandevia & McCloskey 1976b, 1977b, 1977c, Gandevia et al 1980). It should be noted, however, that the classical size-weight illusion, in which a large object feels lighter than a small object of the same mass, excludes the existence of a simple unique one-to-one relation between the descending motor commands (as in the pyramidal tract) and the sensory experience.

The employment of corollary discharges in making judgments of the strength of muscle contraction has been independently demonstrated by Cafarelli & Bigland-Ritchie (1979) in experiments in which they asked subjects to exert isometric forces, using adductor pollicis or biceps brachii, and to make the contractions "feel the same" bilaterally. When, by virtue of its length-tension properties, the effective strength of the muscle on one side was altered by altering the angle of the relevant joint, the subject much more nearly maintained the activation at a constant level, rather than the actual force exerted. This argues that in this situation also corollary discharges were being utilized to perform the judgment, although a contribution from tendon organs was also suspected. The degree of activation of the muscles was assessed both electromyographically and by comparing the force exerted with the maximum strength of the muscle at each length.

Thus on all the evidence it must be concluded that at least one component of Sherrington's muscular sense, namely the sense of effort and of heaviness, depends primarily upon corollary discharges. This is important both for itself, and for removing any general objection to the suggestion that other components of muscle sense may also utilize corollary discharges.

Inability of Corollary Discharges on Their Own to Produce a Sensation of Movement

This inevitably leads on to the question whether corollary discharges on their own might suffice also to give the whole range of kinesthetic sensations, with the muscle receptors providing only a limited contribution by virtue of creating some kind of mismatch between command and performance. But this cannot be accepted on two counts. First, muscle receptors seem quite capable of producing sensory effects in the absence of motor commands. As already detailed, this is amply described both with vibration and on movement of anesthetized digits. Second, corollary discharges on their own, unsupported by muscle afferent feedback, never seem to produce an indubitable sensation of movement and usually there is no awareness of anything happening at all. For example, during progressive anoxic paralysis of the whole arm a stage is passed through, during which the subject can still make a movement with the nearly paralyzed muscles, but entirely without being aware that he has done so (Laszlo 1966, Goodwin et al 1972). Demonstrably, a motor command must still be dispatched from higher centers and presumably with it the usual corollary discharges, but the action is quite without kinesthetic effect; it may be presumed that the muscle afferents which would normally convey the essential information were paralyzed in advance of the motor fibers. Likewise, during local anesthesia of a digit or residual spinal anesthesia, a subject may successfully complete a movement, yet have the experience that he has failed to do so (Goodwin et al 1972, Granit 1972); exclusive reliance on corollary discharges should give the opposite result.

It must be noted, however, that there are two very brief reports stating that subjects with anoxically anesthetized hands may believe that they have successfully performed a finger movement although the movement had in fact been prevented (Merton 1964, Kelso & Holt 1980). Accepted at face value this implies, contrary to the rest of the evidence, that it is the muscle afferents rather than corollary discharges which are without direct sensory action. It might be, however, that the muscle afferent signal was not recognized by these subjects because it was insufficiently intense in the particular circumstances studied. In the absence of such evidence to the contrary, the subjects might then find it most natural to believe, and so to state, that they had completed their intended movement although they had no very definite subjective experience either way. Such matters merit further study.

A classical argument favoring the idea that corollary discharges directly produce kinesthetic sensations in the complete absence of afferent input is that movements can sometimes be experienced for a phantom limb on commanding its phantom parts to move relative to each other. But the evidence for this interpretation has not stood up to recent more detailed scrutiny since some slight muscular contraction, and with it inevitably some afferent discharge, always seems to be present while the ability to "move" and the sensation thereof persists. Henderson & Smyth (1948) found this for the real phantoms occurring chronically after amputation, and Melzack & Bromage (1973) found it for artificial "phantoms" produced acutely by injecting local anesthetic into the brachial plexus.

Rather similarly, after curarization whether local or general, or complete anoxic paralysis of an arm, the attempt to move is completely unaccompanied by any sensation of having done so although the continued existence of motor commands with their corollaries may be presumed (Goodwin et al 1972, McCloskey & Torda 1975, Stevens 1978). The curare experiments are particularly interesting since in these some spindle afferents, whether in flexor or extensor muscles, should have continued to fire tonically, albeit at a low level. Thus there now seems little possibility that corollary discharges can, in their own right, regularly produce a sensation of movement. Moreover, accepting the curare experiments at face value, it would appear that some *change* in afferent input is required for corollary discharges to be brought into play, as by their providing the standard against which the afferent signal may be tested.

Nature of Decoding Mechanisms?

The way in which corollary discharges might be used to decode the spindle messages has as yet hardly been explored. One negative proposition, however, seems defendable, namely that they are not used just in the particular simple subtractive manner proposed by von Holst & Mittelstaedt (1950). Their suggestion had two components. First, that the corollary discharge (though they did not use this term) provides an "efference copy" which is already in, or is readily translated into, the correct symbolic language to provide a direct match with the expected sensory consequences of the programmed motor act (or vice versa). Second, that the efference copy is then compared by subtraction with the actual returning afferent signal and the difference signal immediately and automatically employed to give information about the external world. In the present context, this can at its simplest be taken to mean that the change in spindle firing due to changing fusimotor activity would be subtracted from the actual spindle firing and the residuum used to indicate the change in muscle length and so on. In addition, to monitor the extent and course of any movement, the moment-to-moment calibration of the spindle would have to be continuously calculated on the basis of the prevailing level of fusimotor activity.

A particular difficulty for this view in relation to kinesthesia is that once the corollary discharge is treated precisely like the afferent discharge, the corollary of itself ought to be able to give rise to the sensory experience of movement, whereas on the evidence cited earlier it does not do so. It might be suggested that this is only because in the complete absence of afferent feedback the system is biased into a nonworking region and is unable to transmit signals at all. This could happen if the corollary discharges were conveyed as purely inhibitory signals, since these would pass undetected when there was no excitation to be reduced. But this would not cover the very important situation when a subject is curarized and fails to perceive anything happening when he commands a movement, since the spindle afferents should still be continuing to discharge and the interpretation of this tonic discharge should be altered by a change in the efference copy on attempted movement. But as on present evidence it seems not to be, it must be concluded that the simplest version of the efference copy hypothesis is inapplicable to the kinesthetic component of muscle sense.

A more general explanation has been urged by McKay (1973) who suggests that any change in the afferent input has to be "evaluated" in relation to motor activity, as signaled by corollary discharges, to decide whether it is consistent with having been produced internally (in this case by fusimotor activity) or whether an external change must have occurred (in this case of muscle length). Such evaluation might in some cases be expeditiously performed by a subtractive mechanism, particularly for lower level tasks of motor coordination. If the change is accepted as externally generated, then it is employed to update a separately stored sensory map (in this case of the body image). One advantage of such progressive "evaluation" of the input rather than just forwarding the results of an automatic subtraction of the efference copy as the "sensory" signal is that the latter procedure makes great demands upon the accuracy of the system. A small error in the subtraction of two large signals (or in correctly computing the efference copy) could lead to a finite "sensory" signal when, as under isometric conditions, there should be none.

COMPLEXITY OF DECODING DURING MOVEMENT However, all these matters look more complex from the point of view of the quantitative monitoring of a variety of movements, which presumably is what the system is chiefly concerned with. The isometric situation is a special and limited case; to employ routinely the isometric response of the spindle to changing fusimotor activity as the standard to which the actual spindle discharge should be referred, so as to evaluate whether anything has happened externally, would be a clumsy way of doing things. As already noted, massive computation would still be required subsequently to work out the extent and time course of a movement. A more elegant alternative would be to use the corollary discharge to compute the expected pattern of spindle firing that would occur if a voluntary movement were to follow its planned course. This would involve taking into account the programmed activity of both alpha and gamma motoneurons, and the expected parameters of the mechanical load on the muscle in relation to muscle properties; formidable though such computation might appear, much of it is already a prerequisite for the generation of a motor program that is suitable to perform its particular task. Such a mirroring of the intended movement rather than just of the fusimotor discharge would provide an appropriate reference for evaluating the actual afferent signals, possibly by simple subtraction, to decide whether there has been any additional unintended significant change in the mechanical state of the muscle, and thus also in joint angle. If there is, this could be incorporated into the body image. It must be noted, however, that this line of thought tends to imply that if the movement goes as intended then the corollary discharge itself would appear to be suitable for updating the body image, whereas all the evidence is against the possibility of its being used in this way on its own.

Another problem is raised by the normally occurring coactivation of the alpha and gamma systems. If, as has sometimes been suggested (Matthews 1964, 1972), the fusimotor discharge were to be tailored so as to keep spindle firing approximately constant during a movement, there would be no frank afferent signal of change to initiate sensation, either alone or in combination with corollary discharges. Following the evidence of the curare experiments, a change in afferent input appears to be a prerequisite to experience movement, and a change in corollary discharge alone is not enough. The difficulty does not arise for fast movements since unitary recordings show that the spindles then regularly behave as "stretch receptors" and change their firing with change of muscle length (Goodwin & Luschei 1975, Prochazka et al 1977). This is particularly so when the muscle concerned is acting as antagonist and is being stretched; whether a muscle is being used as agonist or antagonist is immaterial in this respect since, in sensory terms, the CNS uses the muscle spindle signals to tell it what is happening at the joint rather than within the muscle itself. For slow movements, the change in spindle firing may well turn out to be modest or absent, when the dilemma would be firmly with us. But it also seems possible that the subject's belief in what was going on would then be based on visual monitoring of his performance, for example (or from memory of it in the training situation), rather than representing a direct sensory experience elaborated from the interaction of afferent input and corollary discharge.

The fusimotor system still occupies a paradoxical position in relation to our present elementary ideas about efference copies. On the one hand, its existence leads one to postulate the occurrence of corollary discharges at a high level so that the spindle message can be read. On the other hand, the fusimotor discharge itself has some of the characteristics of a corollary discharge in relation to alpha motor firing and its effects on the spindles (see also Granit 1972). Muscle contraction on its own silences the spindle, but this is prevented by intrafusal contraction so the spindle is often said to signal the misalignment between the intrafusal and extrafusal lengths (cf Matthews 1972); in other words, with appropriate adjustment of relative firing, the "efference copy" conveyed by the gamma fibers could ensure that all "reafference" produced by the contraction is removed from the spindle signal, leaving it to transmit purely "exafference." Further thought would be greatly helped if the current wave of experiment in conscious animals and man can succeed in establishing how far the level of fusimotor firing is obligatorily coupled to the alpha motor discharge, and whether the relative activity of the alpha and the two gamma systems can be varied for different types of contraction and for different speeds of movement—as might be expected from their anatomical independence.

A NEW ILLUSION   Discussion of such matters is currently seriously restricted both because we are almost totally lacking in direct experiment to show what is going on inside the black box of the CNS in such situations, and also because the psychophysical study of the behavior of kinesthetic mechanisms during the performance of a normal motor task remains in its infancy. Since muscle sense comprises the sensory side of the motor system it must be presumed to be organized to handle voluntary movements and their deviations from target, rather than to perceive passively imposed displacements; in this respect muscle receptors and their central analytical mechanisms may be presumed to resemble the labyrinth rather than the cutaneous receptors. With luck, the central sensorimotor machinery will operate under constraints of which we are still unaware, but which will force it to provide clues about its internal mechanism. An example of this may perhaps be provided by a newly described illusion: when a subject estimates the distance that a finger moves into flexion at the same time as he is *changing* the voluntary muscle force exerted by the finger, he goes systematically into error over the position estimate, with the greater the final force the further he feels he has moved. This has now been described for two rather different experimental situations, both with and without cutaneous anesthesia, and thus seems unlikely to be an artifact of some detail of the experimental arrangement and to be independent of whether the load on the finger is rigid or yielding. In their spring-matching experiments, Roland & Ladegaard-Pedersen (1977, Roland 1978) found that the stronger the spring the further the subject felt that he had compressed it when asked to reproduce a given displacement against a weaker spring held in the other hand. The difference in displacement was linearly related to the difference in spring strengths, and thus to the difference in the forces involved. Quite independently, Rymer & D'Almeida (1980) showed that if a subject was required to press his finger against an isometric support which moved a certain distance at the same time as the subject was required to increase or decrease the strength of his voluntary contraction, against the support, then the movement was overestimated in proportion to the level of final force. But the subject estimated the movement accurately if he was merely required to produce a steady force throughout the period of movement. Thus, there seems to be a confounding of movement information, presumably derived from spindles, and information on changing contractile force, whether derived from tendon organs or corollary discharges. The effect is the opposite of that to be expected if the fusimotor-induced increase in spindle firing, occurring in conjunction with the contraction, were to be perceived as due change of muscle length; this should give the sensation of movement in the direction of muscle stretch, whereas excess movement in the direction of muscle shortening was what was observed. The illusion remains an enigma and emphasizes that the field is ripe for further exploration, both theoretical and practical.

CONCLUSION

Except for his rejection of corollary discharges Sherrington in 1900 promulgated a view on the origin of "muscle sense" (kinesthetic sensations) to which we can once again subscribe, even though throughout the 1960s it was thought to be largely refuted. Joint afferents may on occasion be presumed to be contributing, since for some joints such as the hip they do indeed provide an appropriate signal. But in other joints, notably the knee, they fail to fire appreciably in the middle of the physiological range. On this and other evidence it must be doubted whether the receptor of these latter joints contribute at all to kinesthesia, thus making it uncertain whether they can do so for any joint. Anyhow, there is no possibility that joint receptors can provide the unique source of kinesthesia.

Cutaneous receptors are demonstrably excited by joint movement, but it is unknown whether they contribute significantly to kinesthesia. For muscle all three of the main mechanoreceptor afferents are now reasonably established as contributing to muscle sense, both on their own and by the interrelationship of their patterns of firing. The spindle primaries, whether from agonist or antagonist, probably contribute mainly movement information; the secondaries (about which there is still little firm evidence), relatively more of positional information, the resulting sensations being referred to the relevant joint and not to the muscle itself; and the tendon organs, sensations of force.

Surprisingly, the discharges of tendon organs are not routinely used for the estimation of the heaviness of lifted objects or of the degree of effort exerted in the performance of a motor task. For this, reliance seems to be placed on corollary discharges sent from motor centers to sensory centers at the same time as the dispatch of the motor command to lower centers. Other corollary discharges have of necessity also to be postulated to explain how the sensorium can distinguish between spindle discharges induced by fusimotor activity and those produced by increase of muscle length. But these latter corollary discharges do not of themselves lead to a sensation in the absence of a changing afferent input. This shows that the higher evaluation of the spindle messages must be more complex than the simple subtractive mechanism sometimes suggested for the visual system, with the difference signal between the corollary and the afferent discharge automatically used to produce a sensation.

Literature Cited

Brodal, A. 1973. Self-observations and neuro-anatomical considerations after a stroke. *Brain* 96:675–94

Brown, M. C., Engberg, I. E., Matthews, P. B. C. 1967. The relative sensitivity to vibration of muscle receptors of the cat. *J. Physiol.* 192:773–800

Burgess, P. R., Clark, F. J. 1969. Characteristics of knee joint receptors in the cat. *J. Physiol.* 203:317–35

Burgess, P. R., Clark, F. J., Simon, J., Wei, J. U. 1982. Signaling of kinesthetic information by peripheral sensory receptors. *Ann. Rev. Neurosci.* 5:171–87

Burke, D., Hagbarth, K.-E., Löfstedt, L., Wallin, B. G. 1976. The responses of human muscle spindle endings to vibration during isometric contraction. *J. Physiol.* 261:695–711

Cafarelli, E., Bigland-Ritchie, B. 1979. Sensation of static force in muscles of different length. *Exp. Neurol.* 65:511–25

Campbell, E. J. M., Edwards, R. H. T., Hill, D. K., Jones, D. A., Sykes, M. K. 1976. Perception of effort during partial curarisation. *J. Physiol.* 263:186–87P

Capaday, C., Cooke, J. D. 1981. The effects of muscle vibration on the attainment of intended final position during voluntary human arm movements. *Exp. Brain Res.* 42:228–30

Carli, G., Farabollini, F., Fontani, G., Meucci, M. 1979. Slowly adapting receptors in cat hip joint. *J. Neurophysiol.* 42:767–78

Clark, F. J. 1975. Information signalled by sensory fibres in medial articular nerve. *J. Neurophysiol.* 38:1464–72

Clark, F. J., Burgess, P. R. 1975. Slowly adapting receptors in cat knee joint: Can they signal joint angle? *J. Neurophysiol.* 38:1448–63

Clark, F. J., Horch, K. W., Bach, S. M., Larson, G. F. 1979a. Contribution of cutaneous and joint receptors to static knee-position sense in man. *J. Neurophysiol.* 42:877–88

Clark, F. J., Matthews, P. B. C., Muir, R. B. 1979b. Effect of the amplitude of muscle vibration on the subjectively experienced illusion of movement. *J. Physiol.* 296:14–15P

Craske, B. 1977. Perception of impossible limb positions induced by tendon vibration. *Science* 196:71–73

Cross, M. J., McCloskey, D. I. 1973. Position sense following surgical removal of joints in man. *Brain Res.* 55:443–45

Eklund, G. 1972. Position sense and state of contraction. *J. Neurol. Neurosurg. Psychiatr.* 35:606–11

Evarts, E. V. 1971. Feedback and corollary discharge: merging of the concepts. *Neurosci. Res. Bull.* 9:86–112

Ferrell, W. R. 1980. The adequacy of stretch receptors in the cat knee joint for signalling joint angle throughout a full range of movement. *J. Physiol.* 299:85–99

Gandevia, S. C. 1978. The sensation of heaviness after surgical disconnection of the cerebral hemispheres in man. *Brain* 101:295–305

Gandevia, S. C., McCloskey, D. I. 1976a. Joint sense, muscle sense, and their combination as position sense, measured at the distal interphalangeal joint of the middle finger. *J. Physiol.* 260:387–407

Gandevia, S. C., McCloskey, D. I. 1976b. Perceived heaviness of lifted objects and effects of sensory inputs from related, non-lifting parts. *Brain Res.* 109:399–401

Gandevia, S. C., McCloskey, D. I. 1977a. Sensations of heaviness. *Brain* 100:345–54

Gandevia, S. C., McCloskey, D. I. 1977b. Changes in motor commands, as shown by changes in perceived heaviness, during partial curarisation and peripheral anaesthesia in man. *J. Physiol.* 272:653–72

Gandevia, S. C., McCloskey, D. I. 1977c. Effects of related sensory inputs on motor performance in man studied through changes in perceived heaviness. *J. Physiol.* 272:673–89

Gandevia, S. C., McCloskey, D. I., Potter, E. 1980. Alterations in perceived heaviness during digital anaesthesia. *J. Physiol.* 306:365–75

Gelfan, S., Carter, S. 1967. Muscle sense in man. *Expl. Neurol.* 18:469–73

Godwin-Austen, R. B. 1969. The mechanoreceptors of the costovertebral joints. *J. Physiol.* 202:737–54

Goodwin, G. M., Luschei, E. S. 1975. Discharge of spindle afferents from jaw-closing muscles during chewing in alert monkeys. *J. Neurophysiol.* 38:560–71

Goodwin, G. M., McCloskey, D. I., Matthews, P. B. C. 1972. The contribution of muscle afferents to kinaesthesia shown by vibration induced illusions of movement and by the effects of paralysing joint afferents. *Brain* 95:705–48

Granit, R. 1972. Constant errors in the execution of movement. *Brain* 95:649–60

Grigg, P. 1975. Mechanical factors influencing response of joint afferent neurons from cat knee. *J. Neurophysiol.* 38: 1473–84

Grigg, P. 1976. Response of joint afferent neurons in cat medial articular nerve to active and passive movements of the knee. *Brain Res.* 18:482–85

Grigg, P., Finerman, G. A., Riley, L. H. 1973. Joint position sense after total hip replacement. *J. Bone Joint Surg. A* 55:1016–25

Grigg, P., Greenspan, B. J. 1977. Response of primate joint afferent neurons to mechanical stimulation of knee joint. *J. Neurophysiol.* 40:1–8

Head, H., Holmes, G. 1911. Sensory disturbances from sensory lesions. *Brain* 34:102–254

Henderson, W. R., Smyth, G. E. 1948. Phantom limbs. *J. Neurol. Neurosurg. Psychiatr.* 11:88–112

Holmes, G. 1917. The symptoms of acute cerebellar injuries due to gunshot injuries. *Brain* 40:461–535

Hulliger, M., Nordh, E., Thelin, A.-E., Vallbo, Å. B. 1979. The responses of afferent fibres from the glabrous skin of the hand during voluntary finger movements in man. *J. Physiol.* 291:233–49

Jones, E. G. 1972. The development of the 'muscular sense' concept during the Nineteenth Century and the work of H. Charlton Bastian. *J. Hist. Med. Allied Sci.* 27:298–311

Juta, A. J. A., van Beekum, W. T., Denier van der Gon, J. J. 1979. An attempt to quantify vibration induced movement sensation. *J. Physiol.* 292:18P

Kelso, J. A. S., Holt, K. G. 1980. Exploring a vibratory systems analysis of human movement production. *J. Neurophysiol.* 43:1183–96

Knibestöl, M. 1975. Stimulus-response functions of slowly adapting mechanoreceptors in the human glabrous skin area. *J. Physiol* 245:63–80

Laszlo, J. 1966. The performance of a simple motor task with kinaesthetic sense loss. *J. Exp. Psychol.* 18:1–8

Loeb, G. E., Bak, M. J., Duysens, J. 1977. Long-term unit recording from somatosensory neurons in the spinal ganglia of the freely moving walking cat. *Science* 197:1192–94

Matthews, P. B. C. 1964. Muscle spindles and their motor control. *Physiol. Rev.* 44:219–88

Matthews, P. B. C. 1972. *Mammalian Muscle Receptors and Their Central Actions.* London: Arnolds

Matthews, P. B. C. 1977. Muscle afferents and kinaesthesia. *Br. Med. Bull.* 33: 137–42

Matthews, P. B. C., Simmonds, A. 1974. Sensations of finger movement elicited by pulling upon flexor tendons in man. *J. Physiol.* 239:27–28P

McCloskey, D. I. 1973. Differences between the sense of movement and position shown by the effects of loading and vibration of muscles in man. *Brain Res.* 61:119–31

McCloskey, D. I. 1978. Kinesthetic sensibility. *Physiol. Rev.* 58:763–820

McCloskey, D. I. 1981. Corollary discharges and motor commands. In *Handbook of Physiology—The Nervous System III, Motor Control,* ed. V. B. Brooks. Bethesda: Am. Physiol. Soc. In press

McCloskey, D. I., Ebeling, P., Goodwin, G. M. 1974. Estimation of weights and tensions and apparent involvement of a "sense of effort." *Exp. Neurol.* 42: 220–32

McCloskey, D. I., Torda, T. A. G. 1975. Corollary motor discharges and kinaesthesia. *Brain Res.* 100:467–70

McIntyre, A. K., Proske, U., Tracey, D. J. 1978. Afferent fibres from muscle receptors in the posterior nerve of the cat's knee joint. *Exp. Brain Res.* 33:415–24

McKay, D. M. 1973. Visual stability and voluntary eye movements. In *Central Processing of Visual Information,* ed. R. Jung, pp. 307–33 Berlin/New York: Springer-Verlag

Melzack, R., Bromage, P. R. 1973. Experimental phantom limbs. *Exp. Neurol.* 39:261–69

Merton, P. A. 1964. Human position sense and sense of effort. *Symp. Soc. Exp. Biol.* 18:387–400

Millar, J. 1975. Flexion-extension sensitivity of elbow joint afferents in cat. *Exp. Brain Res.* 24:209–14

Moberg, E. 1972. Fingers were made before forks. *Hand* 4:201–6

Prochazka, A., Westerman, R. A., Ziccone, S. P. 1977. Ia afferent activity during a variety of voluntary movements in the cat. *J. Physiol.* 268:423–48

Roland, P. E. 1978. Sensory feedback to the cerebral cortex during voluntary movement in man. *Behav. Brain Sci.* 1: 129–47

Roland, P. E., Ladegaard-Pedersen, H. 1977. Sensations of tension and kinaesthesia from musculotendinous receptors in man. Evidence for a muscular sense and sense of effort. *Brain* 100:671–92

Roll, J. P., Gilhodes, J. C., Tardy-Gervet, M. F. 1980. Effects perceptifs et moteurs des vibrations musculaires chez l'homme normal: Mise en evidence d'une reponse des muscle antagonistes. *Arch. Ital. Biol.* 118:51–71

Rose, J. E., Mountcastle, V. B. 1959. Touch and kinesthesis. In *Handbook of Physiology, Section 1, Neurophysiology,* ed. J. Field, 1:387–429. Washington DC: Am. Phys. Soc.

Rymer, W. Z., D'Almeida, A. 1980. Joint position sense: The effects of muscle contraction. *Brain* 103:1–22

Sherrington, C. S. 1900. The muscular sense. In *Textbook of Physiology,* ed. E. A. Schäfer, 2:1002–25. Edinburgh/London: Pentland

Sperry, R. W. 1950. Neural basis of the spontaneous optokinetic response produced by visual neural inversion. *J. Comp. Physiol. Psychol.* 45:482–89

Stevens, J. K. 1978. The corollary discharge: is it a sense of position or a sense of space? *Behav. Brain Sci.* 1:163–65

Teuber, H. 1960. Perception. In *Handbook of Physiology, Section II, Neurophysiology,* ed. J. Field, 3:1595–1668. Washington

DC: Am. Phys. Soc.

Tracey, D. J. 1979. Characteristics of wrist joint receptors in the cat. *Exp. Brain Res.* 34:165–76

Vallbo, Å. B., Hagbarth, K.-E., Torebjörk, H. E., Wallin, B. G. 1979. Somatosensory, proprioceptive, and sympathetic activity in human peripheral nerves. *Physiol. Rev.* 59:919–56

Vallbo, Å. B., Hulliger, M., Nordh, E. 1981. Do spindle afferents monitor joint position in man? *Brain Res.* 204:209–13

von Holst, E. 1954. Relations between the central nervous system and the peripheral organs. *Br. J. Animal Behav.* 2:89–94

von Holst, E., Mittelstaedt, H. 1950. The reafference principle. Interaction between the central nervous system and the periphery. In *Selected Papers of Erich von Holst: The Behavioural Physiology of Animals and Man,* London: Methuen. (From German) 1:139–73.

Ann. Rev. Neurosci. 1982. 5:219–39

MULTIPLE SCLEROSIS

Guy M. McKhann

Department of Neurology, Johns Hopkins Hospital and Johns Hopkins School of Medicine, Baltimore, Maryland 21205 USA

Introduction

Multiple sclerosis is the leading cause of serious neurological disease in young and middle-aged adults in the United States and in Western Europe (Johnson et al 1979). In its classical form, the disease is characterized by acute exacerbations followed by spontaneous remissions. As the disease progresses, many patients have fewer "attacks" and are affected by a slowly progressive pattern of disease. There are variants in the clinical course, ranging from an acute, fulminating disease to a pattern of few attacks at intervals as long as years. The propensity for spontaneous remission and the changing patterns of the disease with time make evaluation of therapeutic attempts extremely difficult.

The gross pathology of the brain in multiple sclerosis is the presence of circumscribed areas, or plaques, of loss of myelin which have irregular borders and occur anywhere in the white matter. The characteristic microscopic features of these plaques are sparing of nerve cell bodies and axons, areas of myelin dissolution, and an accompanying astroglial proliferation. Thus, multiple sclerosis has been considered the prototype of a *demyelinative disease.*

In this review, I present the unique features of multiple sclerosis and then use this information to discuss possible mechanisms of disease and approaches to therapy.

Epidemiology

The clinical onset of multiple sclerosis is rare before the age of 15 and decreases in frequency after the age of 45 years. There is a marked variation in the incidence (number of new patients recorded during a defined period) and prevalence (total number of current cases of the disease) according to latitude (Kurland 1970, Kurtzke 1975, Alter 1980). In Northern latitudes

0147-006X/82/0301-0219$02.00

the prevalence is higher (e.g. Denmark: 64/100,000; Southeastern Norway: 80/100,000; Rochester, Minnesota: 64/100,000). In lower latitudes it is lower (e.g. Parma, Italy: 12/100,000; Israel: 4/100,000). Near the equator the disease is virtually nonexistent or not reported.

This North-South gradient can be observed within a single country. For example, the prevalence in Rochester, Minnesota (latitude 44° N) is 64/100,000, while in New Orleans (latitude 30° N) it is 6/100,000. In Great Britain this latitudinal correlation is even more striking, with Aberdeen (latitude 57°) having a prevalence of 144/100,000 and Cornwall (latitude 50°), 63/100,000. In addition, there are areas such as the Orkney Islands (latitude 59° N) where the prevalence, 125–150/100, is three times the expected rate. There is also an East-West gradient: the disease is far less common in Japanese than in Europeans living at similar latitudes.

A pertinent question is what happens to people who migrate from an area of high incidence to one of low incidence? In two populations, those migrating to an area of high incidence (Great Britain) and those migrating to an area of low incidence (Israel), studies suggest that those who migrate after adolescence carry with them the prevalence rate of their place of origin, whereas those migrating in childhood acquire the prevalence of that in their host country (Dean et al 1977, Alter et al 1978, Dean 1980, Alter 1980).

INTERPRETATION   A possible interpretation is that susceptible individuals are exposed to and acquire the disease well before clinical onset and before age 12 to 15 years. The disease then presents after a significant latent period. This interpretation is an integral part of the "viral hypothesis" of multiple sclerosis.

There are, however, other possible interpretations. For example, early exposure may provide immunity, as is presumably the case with poliomyelitis. Alternatively, some environmental factor, other than an infectious agent, may vary with latitude such as diet, exposure to sunlight, or sociocultural status. This area of multiple sclerosis research has recently been reviewed (Alter 1980, Dean 1980, Kurtzke 1980).

Genetic Studies

The incidence of multiple sclerosis is 15–20 times higher in relatives of patients than in the general population; the highest incidence is in siblings and the next highest in parents. The concordance in identical twins is higher than in fraternal twins. The concept of a genetic factor is strengthened by studies relating to leukocyte functions regulated by the sixth human chromosome. In the United States and Western Europe, the frequency of HLA antigens A3 and B7 are increased, as are B cell antigens DRw2 and DRw3. The strongest association is with DRw2, which occurs three to four times more frequently in patients with multiple sclerosis than in the general population (Oger & Arnason 1980).

INTERPRETATION   The data support the hypothesis that there is a genetic component or a genetically determined susceptibility to some exogenous factor operative in multiple sclerosis.

Pathology

The characteristic pathology of multiple sclerosis is scattered plaques of demyelination of various ages. Histologically, these plaques show (a) sparing of nerve cell bodies and axons, except in older lesions where axons may be lost, (b) areas of myelin dissolution and the presence of macrophages containing myelin breakdown components around the periphery, (c) a loss of oligodendria, particularly in the center of plaques, (d) an intense astroglial response, and (e) perivascular cuffs of lymphocytes. Plaques can occur anywhere in central white matter, but have a predilection for optic nerve, brainstem, spinal cord, and the periventricular areas around the lateral ventricles. The correlation between the distribution and number of plaques and recorded clinical symptoms and signs is often quite poor.

Because of the chronicity of the disease, analysis of older plaques does not reveal the cellular events associated with the development of acute lesions. Light and electron microscopic studies of acute plaques have indicated glial hyperplasia, often in a ring around the plaque. The cells participating in this hyperplasia are astrocytes, microglia, and perhaps oligodendroglia. Inclusion-bearing cells may occur at the edge of acute plaques; these could be either altered oligodendroglia or a form of undifferentiated neuroglial cell. The nature of the inclusion material is not known; it has been suggested by some to represent viral antigen and by others to reflect storage of material by these cells. Another feature of early plaques is the degree of perivascular infiltration with mononuclear cells, often extending some distance away from the plaque. Associated with this infiltration there may be edema of the surrounding brain tissue (Adams 1977, Prineas & Connell 1978).

Biochemical studies of involved areas show the changes one might expect in areas of glial proliferation and myelin breakdown, such as an increase in glial fibrillary astrocyte protein, an increase in lysosomal enzymes, and the presence of cholesterol esters around plaques. There is, however, a disproportionate loss of myelin basic protein and myelin-associated glycoproteins compared to other components of myelin (Itoyama et al 1980).

An important but unresolved question is whether the normal-appearing white matter is really normal. Both histological abnormalities (primarily a diffuse astrocytic proliferation) and biochemical abnormalities (consisting of an increase in lysosomal enzymes, particularly n-acetyl-B-D-gluco-saminidase) have been reported (Allen 1980).

A major discrepancy between the pathology and clinical course of multiple sclerosis is the lack of evidence of a reparative process, specifically remyelination, as a basis for clinical remissions. In most plaques the majority of axons remain demyelinated, with remyelination limited to the periphery of lesions (Prineas & Connell 1979).

INTERPRETATION   The pathological analysis indicates a selective disease process with loss of myelin and oligodendroglia; proliferation of astrocytes, microglia, and plasma cells; and relative sparing of axons. There is no convincing evidence of viral inclusions. The early cellular events are not known. The disease is still considered to be one with multiple foci of discrete demyelination. However, there remains the possibility that there is more generalized involvement of white matter or blood vessels. The latter point is important in terms of research strategy aimed at the study of autopsy tissue.

Cellular and Chemical Pathology

STUDIES WITH ISOLATED OLIGODENDROGLIA   If demyelination is the primary pathological process, then it is logical to study the putative targets: oligodendroglia and myelin. Oligodendroglia can be obtained as relatively pure populations from a variety of species, including man (Poduslo & Norton 1972, Fewster & Blackstone 1975, Snyder et al 1980). These cells can be maintained in vitro and will synthesize lipids enriched in myelin, such as galactocerebrosides (Poduslo et al 1978, Szuchet et al 1980). In multiple sclerosis, fresh autopsy tissue is difficult to obtain under circumstances in which current cell isolation techniques can be applied. In addition, there is no methodology for obtaining oligodendroglia from only the periplaque areas. To date, the few preliminary studies that have been performed, have demonstrated no abnormalities in oligodendroglia isolated from patients who have died with multiple sclerosis.

An alternative approach has been to use oligodendroglia from some other species, such as the rat or cow, or from human abortus material, as a target for possible serum factors in multiple sclerosis. These studies are performed on either oligodendrocytes obtained by bulk isolation or on cultured, dispersed cells from cerebrum or corpus callosum. In the latter experimental situation, cell-specific antibodies, such as anti-galactocerebroside, are used to identify oligodendroglia (Raff et al 1978, 1979, Kennedy et al 1980, Schachner et al 1981). Abramsky et al (1977) reported that patients with multiple sclerosis had serum factors, presumably antibodies, that bound to the surface of oligodendroglia maintained in suspension cultures. Subsequent studies have indicated that sera from control patients are indistinguishable from those from patients with multiple sclerosis (Traugott et al 1979, Kennedy & Lisak 1979). In addition, using indirect immunofluorescence, it was observed that almost all sera contained immunoglobulins that bound weakly to human oligodendrocytes and fibroblasts, and to a lesser degree to Schwann cells and astrocytes (Kennedy & Lisak 1981).

It should be emphasized, however, that the methodologies for obtaining specific cell types from brain and for delineating cell-specific functions are undergoing rapid development and undoubtedly there will be efforts to apply these approaches to multiple sclerosis in the future.

STUDIES OF MYELIN   There have been no consistent reports to indicate that myelin is abnormal in multiple sclerosis. Myelin is a relatively lipid-rich, protein-poor membrane. These characteristics allow it to be isolated by density gradient techniques. There have been several recent reviews of the morphology and biochemistry of myelination (Morell 1977, Palo 1978). Many of the components of myelin have been the subject of studies to determine whether a particular component is released into cerebrospinal fluid or blood at times of demyelination. Prior to the introduction of the radioimmunoassay and high pressure liquid chromatography, the insensitivity of analytical methods was a limiting factor for these studies. There is now evidence, however, that both myelin basic protein and myelin-associated glycoproteins are selectively decreased around plaques (Itoyama et al 1980). These latter observations have led to two lines of investigations: (a) the search in cerebrospinal fluid (CSF) and blood for myelin breakdown products, particularly myelin basic protein and peptides from that protein and (b) the study of proteases that might selectively attack myelin.

Myelin basic protein   Several laboratories have reported the presence in CSF from persons with multiple sclerosis of material that reacts with antibodies raised against myelin basic protein (MBP) or its fragments (Cohen et al 1976, Whitaker 1977, Carson et al 1978, Trotter et al 1978). Studies from our laboratory indicate that this material is present during acute attacks and correlates with the duration of attack (Cohen et al 1980). The nature of the material in CSF is not entirely clear. One group has reported the presence of a molecule larger than basic protein (Carson et al 1980), while others have reported either the whole molecule (Cohen et al 1980) or a specific peptide fragment containing amino acids 43–88 (Whitaker et al 1980). In addition, it is possible to raise antibodies to purified basic protein that do not react with the material in CSF. Some of these discrepancies may be clarified when different laboratories use standardized monoclonal antibodies raised by the hybridoma technique (Kohler & Milstein 1976). Most attempts to demonstrate myelin basic protein or its fragments in blood or serum have been negative, perhaps owing to the rapid clearance from blood of exogenously administered basic protein. However, fragments of myelin basic protein and antibodies to myelin basic protein have been recently demonstrated in sera of normals and of patients during acute exacerbations of multiple sclerosis (Paterson et al 1980). Further, these authors (Paterson et al 1981) have demonstrated that during acute attacks of multiple sclerosis, as in rats during the early stages of acute experimental allergic encephalitis (EAE), antibodies with a high affinity for MBP are found in sera; there is as well a corresponding decrease in free MBP fragments. This pattern of high affinity antibody with reduced levels of free MBP fragments is not seen in normals or in patients with quiescent disease. However, a full survey of patients with neurological diseases that are believed not to be of immunologic origin has not been completed (P. Y. Paterson, personal communication).

Attempts to correlate the presence of other components of myelin with the disease activity have been less successful. In our laboratories, the presence of lipids enriched in myelin, galactocerebroside, and sulfatide, and of a myelin-related enzyme, 2'-3' cyclic nucleotide 3'-phosphodiesterase (CNP), have not correlated with disease activity or with levels of myelin basic protein. Attempts are being made to develop sensitive radioimmunoassay for myelin-associated glycoproteins and proteolipid proteins and to determine possible correlation of these compounds with clinical disease.

Proteases in multiple sclerosis   The study of proteases in multiple sclerosis is based, at least in part, on the sequence of cellular events in experimental allergic encephalomyelitis. In that disease, reactive cells, lymphocytes, and macrophages surround the myelin sheath, followed by a vesicular myelinolysis and a peeling off of myelin lamellae. A possible mechanism is the release from the reactive cells of specific proteases which attack myelin proteins, particularly myelin basic protein and myelin-associated glycoprotein. In multiple sclerosis, active stripping of myelin by macrophages has not been seen; however, close proximity of reactive cells (macrophages and microglia) to areas of myelin loss is seen. A sequence of events has been suggested in which macrophages are attracted to specific areas and then release neutral proteases which attack myelin basic protein (Bloom et al 1978). Recently, Bloom et al (1978) presented evidence that antimacrophage agents, such as silica, or proteinase inhibitors, such as *p*-nitrophenylguanidinobenzoate, pepstatin, or trans-4-aminomethylcyclohexane-1-carboxylic acid, will protect Lewis rats from experimental allergic encephalomyelitis. These findings suggest that macrophages play a role in the clinical and histological expression of this experimental demyelinating disease (Brosnan et al 1980). Others have suggested that acid proteases, which might be of lysosomal origin from macrophages, lymphocytes or astrocytes, also selectively affect myelin basic protein. In CSF, both acidic and neutral proteinases have been found; both may vary with acute exacerbations (Richards & Cuzner 1978).

INTERPRETATION   Several lines of evidence suggest that myelin basic protein and possibly myelin-associated glycoprotein are selectively affected in demyelination. A possible consequence is the release of fragments of basic protein into CSF during acute attacks of multiple sclerosis. Regardless of the mechanism, the presence in cerebrospinal fluid of myelin basic protein-like material provides an indication of activity of disease. The role of proteases in the process of demyelination is not clear.

Mechanism of Disease

The cause of multiple sclerosis is not known. A number of unique features of the disease have been used to suggest both an immunological mechanism and an infectious etiology. Both possible mechanisms are supported by experimental diseases in animals that have some of the features of multiple sclerosis. In recent years, as is discussed below, the distinction between the two mechanisms has become somewhat blurred.

IMMUNOLOGIC ABNORMALITIES IN MULTIPLE SCLEROSIS   An immunological basis for multiple sclerosis has been suggested by the demonstration of a demyelinative disease, experimental allergic encephalomyelitis (EAE), produced by injecting animals with central nervous system antigens, particularly whole myelin, myelin basic protein, and galactocerebroside (Seil 1977). The acute form of EAE differs from multiple sclerosis in its pathology and its single-phase clinical course. Recently, a chronic, relapsing form of this disease has been produced which more closely resembles multiple sclerosis (Raine et al 1980). In this form of the disease, myelin basic protein may not be the responsible antigen (Wisniewski & Lassmann 1980).

Patients with multiple sclerosis have a number of immunologic abnormalities including (a) increased synthesis of immunoglobulins within the brain, (b) the possible presence of circulating factors which can cause demyelination or block neuroelectric transmission, and (c) decreased levels of lymphocyte subgroups, particularly T suppressor cells (Bloom 1980).

Immunoglobulins in CSF   IgG is elevated in 70% of patients with clinically defined multiple sclerosis (MS). Subfractionation of these immunoglobulins by electrophoresis or isoelectric focusing has indicated separate "oligoclonal bands" of immunoglobulin. Oligoclonal bands are present in the CSF in 80–90% of patients with MS, even those in which IgG may not be elevated. Once present, the bands persist. Recent observations suggest that the pattern of oligoclonal bands may change with repeated attacks of the disease (E. J. Thompson, personal communication). However, there is no specific pattern of oligoclonal bands that allows one to diagnose multiple sclerosis; each patient may have his or her own unique pattern (Laterre 1966, Link 1973, Thompson 1977). Analysis of individual plaques from the brains of a single patient indicates that different oligoclonal patterns are obtained from separate plaques, with only a few species of gamma globulin in common (Mattson et al 1980).

The source of the immunoglobulin in CSF of patients with multiple sclerosis is presumably plasma cells derived from B cell clones within brain parenchyma. As mentioned, the pattern of immunoglobulin extracted from individual plaques varies, suggesting that the patterns in cerebrospinal fluid is a composite from different brain regions. It is not at all clear to what antigens these centrally derived immunoglobulins are directed. Patients with multiple sclerosis have elevated titers to a number of viruses, particularly measles. However, an individual patient may have elevated titers to as many as four viruses (Vandvik & Norby 1980). Absorption of the CSF immunoglobulins with viral antigens does not selectively remove oligoclonal bands (Norby 1978). Similarly, antibodies to brain antigens, such as mylein basic proteins, when present, are at low levels and are not a significant proportion of the total CSF immunoglobulin fraction (Shorr et al 1981).

An alternative approach to relating the gamma globulin abnormalities in cereberospinal fluid to the mechanism of disease has been the characterization of CSF immunoglobulins by making anti-idiotypic antibodies (Ebers et al 1979, Naglekerken et al 1980). These antibodies, raised against the CSF IgG, would presumably bind to the combining site of CSF IgG. Possible antigens might then be used to displace the anti-idiotypic antibody. In such a study, anti-idiotypic antibodies have been raised against both CSF and serum IgG from patients with multiple sclerosis. Over a five-year period, up to 20% of the CSF IgG reacted with these anti-idiotypic antibodies. In this particular study, assays of these anti-idiotypic antibodies against CSF immunoglobulins of other patients with multiple sclerosis were negative in 12 of 13 patients, but did cross-react with one other patient (Baird et al 1980, Tachovsky & Baird 1980).

INTERPRETATION   The antigenic determinants for the locally synthesized immunoglobulin in multiple sclerosis have not been defined. In other words, no "multiple sclerosis antigen" has been demonstrated. It is possible that much of the immunoglobulin is the result of nonspecific activation of sequestered B cell clones; however, the analysis of immunoglobulin from isolated plaques and the binding of anti-idiotypic antibodies suggest that some IgG may be more specific. There remain the problems of identification of specific antibodies, determination of the associated antigens, and, finally, the demonstration of similar, if not identical, specificity in other patients with multiple sclerosis. The hope, obviously, is that this approach will indicate the antigenic nature of causal agent(s), as has been possible in chronic viral infections of the central nervous system, such as subacute sclerosing panencephalitis, in which similar immunologic responses occur.

CIRCULATING FACTORS THAT AFFECT MYELIN   One approach to demonstrating a cell-specific abnormality in multiple sclerosis has been the search for circulating factors that affect myelin metabolism or alter oligodendroglial function. A number of laboratories have reported that 60–75% of MS patients have serum factors that will demyelinate myelinating explants (Bornstein 1963, Seil 1977). In some patients, there appears to be a correlation of the presence of serum demyelinative factors and disease activity. In contrast to EAE, the chemical nature of the demyelinating factors in MS sera has been difficult to define. The factor is sensitive to a number of manipulations such as freezing and thawing, heating, exposure to dialysis tubing or ammonium sulfate. In EAE sera, at least a portion of the demyelinating factor consists of IgG. On this basis, it has been assumed that the factor in MS sera is also IgG. However, adsorption of the sera with staphylococcus protein A, which removes IgG_1, IgG_2, and IgG_4, does not significantly decrease the demyelinating factor. Similarly, removal of IgA or IgM has no effect on demyelinating activity. Among immunoglobulins it is possible that the factor is IgG_3 (which is 5–7% of the total IgG fraction) or some other minor immunoglobulin such as IgD or IgE. On the other hand, it is possible that the demyelinating factor may not be an immunoglobulin at all (Grundke-Iqbal & Bornstein 1979).

INTERPRETATION   One problem in interpretation of the significance of demyelinating factors in MS serum is the question of specificity to multiple sclerosis. Ten percent or more of the sera obtained from controls contain demyelinative factors. In amyotrophic lateral sclerosis, a disease in which the "target tissue" is the anterior horn cell, 50% of sera contain demyelinative factors. Thus, at the present time, it is difficult to assign a pathogenic role to an antimyelin or demyelinating factor in MS sera.

NEUROELECTRIC BLOCKING FACTORS   Despite the observed pathology and the resulting historical emphasis on demyelination in multiple sclerosis, there is evidence to suggest an alternative pathophysiological mechanism. The rapidity of appearance and disappearance of some clinical signs and symptoms in multiple sclerosis has suggested that demyelination and possible remyelination occurs too slowly to be the primary pathogenetic mechanism. Thus, the existence of rapidly acting serum factors that block nerve conduction has been proposed. The earliest indication that sera from patients with MS and from animals with EAE had neuroelectric blocking factors as well as a myelinotoxic factor resulted from the studies of Bornstein & Crain (1965). As with myelinotoxic factor, the specificity of possible neuroelectric factors soon came into question. Recent reports suggest, however, that one or more factors capable of blocking neuroelectric activity in invertebrate spinal cords are present in sera from MS patients and from animals with EAE. In MS the blocking factor(s) may correlate with clinical activity and may be within the IgG fraction (Schauf & Davis 1978). Much further work remains to be done in this area—in particular, serial studies in individual patients and further characterization of these factors.

CELLULAR RESPONSES   The cellular responses in peripheral blood or CSF of patients with MS have been analyzed in relation to (*a*) reaction to exogenous agents such as brain components or viruses, (*b*) cytotoxic or demyelinative effects on nervous tissue, and (*c*) changes in endogenous cell populations.

Response to exogenous agents   Many of the reports in this area have been conflicting, with difficulties in assaying cellular responses, presentation of exogenous stimuli (?antigens), and determination of differences between patients with multiple sclerosis and those with other conditions, including normal controls. A major focus has been on responses of lymphocytes to possible brain antigens, such as basic protein, or to viral antigens, particularly those from the measles virus. To this reviewer, the results have been inconclusive in terms of either indicating a response specific to multiple sclerosis or providing information about the mechanism of disease. Attempts have been made with peripheral blood lymphocytes or lymphocytes from CSF to demonstrate cytotoxic effects on tissue explants; these technically difficult studies have not been conclusive. This area of multiple sclerosis research has recently been reviewed (Antel et al 1978).

Two recent approaches to the cellular immunology of multiple sclerosis appear more promising:

1. The observation that T suppressor cells are decreased in both peripheral blood and CSF coincidentally with exacerbation of MS.
2. Fusion studies with lymphocytes.

1. Originally, suppressor cells were assayed by concanavalin A induction (Antel et al (1978). However, a recent report (Reinherz et al 1980) using specific monoclonal antibodies directed against T cells, helper cells, and suppressor cells indicated that most patients had a marked reduction of suppressor cells during an acute attack, with return of this population of T cells during remission. Suppressor cells were not affected in other inflammatory demyelinative diseases such as disseminated encephalomyelitis or the Landry-Guillain-Barre syndrome. In contrast, a similar decrease in suppressor cells has been demonstrated in other possible autoimmune diseases, such as systemic lupus erythematosis, hemolytic anemia, atopic eczema, and inflammatory bowel disease. The precise mechanism of the decrease in suppressor cells is not known, but it may be related to the release of autoantibodies that selectively eliminate the suppressor cell population. The role of inducer and suppressor T lymphocytes in regulation of immune responses has recently been reviewed (Reinherz & Schlossman 1980).

2. The second approach is the fusion of lymphocytes in blood or cerebrospinal fluid with myeloma cell lines to form antibody-generating hybridomas (Sandberg-Wollheim et al 1980). The availability of myeloma cell lines from human may eliminate the interspecies problems that limited earlier attempts with this approach (Croce et al 1980). With this approach it might be possible to determine whether CSF lymphocytes from MS patients generate specific antibodies, particularly during acute exacerbations.

INTERPRETATION   Many of the reported changes in peripheral leukocytes have been difficult to reproduce. However, the decrease in suppressor T cells during acute attacks and a return to normal with remission has been reported from a number of laboratories. The relation of this decrease to the mechanism of disease is unclear. Possibilities include common antigens between oligodendroglia and T suppressor cells, a role for T suppressor cells in limiting an autoimmune response during periods of remission, and a nonspecific response similar to the production of most of the oligoclonal immunoglobulin.

An Infectious Etiology to Multiple Sclerosis

The hypothesis of an infectious agent in MS is not new, just difficult to prove. As attempts to demonstrate or culture more conventional agents such as bacteria, spirochetes, or fungi have failed, proponents of this hypothesis have turned to unconventional agents, or conventional agents acting in unconventional ways. This translates in today's terms as the "viral hypothesis of multiple sclerosis." This hypothesis is based on (*a*) the population migration data (reviewed above), (*b*) the immunological responses in

CSF, which are similar to those seen after known chronic panencephalitis of known viral origin, and (c) the demonstration in animals that viral infections can produce demyelination as the primary pathology (Johnson 1980).

The strongest argument against the viral hypothesis is that no agent has been reproducibly seen, cultured, rescued, or biochemically identified from brain or other tissue from patients with multiple sclerosis. Even the research strategies used successfully to analyze the viral or transmissible nature of slow viruses, or latent viral infections in the human, such as subacute sclerosing panencephalitis, progressive multifocal leukoencephalopathy, Jakob-Kreutzfeld disease, or Herpes Simplex infection, have been negative. For example, intracerebral injection of MS brain into primates (Sibley et al 1980), as was used in kuru and Jakob-Kreutzfeld disease (Gajdusek & Gibbs 1975), cocultivating to rescue virus, as was done with progressive multifocal leukoencephalopathy (Weiner et al 1972), and the use of bio-chemical probes for viral nucleic acids (Dorries & ter Meulen 1980), as has been done in herpes simplex encephalitis (Marsden 1980, Cabrera et al 1980) have all been negative.

Analysis of viral infection in animals suggests several possible mecha-nisms by which viruses might induce a demyelinative disease (Brooks et al 1979, Fields & Weiner 1981). Direct cytotoxic effect on oligodendroglia occurs with the JHM strain of mouse hepatitis virus and with the JC papavovirus in progressive multifocal leukoencephalopathy. Viral infection might not necessarily be acutely cytotoxic but rather, in a chronic process, might interfere with cellular functions required for maintenance of myelin. In the human disease, this possibility may be difficult to prove if the latent infection is limited to oligodendroglia around plaques. However, if the viral process is more generalized, despite the focal pathology, then current tech-niques for obtaining oligodendroglia from brain and characterizing their biochemical properties may be applicable. Alternatively, more accessible tissue, such as lymphocytes, might contain the viral genome.

An alternative to direct effect on cellular function by a virus is the possibility that viral infection induces immunologic responses in the host that are damaging to the nervous system. In lymphocytic choriomeningitis infection of mice, the CNS involvement can be prevented by prior im-munosuppression of the animals. In addition, the disease can be transferred by immune T cells but not by immune sera (Nathanson et al 1975). In murine encephalomyelitis (Theiler's Virus) there is a two-stage disease, an acute encephalitic involvement of gray matter, and a chronic, delayed in-volvement of white matter. Immunosuppression enhances the acute phase and abolishes the chronic phase (Lipton & Dalcanto 1976). These observa-tions suggest two possible mechanisms: (a) viral infection alters neural cell membranes to produce an autoimmune type of disease and/or (*b*) there are similar receptors for the virus on both neural cells and peripheral blood lymphocytes. This latter mechanism seems to be operative in recognition of the hemagglutin of reovirus type 3 in which both fatal encephalitis and generation of suppressor cells occur in the course of the disease (Weiner et al 1980a,b).

Repeated attacks of demyelination have not been reported in virally induced demyelinative diseases. However, in visna, a virally-induced demyelinative disease in sheep, the RNA virus persists as a DNA provirus intermediate. The infected animals initially develop a cellular and humoral response to the agent. However, in chronically infected animals, virulent viruses can be recovered by co-cultivation that are not neutralized by the antibodies formed against the initial infectious agent (Narayan et al 1978). This antigenic shift caused by spontaneous mutations of the virus within the host can lead to serial exacerbations of disease with the cumulative effect appearing as progressive neurological disease (Johnson 1980). It is conceivable that a similar mechanism might result in an exacerbating and remitting form of disease.

INTERPRETATION   At present there is no hard evidence to substantiate a role for viruses or some other form of transmissible agent in either the onset or recurrent attacks of multiple sclerosis. However, the methods for detecting latent viral infection such as the use of nucleic acid hybridization, monoclonal antibodies, or the use of anti-idiotypic antibodies to determine antigenic determinants on centrally derived immunoglobulin have only recently been applied to multiple sclerosis. Those who subscribe to the viral hypothesis of multiple sclerosis hope these advances will provide positive evidence to substantiate the role of an infectious agent.

Diagnosis of Multiple Sclerosis

The diagnosis of MS rests primarily on the interpretation of the clinical signs and symptoms, with emphasis on the presence of lesions which are spread geographically in the nervous system and spread temporally in terms of exacerbations and remissions.

The symptoms reflect the involvement of white matter. Thus, disruption of tracts presenting as unilateral visual loss (optic nerve), hemiparesis (corticospinal tracts), ataxia (cerebellar outflow systems), internuclear ophthalmoplegia (medial longitudinal fasciculus) are common. In contrast, symptoms relative to neuronal involvement such as seizures, focal deficits in high cortical function, and dementia are uncommon, particularly early in the course of the disease.

The clinical course of the disease is highly variable. Some patients have a fulminant, monotropic disease with inexorable progression. More common is the pattern of exacerbations and remissions followed by a decrease in exacerbations and the appearance of slow progression. Finally, some patients may have one or two episodes and then be symptom-free for many years. These clinical patterns among patients are so different that one wonders if multiple sclerosis is a single disease.

Contrary to the image most people have of multiple sclerosis, the disease carries a reasonable prognosis in most patients. Data from follow-up of patients seen at the Mayo Clinic (Percy et al 1971) and in follow-up of US Army personnel (Kurtzke 1970) indicate that 75% of patients are alive 25 years after the onset of the disease. Of these survivors, 55% are without significant disability. Even these figures may be falsely skewed toward more severely involved patients because of the diagnosis requiring rigid clinical criteria. Thus, early cases and milder cases may be excluded. The true clinical spectrum and prognosis of MS will not be known until a reliable laboratory diagnostic test for multiple sclerosis is available.

LABORATORY AIDS TO DIAGNOSIS   Aids to diagnosis are listed in Table 1. The changes in CSF are discussed above. The physiologic tests involving evoked potential recording in the visual, auditory, or somatosensory systems are useful in demonstrating the existence of a second, separate lesion which may be subclinical in its presentation. Evoked responses have also been useful for studying the physiology of recovery from an acute attack, particularly in the visual system (McDonald & Halliday 1977).

INTERPRETATION   A laboratory test that is unequivocally diagnostic for multiple sclerosis does not exist. Thus, it is virtually impossible to make the diagnosis early in the course of the disease for many patients; hence the use of such terms as "possible multiple sclerosis," or "probable multiple sclerosis" (this problem in diagnostic classification is well reviewed in Brown et al 1979). Such a diagnostic test is badly needed, not only for individual patients, but also for epidemiological and genetic studies.

Mechanisms of Recovery

A striking feature of multiple sclerosis, particularly early in the course of the disease, is the degree of recovery patients can achieve during remissions. This recovery takes place under circumstances where abnormal physiological responses, such as altered visual evoked potentials, may persist (McDonald & Halliday 1977). Remyelination of plaques is rarely seen, and then only around the edges (Prineas & Connell 1979). Thus, explanations of recovery have centered not on remyelination but on restoration of conduc-

Table 1   Abnormalities in clinically definite multiple sclerosis

Test	Remarks
Visual evoked responses	Indication of topical disease.
Auditory evoked responses	
Somatosensory evoked responses	
Cerebrospinal fluid:	
IgG concentration	Elevated in approximately 70% of patients. No correlation with disease activity.
IgG Index[a]	Elevated in approximately 70–90% of patients. Possible correlation with disease activity.
Oligoclonal immunoglobulin pattern	Present in 80–90% of patients. No definite correlation with disease activity. Onset of abnormality in early disease not known.
Myelin basic protein	Present during exacerbation. Decrease may correlate with clinical improvement.

[a] CSF IgG index: CSF-IgG/CSF-Albumin/(Serum IgG/Serum albumin.

tion in demyelinated nerve (Bostock & McDonald 1981). The model for this process has been the recovery of conduction in the peripheral nervous system following demyelination with either diphtheria toxin or lysolecithin. In these models, recovery of conduction in persistently demyelinated areas is, in part, related to the spread of sodium channels along the nerve from their normal site at internodes (Ritchie & Rogart 1977, Foster et al 1980, Bostock & McDonald 1981). This type of conduction is slower than normal saltatory conduction (Sears et al 1978)—a finding compatible with the observation in multiple sclerosis of restoration of function despite persistently slowed central conduction (Halliday & McDonald 1977).

INTERPRETATION   It is not clear why remyealination does not occur in multiple sclerosis. Possibilities include the following:

1. Adult oligodendroglia do not have the capacity to divide and migrate into the areas of myelin dissolution and loss of oligodendroglia.
2. There is, in the adult nervous system, no pool of immature oligodendroglia capable of migrating to the site of injury and differentiating into reparative, remyelinating cells.
3. Some factors, such as astrocytic processes, altered axonal membrane, or persistance of proteases, interfere with the normal recognition of an unmyelinated axon by a myelinogenic oligodendroglia.

These possibilities are being explored in animal models of demyelination using cell-specific markers.

Therapy

There is no specific therapy for MS. Attempts have been directed at altering immunological responses by immunosuppression, introduction of antigens, and dietary alteration. The unpredictability of the disease, the changing clinical course from one of exacerbations and remissions to one of progression, and the lack of proven laboratory tests to demonstrate clinical activity make therapeutic trials extremely difficult to design and carry out. Guidelines for such clinical trials have recently been published (Brown et al 1979).

Immunosuppression with high doses of steroids or ACTH results in a shortening of the duration of acute attacks (Rose et al 1970). There is no evidence that this form of therapy prevents attacks or alters the progressive form of the disease. The use of other forms of immunosuppression such as azathioprine or cyclophosphamide may reduce the relapse rate, but apparently does not benefit the chronic-progressive forms. Several controlled clinical trials with these agents as well as with antilymphocyte sera are in progress (Bauer 1980).

The use of myelin basic proteins or related copolymers in therapy of multiple sclerosis is based on the observation that these agents can alter the course of the experimental disease, EAE. Preliminary trials with myelin basic protein have not been reported to alter the course of the disease. Dietary therapy has been based on a possible role of unsaturated fatty acids in the disease. Reports of the use of these diets have been inconclusive (Bauer 1980).

Future Directions

It is extremely hazardous to make predictions about a disease with unknown etiology, unknown mechanisms, and unknown therapy. Nevertheless, advances in recent years do suggest some promising leads.

ETIOLOGY   The emphasis continues to be on the demonstration of a transmissible agent. Future work in this area may be directed at the demonstration of altered cellular responses induced by a virus. The possible use of peripheral blood leukocytes or lymphocytes in CSF may aid in these investigations, because brain tissue from patients in the acute phases of the disease is rarely available. Thus, studies dependent on demonstration or isolation of an agent from brain are limited by availability of tissue from both patients with multiple sclerosis and appropriate controls.

MECHANISM   The immunologic abnormalities in MS have not provided an understanding of the basic mechanism of the disease. However, as mentioned above, the use of anti-idiotypic antibodies as well as more highly specific monoclonal antibodies may provide significant new information. Such information may be useful not only in understanding the mechanisms of disease, but also in providing a definitive diagnostic test.

THERAPY   There are numerous examples in neurology where therapy has been successfully introduced without a clear understanding of the underlying disease process. Such is the case in epilepsy, Parkinson's Disease and manic-depressive illness. A number of new approaches to therapy have been suggested or are under investigation.

Anti-viral agents   Despite the absence of evidence implicating a specific virus, attempts at therapy with potential anti-viral agents have been attempted. The most recent trial in this area involves the use of interferon in a controlled clinical trial.

Removal of circulating factors   Based on the successful use of plasmapheresis in myasthenia gravis (Dau et al 1977) and chronic relapsing polyneuritis (Server et al 1979), the removal of circulating factors is being evaluated in a controlled trial of therapy during acute attacks of MS.

Modification of properties of demyelinated nerve   Assuming that remyelination is not the primary mechanism of recovery in MS, attempts to use agents which might prolong the action potential are being considered. Clinical trials are underway with 4-aminopyridine, a drug which blocks mammalian potassium channels and aids conduction in experimental demyelination (Sherratt et al 1980, Bostock & McDonald 1981).

Alteration of protease actions   Attempts to neutralize extracellular effects of such proteases are being considered on the assumption that macrophage-derived neutral proteases selectively attack myelin basic protein.

Immunopharmacologic therapy   The current approaches using high doses of steroids, ACTH, or other immunosuppressive drugs have not yielded any striking improvements in the clinical course of the disease, particularly in the chronic progressive form. Similarly, trials with transfer factor have been inconclusive. Agents that might augment suppressor cell functions are under study.

Speculative Overview

Multiple sclerosis continues to be a baffling disease, although new observations continue to be made. Most of those in recent years have been in the immunologic area. What is lacking, however, is a consistent thread that will tie the epidemiological, genetic, and immunological data together. The most attractive hypothesis continues to be that some form of transmissible agent, acquired early in life by susceptible individuals, alters the host's immune system. In the central nervous system this results in the activation of B cell clones and in the alteration of cellular membranes of oligodendroglia and vascular endothelium. The repeated attacks may be related to a nonspecific immunologic challenge or to antigenic shifts of the latent transmissible agent. So much for theory. How to prove this or other possible hypotheses, and how to translate that proof into an alteration of the disease remains the challenge.

ACKNOWLEDGMENTS

Personal work referred to in this review was supported by grants from the National Multiple Sclerosis Society and from the National Institute of Neurological Diseases, Communicative Disorders, and Strokes.

Literature Cited

Abramsky, O., Lisak, R. P., Silberberg, D. H., Pleasure, D. E. 1977. Antibodies to oligondendroglia in patients with multiple sclerosis. *N. Engl. J. Med.* 297:1207–11

Adams, C. W. M. 1977. Pathology of multiple sclerosis: Progression of the lesion. *Br. Med. Bull.* 33:15–20

Allen, I. U. 1980. A histological and histochemical study of the macroscopically normal white matter in multiple sclerosis. In *Progress in Multiple Sclerosis Research*, ed. H. J. Bauer, S. Poser, G. Ritter, pp. 340–47. Berlin/Heidelberg/New York: Springer. 677 pp.

Alter, M. 1980. The geographic distribution of multiple sclerosis: New concepts. See Allen 1980, pp. 495–502

Alter, M., Kahana, E., Loewenson, R. 1978. Migration and risk of multiple sclerosis. *Neurology* 28:1089–98

Antel, J. P., Arnason, B. G. W., Medof, M. E. 1978. Suppressor cell function in multiple sclerosis: Correlation with clinical disease activity. *Ann. Neurol.* 5:338–42

Antel, J. P., Richman, D. P., Medof, M. E., Arnason, B. G. W. 1978. Lymphocyte function and the role of regulator cells in multiple sclerosis. *Neurology* 28(2):106–10

Baird, L. G., Tachovsky, T. G., Sandberg-Wollheim, M., Kaprowski, H., Nisonoff, A. 1980. Identification of a unique idiotype in cerebrospinal fluid and serum of a patient with multiple sclerosis. *J. Immunol.* 124:2324–28

Bauer, H. J. 1980. Recent trends in the therapy of multiple sclerosis. See Allen 1980, pp. 653–56

Bloom, B. R. 1980. Immunological changes in multiple sclerosis. *Nature* 287:275–76

Bloom, B. R., Ju, G., Brosnan, C., Cammer, W., Norton, W. 1978. Notes on the pathogenesis of multiple sclerosis. *Neurology* 28(2):93–101

Bornstein, M. B. 1963. A tissue culture approach to demyelinative disorders. *Natl. Cancer Inst. Monogr.* 11:197–214

Bornstein, M. B., Crain, S. M. 1965. Functional studies of cultured brain tissues as related to demyelinative disorders. *Science* 148:1242–44

Bostock, H., McDonald, W. I. 1981. Recovery of function after demyelination. In *Neuronal-Glial Cell Interrelationships*, ed. T. A. Sears. Dahlen Konferenzen. Weinheim/Deerfield Beach/Basel: Verlag Chemie. In press

Brooks, B. R., Jubelt, B., Swarz, J. R., Johnson, R. T. 1979. Slow viral infections. *Ann. Rev. Neurosci.* 2:309–40

Brosnan, C. F., Cammer, W., Norton, W. T., Bloom, B. R. 1980. The role of activated macrophages in inflammatory demyelinating diseases. *4th Int. Cong. Immunol.* 18.6.05

Brown, J. R., Beebe, G. W., Kurtzke, J. F., Loewenson, R. B., Silberberg, D. H., Tourtelotte, W. W. 1979. The design of clinical studies to assess therapeutic efficacy in multiple sclerosis. *Neurology* 29(2):1-23

Cabrera, C. V., Wohlenberg, C., Openshaw, H., Ray-Mendez, M., Puga, A., Notkins, A. L. 1980. Herpes simplex virus DNA sequences in the CNS of latently infected mice. *Nature* 188:288-90

Carson, J. H., Barbarese, E., Braun, P. E., McPherson, T. A. 1978. Components in multiple sclerosis cerebrospinal fluid that are detected by radioimmunoassay for myelin basic protein. *Proc. Natl. Acad. Sci. USA* 75:1976-78

Cohen, S. R., Brooks, B. R., Herndon, R. M., McKhann, G. M. 1980. A diagnostic index of active demyelination—myelin basic protein in cerebrospinal fluid. *Ann. Neurol.* 8:25-31

Cohen, S. R., Herndon, R. M., McKhann, G. M. 1976. Radioimmunoassay of myelin basic protein in spinal fluid: An index of active demyelination. *N. Engl. J. Med.* 295:1455-57

Croce, C. M., Linnenbach, A., Hall, W., Steplewski, Z., Koprowski, H. 1980. Production of human hybridomas secreting antibodies to measles virus. *Nature* 288:488-89

Dau, P. C., Lindstrom, J. M., Cassel, C. K., Denys, E. H., Sher, E. F., Spitler, L. E. 1977. Plasmapheresis and immunosuppressive drug therapy in myasthenia gravis. *N. Engl. J. Med.* 297:1134-40

Dean, G. 1980. Multiple sclerosis among immigrants to Britain and in the Islands of Scilly and Malta. See Allen 1980, pp. 519-24

Dean, G., Brady, R., McLoughlin, H., Elian, M., Adelstein, A. M. 1977. Motor neurone disease and multiple sclerosis among immigrants to Britain. *Br. J. Prev. Soc. Med.* 31:141-46

Dorries, K., ter Meulen, V. 1980. Search for viral nucleic acids in multiple sclerosis brain. See Allen 1980, pp. 47-52

Ebers, G. C., Zabrieske, J. B., Kunkel, H. G. 1979. Oligoclonal immunoglobulins in subacute sclerosing panencephalitis and multiple sclerosis: A study of idiotypic determinants. *Clin. Exp. Immunol.* 35:67-75

Fewster, M. E., Blackstone, S. 1975. In vitro study of bovine oligodendroglia. *Neurobiology* 5:316-28

Fields, B. N., Weiner, H. L. 1981. Mechanism of viral injury to the nervous system. See Bostock & McDonald 1981, in press

Foster, R. E., Whalen, C. C., Waxman, S. G. 1980. Reorganisation of the axon membrane in demyelinated peripheral nerve fibers: Morphological evidence. *Science* 210:661-63

Gajdusek, D. G., Gibbs, C. J. Jr. 1975. Slow virus infections of the nervous system and the laboratories of slow, latent and temperate virus infections. In *The Nervous System*, ed. D. B. Tower, 2:113-35. New York: Raven. 437 pp.

Grundke-Iqbal, I., Bornstein, M. B. 1979. Multiple sclerosis: Immunological studies on the demyelinating serum factor. *Brain Res.* 160:489-503

Halliday, A. M., McDonald, W. I. 1977. Pathophysiology of demyelinating disease. *Br. Med. Bull.* 33:21-27

Itoyama, Y., Sternberger, H. H., Webster, H. de F., Quarles, R. H., Cohen, S. R., Richardson, E. P. 1980. Immunocytochemical observation on the distribution of myelin-associated glycoprotein and myelin basic protein in multiple sclerosis lesions. *Ann. Neurol.* 7:167-77

Johnson, R. T. 1980. Selective vulnerability of neural cells to viral infections. *Brain* 103:447-72

Johnson, R. T., Katzmann, R., McGeer, E., Price, D., Shooter, E. M., Silberberg, D. 1979. Report of the panel on inflammatory, demyelinating and degenerative diseases. NIH Publ. No. 79-1916. Washington DC: US Dept. Health, Educ., Welfare

Kennedy, P. G. E., Lisak, R. P. 1979. A search for antibodies against glial cells in the serum and cerebrospinal fluid of patients with multiple sclerosis and Guillain-Barre syndrome. *J. Neurol. Sci.* 44:125-33

Kennedy, P. G. E., Lisak, R. P. 1981. Do patients with demyelinating diseases have antibodies against human glial cells in their sera? *J. Neurol. Neurosurg. Psychiatr.* In press

Kennedy, P. G. E., Lisak, R. P., Raff, M. C. 1980. Cell type specific markers for human glial and neuronal cells in culture. *Lab. Invest.* 43:342-51

Kohler, G., Milstein, C. 1976. Derivation of specific antibody-producing tissue culture and tumor lines by cell fusion. *Eur. J. Immunol.* 6:511-19

Kurland, L. T. 1970. The epidemiological characteristics of multiple sclerosis. In *Handbook of Clinical Neurology*, ed. P. J. Vinken, G. W. Bruyn, 9:63-84. Amsterdam: North-Holland. 706 pp.

Kurtzke, J. F. 1970. Clinical manifestations of multiple sclerosis. See Kurtzke 1970, pp. 161-216

Kurtzke, J. F. 1975. A reassessment of the distribution of multiple sclerosis. *Acta Neurol. Scand.* 51:110-57

Kurtzke, J. F. 1980. Clustering and epidemic occurrence of multiple sclerosis. See Allen 1980, pp. 519–24

Laterre, E. C. 1966. Les gammaglobulins du liquide cephalo-rachidier dans la sclerose en plaque. *Acta Neurol. Belg.* 66:305–18

Link, H. 1973. Comparison of electrophoresis of agar gel and agarose gel in the evaluation of fluid gamma-globulin abnormalities in cerebrospinal fluid and serum in multiple sclerosis. *Clin. Chim. Acta* 46:383–89

Lipton, H. L., Dalcanto, M. C. 1976. Theiler's virus-induced demyelination: Prevention by immunosuppression. *Science* 192:62–64

Marsden, H. 1980. Herpes simplex virus in latent infection. *Nature* 288:212–13

Mattson, D. H., Roos, R. P., Arnason, B. G. W. 1980. Isoelectric focusing of IgG eluted from multiple sclerosis and subacute sclerosing panencephalitis brains. *Nature* 287:335–37

McDonald, W. I., Halliday, A. M. 1977. Diagnosis and classification of multiple sclerosis. *Br. Med. Bull.* 33:4–9

Morell, P. 1977. *Myelin.* New York/London: Plenum. 531 pp.

Nagelkerken, L. M., Aalberse, R. C., van Walbeck, H. K., Out, T. A. 1980. Preparation of antisera directed against the idiotype(s) of immunoglobulin G from the cerebrospinal fluid of patients with multiple sclerosis. *J. Immunol.* 125: 384–89

Narayan, O., Griffin, D. E., Clements, J. E. 1978. Virus mutation during "slow infection." Temporal development and characterisation of mutants of visna virus recovered from sheep. *J. Gen. Virol.* 41:343–52

Nathanson, N., Monjan, A. A., Panitch, H. S., Johnson, E. D., Petursson, G., Cole, G. A. 1975. Virus-induced cell-mediated immunopathological disease. In *Viral Immunology and Immunopathology,* ed. A. Notkins, pp. 357–91. New York: Academic

Norby, E. 1978. Viral antibodies in multiple sclerosis. *Prog. Med. Virol.* 24:1–39

Oger, J. J. F., Arnason, B. G. W. 1980. HLA patterns in multiple sclerosis. See Allen 1980, pp. 460–64

Palo, J. 1978. *Myelination and Demyelination.* New York/London: Plenum. 651 pp.

Paterson, P. Y., Day, E. D., Varitek, V. A., Peterson, D. J., Whitacre, C. C., Berenberg, R. A., Harter, D. H. 1980. Circulating myelin basic protein (MBP) fragments: Role in neuroimmunologic diseases. *4th Int. Cong. Immunol.* 18.-6.22

Paterson, P. Y., Day, E. D., Whitacre, C. C., Berenberg, R. A., Harter, D. H. 1981. Endogenous myelin basic protein-serum factors (MBP-SFs) and anti-MBP antibodies in humans. Occurence in sera of clinically well subjects and patients with multiple sclerosis. *J. Neurol. Sci.* In press

Percy, A. K., Nobrega, F. T., Okazaki, H., Glattre, E., Kurland, L. T. 1971. Multiple sclerosis in Rochester, Minn. *Arch. Neurol.* 25:105–11

Poduslo, S. E., Miller, K., McKhann, G. M. 1978. Metabolic properties of maintained oligodendroglia. *J. Biol. Chem.* 253:1592–97

Poduslo, S. E., Norton, W. T. 1972. Isolation and some chemical properties of oligodendroglia from calf brain. *J. Neurochem.* 19:727–36

Prineas, J. W., Connell, F. 1978. The fine structure of chronically active multiple sclerosis plaques. *Neurology* 28(2): 68–75

Prineas, J. W., Connell, F. 1979. Remyelination in multiple sclerosis. *Ann. Neurol.* 5:22–31

Raff, M. C., Fields, K. L., Hakomori, S.-I., Mirsky, R., Pruss, R. M., Winter, J. 1979. Cell-type specific markers for distinguishing and studying neurones and the major class of glial cells in culture. *Brain Res.* 174:283–308

Raff, M. C., Mirsky, R., Fields, K. L., Lisak, R. P., Dorfman, S. H., Silberberg, D. H., Gregson, N. A., Leibowitz, S., Kennedy, M. 1978. Galactocerebroside is a specific cell surface antigenic marker for oligodendrocytes in culture. *Nature* 274:813–16

Raine, C. S., Traugott, H. U., Stone, S. H. 1980. Applications of chronic relapsing experimental allergic encephalomyelitis to the study of multiple sclerosis. See Allen 1980, pp. 3–10

Reinherz, E. L., Schlossman, S. F. 1980. Regulation of the immune response—inducer and suppressor T lymphocyte subsets in human beings. *N. Engl. J. Med.* 303:370–73

Reinherz, E. L., Weiner, H. L., Hauser, S. L., Cohen, J. A., Distaso, J. A., Schlossman, S. F. 1980. Loss of suppressor T cells in active multiple sclerosis. Analysis with monoclonal antibodies. *N. Engl. J. Med.* 303:125–29

Richards, P. T., Cuzner, M. L. 1978. Proteolytic activity in cerebrospinal fluid. *Adv. Exp. Med. Biol.* 100:521–27

Ritchie, J. M., Rogart, R. B. 1977. The density of sodium channels in mammalian myelinated nerve fibres and the nature of the axonal membrane under the myelin sheath. *Proc. Natl. Acad. Sci. USA* 74:211–15

Rose, H. S., Kuzma, J. W., Kurtzke, J. F., Sibley, W. A., Tourtelotte, W. W. 1970. Cooperative study in the evaluation of therapy in multiple sclerosis: ACTH as placebo in acute exacerbations. Final report. *Neurology* 20(2):1–59

Sandberg-Wollheim, M., Shander, M., Croce, C. M., Kaprowski, H. 1980. Production of interspecies hybridomas between CSF or blood lymphocytes from patients with neurological diseases and mouse myeloma cells. See Allen 1980

Schachner, M., Commer, I., Lagenaur, C., Schnitzer, J. 1981. Developmental expression of antigenic markers in glial subclasses. See Bostock & McDonald 1981, in press

Schauf, C. L., Davis, F. A. 1978. The occurrence, specificity and the role of neuroelectric blocking factors in multiple sclerosis. *Neurology* 28(2):34–39

Sears, T. A., Bostock, H., Sherratt, R. M. 1978. The pathophysiology of demyelination and its implication for the symptomatic treatment of multiple sclerosis. *Neurology* 28(2):21–26

Seil, F. J. 1977. Tissue culture studies in demyelinating diseases: A critical review. *Ann. Neurol.* 2:345–55

Server, A. C., Lefkowith, J., Braine, H., McKhann, G. M. 1979. Treatment of chronic relapsing inflammatory polyradiculneuropathy by plasma exchange. *Ann. Neurol.* 6:258–61

Sherratt, R. M., Bostock, H., Sears, T. A. 1980. Effects of 4-aminopyridine on normal and demyelinated mammalian nerve fibres. *Nature* 283:570–72

Shorr, J., Roström, B., Link, H. 1981. Antibodies to viral and non-viral antigens in subacute sclerosing panencephalitis and multiple sclerosis demonstrated by thin-layer polyacrylamide gel isoelectric focusing, antigen immunofixation and autoradiography. *J. Neurol. Sci.* 49:99–108

Sibley, W. A., Kalter, S. S., Laguna, J. F. 1980. Attempts to transmit multiple sclerosis to newborn and germ-free non-human primates: A ten-year interim report. See Allen 1980, pp. 80–85

Snyder, D. S., Raine, C. S., Faroog, M., Norton, W. T. 1980. The bulk isolation of oligodendroglia from whole rat forebrain: A new procedure using physiologic media. *J. Neurochem.* 34:1614–21

Szuchet, S., Stefansson, K., Wallmann, R. L., Dawson, G., Arnason, B. G. W. 1980. Maintenance of isolated oligodendrocytes in long-term culture. *Brain Res.* 200:151–64

Tachovsky, T. G., Baird, L. G. 1980. Unique idiotypes in CSF of MS patients. *4th Int. Congr. Immunol.* 18.6.28

Thompson, E. J. 1977. Laboratory diagnosis of multiple sclerosis: Immunological and biochemical aspects. *Br. Med. Bull.* 33:28–33

Traugott, U., Snyder, D. S., Raine, C. S. 1979. Oligodendrocyte staining by multiple sclerosis serum is non-specific. *Ann. Neurol.* 6:13–20

Trotter, J. L., Huss, B., Blank, W. P., O'Connell, K., Hagen, S., Shearer, W. T., Agrawal, H. C. 1978. Myelin basic protein and CSF in normal and pathological brains. *Trans. Am. Soc. Neurochem.* 9:59

Vandvik, B., Norby, E. 1980. Viral antibody responses in the central nervous system of patients with multiple sclerosis. See Allen 1980

Weiner, H. L., Ault, K. A., Fields, B. N. 1980a. Interaction of reovirus with cell surface receptors. I. Murine and human lymphocytes have a receptor for the hemagglutination of reovirus type 3. *J. Immunol.* 124:2143–48

Weiner, H. L., Powers, M. L., Fields, B. N. 1980b. Reovirus virulence and central nervous system cell tropism: Absolute linkage to the viral hemagglutination. *J. Inf. Dis.* 141:609–16

Weiner, L. P., Hendon, R. M., Narayan, O., Johnson, R. T. 1972. Further studies of an SV40-like virus isolated from human brain. *J. Virol.* 10:147–52

Whitaker, J. N. 1977. Myelin encephalitogenic protein fragments in cerebrospinal fluid of persons with multiple sclerosis. *Neurology* 27:911–20

Whitaker, J. N., Bashir, R. M., Jen Chou, C-H., Kibler, R. F. 1980. Antigenic features of myelin besic protein-like material in cerebrospinal fluid. *J. Immunol.* 124:1148–53

Wisniewski, H. M., Lassmann, H. 1980. Chronic relapsing EAE: Its application to study of human inflammatory demyelinating diseases. See Allen 1980

Ann. Rev. Neurosci. 1982. 5:241–273

THE NEUROPSYCHOLOGY
OF HUMAN MEMORY*

Larry R. Squire

Veterans Administration Medical Center, San Diego, California 92161 and the Department of Psychiatry, University of California at San Diego, School of Medicine, La Jolla, California 92093

INTRODUCTION

The biology of memory is presently being investigated at a variety of levels from the cellular to the neuropsychological. The interdisciplinary nature of neuroscience has encouraged the belief that superficially dissimilar phenomena—e.g. collateral sprouting, receptor adaptations, synaptic depression and facilitation, dendritic growth in response to enriched environment, and recovery of function after brain injury—are all potentially relevant to questions about memory because they all reflect the nervous system's capacity for plasticity. Yet, a satisfying account of the biology of memory must include not only information about the details of synaptic change but also a description of the learning processes and memory systems whose neurobiological mechanisms we wish to understand, plus information about how memory is organized in the brain, how memory changes with time, and which brain regions are involved.

A favorable strategy for addressing these questions has been to study amnesia. Most often, disorders of memory occur in the context of impairment in other aspects of intellectual function, as in depression or dementia. Nonetheless, amnesia can occur as a relatively circumscribed disorder in the absence of other cognitive impairment. As in many areas of biology where disorders of function have taught us about normal function, so the analysis of memory disorders can provide insights into the structure and organization of normal memory (for other recent reviews, see Piercy 1977, Weiskrantz 1978, Squire & Cohen 1982).

This review considers four aspects of memory and amnesia—*anterograde amnesia* or loss of memory for events that occurred after the onset of amnesia, *retrograde amnesia* or loss of premorbid memory, *preserved learning capacity*, and the *anatomy of amnesia*. Each of these topics is organized around the specific issues that have guided the past decade of experimental work, with the intention of summarizing what is presently known about the organization of memory and its neurological substrate.

THE AMNESIAS

The globally amnesic patient can appear normal to casual observation. Such a patient may have normal intellectual capacity, normal digit span and intact social skills, and may retain knowledge acquired in early life. The defect lies in acquiring new memories and in recalling some memories that had been acquired prior to becoming amnesic. This defect often occurs in the absence of confabulation or confusion and with awareness by the patient of his condition. Note that the amnesic syndrome has no connection to hysterical memory loss of psychogenic orgin. By far the most frequently studied example is found in Korsakoff's syndrome. First described in 1887 (Korsakoff 1887), this syndrome has been studied extensively during the past one hundred years (Talland 1965, Victor et al 1971, Butters & Cermak 1980). The disease develops after chronic alcohol abuse and is characterized by symmetrical brain lesions along the walls of the third and fourth ventricles as well as in the cerebellum and cerebral cortex. After the acute stage has passed, the patient with Korsakoff syndrome is alert and responsive and has normal intellectual capacity, as assessed by conventional tests. The disease produces a spectrum of cognitive deficits, but amnesia occurs out of proportion to other neuropsychological findings.

Additional information about global amnesia has come from the rare individual case in which amnesia occurs as a strikingly circumscribed entity. The best-known and most thoroughly studied of these is case H. M. (Scoville & Milner 1957). In 1953, H. M. sustained bilateral resection of the medial temporal region in an attempt to relieve severe and intractable epilepsy. The resection included the anterior two-thirds of the hippocampal formation, parahippocampal gyrus, amygdala, and uncus. Following surgery, H. M. exhibited a profound amnesic syndrome in the absence of any detectable change in general intellectual ability. H. M. has been carefully studied by Brenda Milner and her colleagues at the Montreal Neurological Institute, and the findings from this case alone have provided an enormous amount of information about memory. A second well-studied patient is case

N. A. (Teuber et al 1968, Kaushall et al 1981), who became amnesic in 1960 following a penetrating brain injury with a miniature fencing foil. Recent CT scans have identified a lesion in the region of the left dorsomedial thalamic nucleus (Squire & Moore 1979). Consistent with these radiographic findings, neuropsychological studies have demonstrated that the amnesia is more pronounced for verbal material than for nonverbal material (Teuber et al 1968, Squire & Slater 1978). For example, he forgets lists of words and connected prose more readily than faces or spatial locations. In this sense, N. A.'s memory deficit is material-specific, rather than global (Milner 1968). Like the global amnesia of H. M., N. A.'s deficit occurs in a bright individual (N. A.'s IQ is 124) without notable neuropsychological findings other than amnesia.

A final type of global amnesia is that produced by bilateral electroconvulsive therapy (ECT) (Squire 1981a). ECT is sometimes prescribed for the treatment of depressive illness, and amnesia is its prominent side effect. Since ECT is a scheduled event, it is unique among the better studied amnesias because it provides an opportunity to use each patient as his own control in before-and-after studies. The amnesia recovers to some extent after each treatment in a series and cumulates across treatments. Treatments are usually scheduled every other day, three times a week, and a series of treatments typically consists of six to twelve treatments. By one hour after treatment, amnesia appears as a rather circumscribed deficit in the absence of gross confusion or general intellectual impairment.

The amnesias outlined above do not of course exhaust the list of causes or types of amnesia. For example, amnesia can also occur after head injury, anoxia, encephalitis, tumor, or vascular accident. But the amnesias described above have been studied the most extensively in recent years, particularly from the point of view of the neuropsychology of memory.

ANTEROGRADE AMNESIA

Amnesia as a Non-Unitary Disorder

Comparison of different amnesias clearly indicates that they do not all take the same form. Moreover, some behavioral deficits exhibited by amnesic patients have no necessary relationship to amnesia at all. This point can be made most clearly by comparing the Korsakoff syndrome to other examples of amnesia. Thus, patients with Korsakoff syndrome have cognitive and other clinical deficits not shared by other amnesic groups. These individuals commonly exhibit apathy, blandness, lethargic indifference, or vacuity of expression, loss of initiative, placidity (Talland 1965, pp. 19, 29). Yet there is no necessary relationship between amnesia and these features. Case N. A., for example, is energetic and alert, easily initiates social contact, and interacts with people in an agreeable, friendly way (Kaushall et al 1981). Zangwill also has commented on the very different appearance of the Korsakoff syndrome in comparison to the more "pure amnesic syndromes . . . [possibly] involving the hippocampal region" (Zangwill 1977). Thus, the patient with Korsakoff syndrome performs poorly on a variety of cognitive tests, such as those requiring rapid switching of strategies (Talland 1965, Oscar-Berman 1973, Glosser et al 1976, Butters & Cermak 1980). Case H. M.'s amnesia, by comparison, is well circumscribed, and he performs well on many such tests (Milner 1963, Milner et al 1968).

A particularly clear example of how the memory impairment associated with the Korsakoff syndrome is different from other examples of amnesia comes from a study of proactive interference. Proactive interference (PI) refers to the interfering effects of having learned a first task on the learning of a second task. Like normal subjects, patients with Korsakoff syndrome exhibit a gradual decline in recall due to PI when learning successive groups of words that all belong to the same category (e.g. animal names). Normal subjects, but not patients with Korsakoff syndrome, exhibit an improvement in recall (or release from PI) when words are presented that belong to a new category (e.g. vegetable names) (Cermak et al 1974).

Recently, these results have been placed in clearer perspective by the finding that failure to release from PI is a sign of frontal lobe dysfunction, with no obligatory link to amnesia (Moscovitch 1981). Patients with material-specific memory disorders (Milner 1968), who had sustained left or right unilateral temporal lobectomy, exhibited normal release from proactive inhibition. Moreover, patients who had sustained surgical removal of portions of the frontal lobe, and who were not amnesic, failed to release from proactive inhibition. These findings suggest that frontal lobe dysfunction determines some of the features of the Korsakoff syndrome. They also indicate the value of comparisons between amnesic groups for understanding the nature of memory dysfunction. Clearly, generalizations about amnesia or the amnesic syndrome are not appropriate when speaking about one kind of amnesia.

A final example that makes the same point comes from studies of information-encoding by patients with Korsakoff syndrome, case N. A., and patients receiving ECT (Cermak & Reale 1978, Wetzel & Squire 1980). An orientation procedure was employed to assess how well these patients were capable of three kinds of encoding (Craik & Tulving 1975). By this procedure, encoding during learning is controlled by the experimenter through orienting questions that direct the subject's attention to the superficial appearance of a word, to its sound, or to the semantic category to which it belongs. On such a task, normal subjects exhibit superior recall of semantically encoded words. Case N. A. and patients receiving ECT exhibited the normal superiority of recall for words that had been encoded semantically; and their retention scores, though considerably lower than normal, paralleled the normal pattern of performance. Patients with Korsakoff syndrome failed to exhibit superior retention of semantically encoded words.

These findings are relevant to ideas about the nature of the disorder in amnesia. For example, theories that emphasize a semantic encoding deficit (e.g. Butters & Cermak 1980) seem applicable only to the Korsakoff syndrome. Since N. A.'s amnesia (which did not include this semantic encoding deficit) is thought to result from damage in the region of the dorsomedial thalamic nucleus (Squire & Moore 1979), and since this nucleus is also prominent in the neuropathology of Korsakoff syndrome (Victor et al 1971), some characteristics of the amnesia exhibited in Korsakoff syndrome must reflect lesions outside of the dorsal thalamic region. The possible contribution of frontal lobe dysfunction to the semantic encoding deficit described above has not been evaluated, but it is well known that the frontal lobes have a role in the maintenance and flexible use of central "sets" (Teuber 1964, Luria 1960). The term "set" refers to a temporary domination of behavior by a particular strategy or hypothesis or by the direction of attention to a particular aspect of the environment.

The evidence reviewed so far has tended to place the Korsakoff syndrome apart from other types of amnesia, primarily because this syndrome is associated with various cognitive deficits that occur together with amnesia and in some cases seem to determine its character. Thus, one could suppose that a common core of amnesia does exist in all instances of the amnesic syndrome, and that this core can sometimes be obscured by the effects of additional lesions and superimposed cognitive deficits. In the next section, however, I suggest that the basic amnesic disorder can take different forms.

Two Forms of Amnesia

Recent studies of forgetting provide strong evidence that there are at least two distinct amnesic syndromes and that diencephalic amnesia and bitemporal amnesia are fundamentally different. This idea was first advanced by Lhermitte & Signoret (1972) and is supported by recent findings (Huppert & Piercy 1978, 1979). In brief, these studies used a procedure designed to minimize the problem of comparing forgetting rates in two groups that already differ in levels of initial learning. For learning, patients viewed 120 colored slides. Retention was assessed by a yes/no recognition procedure at three different intervals after learning (10 min, 1 d, and 7 d). Forty new slides and 40 of the original slides were presented at each interval. To equate retention at 10 min after learning, amnesic patients were given longer to view the slides during learning than control subjects ($4x$ or $8x$ for Korsakoff patients, 15x for case H. M.). Having equated performance at 10 min after learning, it was then possible to ask whether the rates of forgetting during the next seven days were normal or abnormal. Huppert & Piercy found that patients with Korsakoff syndrome exhibited a normal rate of forgetting for material they were able to learn, whereas H. M. exhibited an abnormally rapid rate of forgetting. Yet, H. M. required more exposure during learning to raise his 10 min retention score to control levels than did the Korsakoff patients, presumably because his amnesia was more severe than that of the Korsakoff patients. Accordingly, one might worry that severity of amnesia is related to forgetting rate.

This possibility, however, has been ruled out by more recent work that extends the analysis of forgetting rates to case N. A., patients receiving ECT, and to the San Diego Korsakoff population (Squire 1981b). All amnesic patients viewed the material to be learned for a total of 8 sec. Despite similarity in the level of retention achieved at 10 min after learning, Korsakoff patients and case N. A. exhibited a normal rate of forgetting during the next 32 hr, and patients receiving ECT forgot at an abnormally rapid rate. These findings support the idea that diencephalic amnesia, of which Korsakoff patients and case N. A. are clear examples, is character-ized by a normal rate of forgetting. By contrast, bitemporal amnesia of which H. M. is an example, and the amnesia associated with ECT are characterized by rapid forgetting. Although it is obviously not possible to speak with certainty about the anatomy of the amnesia associated with bilateral ECT, the sensitivity of the hippocampal formation to seizures suggest that this amnesia might primarily reflect medial temporal dysfunc-tion (Inglis 1970).

SUMMARY   The experimental study of amnesia has raised the possibility that diencephalic and bitemporal amnesia are distinct entities, and supports the related idea that these two brain regions contribute in different ways to normal memory functions. Patients with Korsakoff syndrome and case N. A. appear to belong to one category, and case H. M. and patients receiving ECT belong to another. This conclusion seems particularly strong in view of the fact that patients with Korsakoff syndrome, who have a variety of cognitive deficits in addition to amnesia, nevertheless had a normal forget-ting rate; whereas H. M. and patients receiving ECT, who exhibit rather circumscribed amnesias by comparison, had abnormal forgetting rates. In addition, the Korsakoff syndrome appears to include features not shared by these other examples of amnesia. These ideas support a scheme for classify-ing amnesia based on behavior and on the presumed nature of the underly-ing neuropathology, a proposal that can be rigorously tested as additional examples of amnesia are submitted to an analysis of forgetting and as animal models of human global amnesia are perfected (Thompson 1981, Mishkin et al 1981, Squire & Zola-Morgan 1982). This discussion of how amnesias are similar and different sets the stage for considering the nature of the underlying deficit and the role of the affected brain regions in normal memory functions. I next consider these issues.

The Nature of the Deficit

Understanding the basic memory defect in amnesia is fundamental to the neurology of memory. Yet, after decades of experimental work, disagreement remains about how best to characterize it. In the next section I consider some reasons that these matters have been so hard to settle and I summarize the current literature.

BACKGROUND: STORAGE VS RETRIEVAL   Usually amnesia has been considered to reflect a disorder in some particular stage of information processing, such as storage, consolidation, or retrieval. This kind of analysis has been applied to amnesia in man (Squire 1980a, Wicklegren 1979) and in laboratory animals (Miller & Springer 1973, Gold & King 1974). Both efforts, following relatively separate courses, have become mired in the same polarization of opinion: Is amnesia a deficit of storage or retrieval?

This issue is connected in turn to a long-standing debate about normal brain function: Is material permanently stored in memory such that normal forgetting reflects only a change in accessibility or retrievability, or is forgetting actually associated with loss of information from the brain? Commitment to the idea of permanent memory storage leads quite naturally to thinking about memory dysfunction in terms of retrieval failure. Popular ideas about hypnosis, repression, and psychoanalytic theory have apparently led the majority of individuals to believe that most, if not all, memories are stored permanently (Loftus & Loftus 1980). This idea has deep historical roots in Freud's view that "in mental life nothing which has once been formed can perish" and that forgetting is motivated in part by repression (Freud 1930).

In fact, it is simply not yet possible to determine directly whether the neural substrate of mammalian information storage is maintained permanently or whether it dissipates with time. Disagreement will probably remain about this fundamental issue until decisive biological evidence can be obtained. Nevertheless, it is worth pointing out that the bias for considering representation of memory in the brain to be permanent has been reviewed recently from the perspective of cognitive psychology (Loftus & Loftus 1980), with the conclusion that there are no psychological data that would require such an idea. The common experience that we have available in memory far more than we are able to recall on any particular occasion should not convince us that all past experiences are permanently represented in storage.

The issue need not be so simple as whether a particular past experience is or is not in memory storage. Memory of a particular face, word, or event could dissipate from storage but could nevertheless influence behavior in a less literal, but long-lasting way by virtue of having been incorporated into more generic representations or schemata (Bartlett 1932, Rumelhart & Norman 1978). Schemata are knowledge structures that constitute procedures for interpreting new information and operating in the world. An experience of a particular face might leave no neural trace that would aid later recognition of the same face; yet it could tune or elaborate neural mechanisms (schemata) specialized for facial recognition.

Not only does there not seem to be any compelling reason for supposing that memory storage in the brain is always permanent, but available information about the nervous system's capacity for plasticity makes it easy to envision ways that some information could be erased or absorbed with time. For example, areas of mammalian cortex are capable of considerable reorganization in dendritic architecture as a function of experience during adulthood (Greenough et al 1979, Juraska et al 1980), and in man growth of these elements normally continues throughout life (Buell & Coleman 1979). During development, where synaptic reorganization has been best studied, growth and elimination of synaptic elements appear to follow a principle of competition, whereby loss of elements leads to an increase in innervation by surviving elements (Purves & Lichtman 1980). This principle might also apply to the changes produced after visual deprivation, and to certain aspects of memory and forgetting (Squire & Schlapfer 1981). These dynamic features of the nervous system provide a way of understanding how the neural substrates of information could change with time. Finally, and perhaps most relevant, in invertebrates, where the cellular correlates of information storage for simple forms of behavioral plasticity can be investigated rather directly, the relevant neural changes gradually disappear over a period of days and weeks (Kandel 1976).

Yet the difficulty in discovering the nature of amnesia reflects more than an historic debate about the permanence of memory storage; it reflects confusion over the use of terms like "storage" and "retrieval." Consider the problem of distinguishing a storage deficit from a retrieval deficit. Any deficit in the initial storage of information will necessarily result in a deficit in its retrieval. Moreover, unless a deficit can be fully reversed, any improvement in recall after special cueing or prompting procedures can be interpreted either as evidence for the reduction of a retrieval deficit or as evidence for incompletely stored information that can be expressed better when prompts or cues are provided. Notwithstanding these difficulties, some progress has been made in understanding amnesia by experiments framed around storage and retrieval issues.

RETRIEVAL EXPLANATIONS   The view that amnesia might reflect a deficit in retrieval of information that had been adequately stored seems to have developed as a way of understanding two prominent features of amnesia:

1. A variety of cues and other techniques seem remarkably effective in eliciting otherwise unavailable information from amnesic patients.
2. When recalling successive lists of words, many amnesic patients make intrusion errors, i.e. they produce words presented on previous lists, directly demonstrating some storage of recent information.

Facilitatory effects on the retention of amnesic patients have been demonstrated with a variety of techniques, including fragments of pictures or words, the initial letters of words, yes-no recognition, and two-choice recognition. Many of these procedures prove to be very effective in improving the performance of amnesic patients. Yet, the critical issue is not whether performance can be improved at all by such techniques but whether they disproportionately improve the performance of amnesic patients, or are simply effective ways to elicit information from all subjects, amnesic patients and normal subjects alike.

The evidence suggests that cues improve the performance of all subjects (Squire 1980a, Squire & Cohen 1982). Tests based on fragmented drawings have been given to H. M. (Milner et al 1968), and tests involving either fragmented drawings, fragmented words, or the initial letters of words have been given to patients with Korsakoff syndrome (Warrington & Weiskrantz 1968, Weiskrantz & Warrington 1970). These patients exhibited considerable retention over intervals of one hour or more, but nevertheless failed to attain the level of performance exhibited by control subjects. The control subjects also benefitted from these cues and maintained a marked advantage over the amnesic patients.

In verbal learning tests, patients receiving bilateral ECT exhibited a marked impairment in retention even when tested by procedures known to benefit performance: yes-no recognition, two-choice recognition, or cueing by the initial three letters of a word (Squire et al 1978). In addition, case N. A. improved his recall for past public events when his task was to discriminate true and false statements about these events (Squire & Slater 1977). But control subjects improved also and maintained their advantage over him. Thus, recognition is easier than recall for amnesic patients as well as for controls. "Like other men and women [amnesics] too are more likely to succeed in recognition than in unaided recall, but in all tests of memory their capacity and reliability are abnormally small" (Talland 1965, p. 231). Taken together, the results with cues of various types provide no basis for supposing that amnesia reflects a failure to retrieve information that is adequately stored.

Another body of work sometimes taken in support of a retrieval explanation of amnesia concerns the finding that amnesic patients, when trying to recall a list of words, sometimes produce words learned on previous lists. Originally, this finding had been interpreted in the context of a retrieval explanation of amnesia to mean that response competition at the time of retrieval interferes with performance (Warrington & Weiskrantz 1974). However, this view has encountered difficulties and is now regarded as inadequate (Warrington & Weiskrantz 1978). For example, for patients with Korsakoff syndrome, retention scores and frequency of intrusions from a just learned list of words can vary independently of each other (Kinsbourne & Winocur 1980). Whatever the correct interpretation of intrusion errors, it should be emphasized that the study of these errors has been limited to the Korsakoff syndrome. Since patients with Korsakoff syndrome have cognitive deficits superimposed on amnesia, including symptoms of frontal lobe dysfunction that affect their performance on some memory tests (Moscovitch 1981), it is reasonable to wonder how typical intrusion errors are in amnesia. Moreover, since frontal lobe damage can cause perseverative errors (Milner 1963), questions can be raised even in the case of the Korsakoff syndrome as to whether intrusion errors constitute a part of the amnesia or reflect superimposed frontal lobe pathology.

STORAGE OR CONSOLIDATION EXPLANATIONS   Another body of work has suggested that amnesia might reflect a deficit of information input or retention such that memory is not encoded normally during learning or is not consolidated or elaborated normally with the passage of time. One hypothesis, applied to the Korsakoff syndrome, is that amnesia reflects an encoding deficit (Butters & Cermak 1980). This view holds that the patient with Korsakoff syndrome has difficulty engaging in deeper (e.g. semantic) levels of information processing and as a result develops a representation of information based on superficial analysis.

Although patients with Korsakoff syndrome can differ from normal subjects, and even from other amnesic patients, with respect to their ability to use semantic information (Wetzel & Squire 1980, Cermak & Reale 1978, Wetzel & Squire 1981), it is also clear that such a deficit does not always appear, especially when the task is simplified (Cermak & Reale 1978, Mayes et al 1978, Mayes et al 1980). The results of information-encoding studies thus provide a mixed picture of the abilities of the Korsakoff patient. In any case, Korsakoff syndrome affects the speed and efficiency of information processing as measured by divided attention and concept formation tasks (Glosser et al 1976, Oscar-Berman 1973)—cognitive deficits that could be expected to affect the ability to register information in a normal way. These findings and others have led Huppert & Piercy (1978) to argue compellingly that the Korsakoff syndrome involves at least a deficit in an early stage of information processing.

In addition to an encoding deficit, recent studies also raise the possibility that patients with Korsakoff syndrome in particular might also have cognitive deficits that could affect their ability to reconstruct information from memory in some circumstances. In one study, patients exhibited impaired retrieval from semantic memory (Cermak et al 1978). In a second study, patients exhibited impaired memory search strategies (Cermak et al 1980). These findings complicate attempts to define the deficit in Korsakoff syndrome in any simple way. They could have some specific cognitive deficit that affects both encoding and retrieval, as suggested by Kinsbourne & Winocur (1980), or their encoding and reconstructive deficits could be separate disorders.

Although the bulk of experimental work on anterograde amnesia has involved the Korsakoff syndrome, useful information about the nature of the deficit has also come from study of case H. M., case N. A., and patients receiving bilateral ECT. Whereas work with the Korsakoff syndrome has identified deficits in information-encoding that selectively interfere with deeper levels of processing, and possibly with the reconstruction of information, study of H. M. and patients receiving ECT has suggested that their amnesia reflects a deficit limited to the formation of new memory.

This view can be illustrated by a study of anterograde amnesia and memory for temporal order (Squire et al 1981). Amnesic patients are frequently confused about when an event occurred, although they might remember something about the event itself. This could mean that information about temporal order was represented weakly in storage, or that it was stored but particularly difficult to retrieve. In the study, case N. A. and ECT patients were tested after a short learning-retention interval. Control subjects were tested at various learning-retention intervals to determine when in the course of normal forgetting they were as poor as the patients at remembering two lists of 12 simple sentences. At this stage of forgetting amnesic patients were no worse than the control subjects at remembering whether sentences belonged to the first or second list. The same kind of result has been obtained for partial information (Squire et al 1978, Mortensen 1980), retention of semantically encoded words (Wetzel & Squire 1980), and cued recall of words (Wetzel & Squire 1981). In all cases the pattern of performance exhibited by amnesic patients (N. A. or patients receiving ECT) was recapitulated by normal subjects during the course of normal forgetting. These results seem easiest to understand by supposing that these amnesias result in a weak representation in memory. A retrieval explanation is not needed to understand such results, except in the sense that such an explanation would apply to normal forgetting.

SUMMARY   This consideration of the nature of anterograde amnesia was begun with the demonstration that there appear to be two classes of amnesia —one associated with a normal rate of forgetting and one associated with an abnormally rapid rate of forgetting. These two forms of amnesia would appear to require different explanations to account for the underlying disorder. The evidence just reviewed is consistent with the proposal that diencephalic amnesia reflects a deficit in the initial encoding of information, together with a normal retentive ability for information that can be acquired. In this respect the two kinds of diencephalic amnesia considered here, case N. A. and the Korsakoff patient, are similar. However, the comparatively widespread neuropathology of Korsakoff syndrome may be responsible for two features of the syndrome not observed in case N. A. and which seem to set the Korsakoff syndrome apart from some of the other amnesias:

1. A qualitative deficit in encoding that affects deeper, semantic, more elaborative aspects of information processing.
2. Reconstructive deficits and an extensive, possibly related, impairment in remote memory (Squire & Cohen 1981, Cohen & Squire 1981).

Patients receiving ECT and case H. M., who exhibit an abnormally rapid rate of forgetting, may be deficient in post-encoding processes involved in the consolidation or elaboration of memory. This idea was put forward several years ago to account for H. M.'s amnesia (Milner 1966). If a system involved in such a process were damaged, it seems reasonable that the deficit would be larger with an increasing learning-retention interval. This view of the amnesias, derived from analysis of anterograde amnesia, emphasizes a deficiency in initial registration or consolidation rather than in retrieval. Even stronger support for this view comes from study of retrograde amnesia: the loss of memory for events that occurred prior to the onset of amnesia.

RETROGRADE AMNESIA AND REMOTE MEMORY IMPAIRMENT

As recently as the early 1970s, opinion was divided about the nature of retrograde amnesia. On the one hand, clinical descriptions of case H. M. had consistently stated that this individual exhibits a brief retrograde amnesia covering a period of perhaps 1 to 3 yr prior to the onset of his amnesia in 1953 (Scoville & Milner 1957, Milner 1966). Such clinical descriptions are not unusual, having been recorded for two other patients who sustained medial temporal resections (Penfield & Milner 1958) and for patients experiencing severe head injury (Russell & Nathan 1946). On the other hand, the first formal assessment of remote memory functions in amnesia, involving a mixed group of three Korsakoff patients and two other patients, revealed an extensive remote memory impairment affecting the entire period of time covered by the test, 1930–1970 (Sanders & Warrington 1971). These two views of retrograde amnesia turn out to be reconcilable in that they apply to etiologically distinct forms of the amnesic syndrome.

Premorbid vs Postmorbid Memory: Cases H. M. and N. A.

Formal testing of remote memory in two well-circumscribed cases of chronic memory dysfunction (case H. M. and case N. A.) has demonstrated that premorbid memory can be less affected than postmorbid memory and has confirmed clinical descriptions of these cases. H. M. performed as well as control subjects in recognizing famous faces from his premorbid period, 1920–1950, but identified fewer than 20% of the faces of individuals who became prominent after his surgery in 1953 (Marslen-Wilson & Teuber 1975). The status of N. A.'s remote memory has been evaluated with tests of past public events, former one-season television programs, and famous faces (Squire & Slater 1978, Cohen & Squire 1981). As assessed by six of seven tests, his remote memory appears unaffected for the period before his accident in 1960.

Brief Retrograde Amnesia: Temporally Limited Retrograde Amnesia

Until recently, the idea that retrograde amnesia could affect memories acquired as much as a few years previously rested entirely on clinical impressions of the sort described above. There has even been reason to wonder about the accuracy of these impressions. First, the findings of Sanders & Warrington (1971) raised the possibility that when retrograde amnesia does occur it is extensive rather than temporally limited. Second, a larger literature concerning experimental amnesia in laboratory animals had suggested that graded retrograde amnesia is typically brief (minutes or hours) and that long-term memory is rather invulnerable to amnesia (McGaugh & Herz 1972, Squire 1975). Accordingly, in trying to understand the clinical descriptions, some forcibly argued that reports of temporally limited retrograde amnesia extending across a year or more might be explained by sampling bias (Coons & Miller 1960). When interviewing a patient about his past memories, questions about remote events tend to be more general and to cover a broader period of time than questions about recent events. Loss of memory for recent events is therefore easier to detect than loss of memory for more remote events.

A series of studies of patients receiving bilateral ECT, using specially designed tests based on former one-season television programs (Squire & Slater 1975, Squire & Fox 1981), has established that temporally limited retrograde amnesia is real and not an artifact of sampling bias. Patients were tested prior to the first treatment of their series and then again an hour or longer after the fifth treatment. Before ECT, patients performed similarly to control subjects. After ECT, memory was selectively affected for events that occurred just a few years prior to ECT but was normal for events that occurred prior to that time (Squire et al 1975, Squire et al 1976, Squire & Cohen 1979). A temporally limited retrograde amnesia has also been demonstrated using a test that asked patients to recall details about past public events (Cohen & Squire 1981). These retrograde amnesias gradually resolve during the months following treatment (Squire et al 1981). These findings, as well as those cases in which temporally limited retrograde amnesia is not observed following ECT, are considered more fully elsewhere (Cohen & Squire 1981).

These findings have interesting implications for the neuropsychology of memory. That memory can be lost for events of the past one to two years without affecting more remote memories has been taken as confirmation of Ribot's Law, that in the dissolution of memory the "new perishes before the old" (Ribot 1882). This finding also implies a normal process by which memory gradually becomes more resistant to disruption with the passage of time.

This phenomenon of increasing resistance could constitute the basis for gradual reorganization and restructuring in memory (Norman & Rumelhart 1975) and for the development of schemata (Bartlett 1932) that have been suggested to appear in normal memory with the passage of time. These considerations suggest that there develops gradually after learning a representation of the original experience that has lost detail through forgetting but that has become reorganized, schematized, and more resistant to disruption. Freud also proposed a relationship between forgetting and constructive changes in the representation of information. "Normal forgetting takes place by way of condensation. In this way it becomes the basis for the formation of concepts" (Freud 1901).

Little or nothing is known about the neurobiological events that might underlie these gradual changes in the development of memory. The nervous system is capable of gradual morphological change in response to rearing in enriched environments or to daily maze training (Greenough et al 1979, Juraska et al 1980, Rosenzweig 1979, Bennett et al 1979). The relationship between forgetting and resistance is formally similar to the competition that can be observed during development of the nervous system (Purves & Lichtman 1980). In the case of development, as inputs to a cell are eliminated, the remaining inputs increase their influence on that cell. In the case of memory, as some information is forgotten, the remainder becomes more resistant to disruption.

Another implication of these findings is that permanent memory storage normally occurs outside the brain regions affected in amnesia. It appears that these brain regions constitute an essential neuronanatomical substrate for the formation of new memories and for their maintenance and elaboration after learning. The finding with objective tests that temporally limited retrograde amnesia can cover a time span of a few years also corroborates clinical impressions that case H. M. has retrograde amnesia for one to three years prior to his surgery. Taken together, these observations suggest that the medial temporal region plays a role in the development and consolidation of memory and that this role can continue for up to a few years after initial learning.

The findings with ECT, and the facts of brief retrograde amnesia in general, speak directly to questions about the nature of amnesia, as just reviewed in the context of anterograde amnesia. The marked discontinuity between premorbid and postmorbid memory, noted for cases H. M. and N. A., and the finding that the retrograde amnesia associated with ECT can be temporally limited, provide strong evidence against a general retrieval interpretation of these amnesias. A retrieval deficit should involve memory for all past events and not just memory for events that have occurred recently. Thus, for case H. M., case N. A., and the amnesia associated with ECT, the experimental facts lead to the consistent view that these amnesias reflect a deficit in the establishment and elaboration of memory.

It should be noted that retrieval explanations of amnesia (e.g. Warrington & Weiskrantz 1970, Kinsbourne & Wood 1975) derive largely from study of patients with Korsakoff syndrome whose anterograde amnesia is different in certain ways from that of other amnesic patients. The next section illustrates that, unlike the other amnesic patients considered here, patients with Korsakoff syndrome also exhibit an extensive impairment of remote memory.

Extensive Remote Memory Impairment: The Korsakoff Syndrome

Beginning with the seminal study of Sanders & Warrington (1971) demonstrating that amnesia can affect a large portion of remote memory, the status of remote memory in the Korsakoff syndrome has come to be rather well understood. Patients with Korsakoff syndrome exhibit a severe and extensive impairment of remote memory that affects most of their adult lives. This impairment is temporally graded, affecting memory of recent periods of time to a greater extent than more remote periods, and has been consistently demonstrated in different patient populations using a variety of remote memory tests (Marslen-Wilson & Teuber 1975, Seltzer & Benson 1974, Albert et al 1979, Cohen & Squire 1981, Squire & Cohen 1981, Meudell et al 1980).

In considering this deficit, the term "remote memory impairment" is preferable to "retrograde amnesia." The Korsakoff syndrome develops gradually so that it is difficult to know what portion of memory for past events reflects anterograde amnesia and what portion reflects retrograde amnesia. Nevertheless, available information suggests how the combined effects of anterograde and retrograde amnesia might explain the remote memory findings. Anterograde amnesia may help to understand the graded aspect of the impairment. That chronic alcoholic patients have information-processing deficits (Ryan et al 1980, Parker & Noble 1977) and that they exhibit impairment on the more recent questions of remote memory tests (Cohen & Squire 1981, Albert et al 1980) suggest that with continued drinking alcoholics should gradually lose ground to nonalcoholic control subjects with respect to their ability to recall recent public events. Accordingly, after years of alcohol abuse the patient eventually diagnosed as having Korsakoff syndrome would be expected to demonstrate an amnesia for public events that is more severe for recent events than for remote events. By this view the gradient of remote memory impairment exhibited in Korsakoff syndrome reflects progressive anterograde amnesia.

The presence of certain cognitive deficits in Korsakoff syndrome might explain the extensiveness of the remote memory impairment. These patients have difficulty in tests of concept formation, problem solving, and rapid switching of cognitive set (Talland 1965, Butters & Cermak 1980), and recently they have been reported to have difficulty in certain tasks that require retrieval from semantic memory (Cermak et al 1978). Although the relationship between remote memory functions and other cognitive skills is far from clear, it seems reasonable that impairment involving retrieval from semantic memory or the deficient use of problem-solving strategies could affect remote memory and contribute to its impairment across all time periods.

Taken together, these considerations suggest that the extent of remote memory impairment in Korsakoff syndrome and its graded pattern reflect the operation of two factors:

1. Progressive anterograde amnesia, which exerts a greater effect on memory for events from recent years than events from remote years, and

which could explain the temporally graded pattern of remote memory impairment.

2. Cognitive deficits that affect in a uniform way the ability to reconstruct past memories, and which could explain the extensiveness of remote memory impairment.

Several lines of evidence indicate that the remote memory impairment, as observed in Korsakoff syndrome, is a distinct entity dissociable from and unrelated to anterograde amnesia. The patient H. M., for example, has profound anterograde amnesia for events that have occurred since his surgery in 1953, without measurable loss of remote memory from his premorbid period. In one study comparing H. M. and patients with Korsakoff syndrome on a remote memory test of famous faces (Marslen-Wilson & Teuber 1975), H. M.'s score for the 1950s and 1960s period was worse than that of patients with Korsakoff syndrome, but his score for the 1930s and 1940s was better. In another comparison involving case N. A., patients tested 1 to 2 hr after their fifth bilateral ECT, and patients with Korsakoff syndrome, all groups scored comparably on four tests of new learning capacity; yet the Korsakoff patients, and not the other patients, scored poorly on seven different tests of remote memory (Cohen & Squire 1981). Accordingly, the remote memory loss of Korsakoff syndrome is not inextricably coupled to anterograde amnesia.

This view of remote memory dysfunction as a distinct entity dissociable from anterograde amnesia in no way contradicts the common view that anterograde amnesia is closely related to brief, temporally limited retrograde amnesia. Indeed, a link between these two entities has been amply demonstrated in studies of experimental amnesia in animals (McGaugh & Herz 1972), and traumatic amnesia in man (Russell & Nathan 1946), and forms the basis of recent discussions of memory and brain function (e.g. Wicklegren 1979). More severe anterograde amnesia appears to be correlated with more prolonged retrograde amnesia. If the facts of retrograde amnesia are taken to imply a normal process whereby recently acquired memories gradually become resistant to disruption for a period up to a year or two after learning, then it is easy to imagine why a correlation between the severity of anterograde amnesia and the temporal extent of retrograde amnesia should be observed. Any brain injury that interfered with the consolidation and elaboration of memory would result in both a deficit in the formation of new, enduring memories (anterograde amnesia) and a deficit in recently acquired memories that were undergoing consolidation and elaboration when the injury occurred (brief retrograde amnesia).

Summary

The available data concerning retrograde amnesia and remote memory dysfunction now appear to resolve much of the confusion that has clouded these issues. Brief retrograde amnesia and extensive remote memory dysfunction appear to be distinct entities. Brief retrograde amnesia has been demonstrated most clearly in the amnesia associated with ECT, but it seems reasonable to suppose that it is present as well in case H. M. Extensive remote memory impairment has been clearly demonstrated in the Korsakoff syndrome. Remote memory dysfunction is presumably related to the involvement of brain regions in addition to those that are affected in more circumscribed amnesias (e.g. the medial temporal region for case H. M. and the dorsal thalamus for N. A.) and might be related to certain cognitive deficits superimposed on amnesia. Based on the available data, one might conjecture that for amnesias presumably caused by temporal lobe dysfunction, remote memory impairment implies damage in areas beyond the medial temporal region. The post-encephalitic patient who exhibits remote memory impairment (Albert et al 1980) in the context of temporal lobe and frontal lobe damage might be an example of this circumstance. For diencephalic amnesia, one might conjecture that severe remote memory loss implies damage in areas beyond the dorsal thalamus. The Korsakoff patient is presumably an example of this circumstance.

PRESERVATION OF LEARNING AND MEMORY IN AMNESIA

The fact that amnesic patients seem to perform well under some conditions has generated considerable interest because of the clear relevance of such findings to the nature of amnesia and the organization of normal memory.

Perceptual-Motor Skills

The best known examples of preserved learning and memory lie in the domain of perceptual-motor skills. Case H. M., for example, exhibited progressive learning of a mirror-tracing task across three days of testing, despite reporting on each day that he had no memory of having performed the task before (Milner 1962). Similarly, H. M. (Corkin 1968) as well as post-encephalitic patients (Brooks & Baddeley 1976) and patients with Korsakoff syndrome (Cermak et al 1973) were able to learn and remember over days, sometimes at a rate comparable to that of control subjects, the hand-eye coordination skills needed for a pursuit-rotor task. This task requires the tracking of a revolving target with a hand-held stylus. In the last few years, additional testing procedures have been cataloged, like those involving fragmented drawings or certain kinds of cues, that also can elicit signs of retention in patients who are by other indications profoundly amnesic (Weiskrantz 1978).

There have been two general views of these matters. On the one hand, the finding that patients can sometimes demonstrate learning or retention behaviorally without reflecting it in their verbal reports has suggested that amnesic patients do not have access to their memories. By this view the ability to demonstrate certain kinds of learning and retention does not require this sort of access, and amnesia is fundamentally a retrieval deficit (Weiskrantz 1978). On the other hand, amnesia has been considered to reflect essentially a deficit in the formation of memory. This view recognizes that amnesic patients may perform well or even normally at certain tasks, but takes these data as a suggestion that different forms of memory are differently organized in the nervous system and that the amnesic syndrome affects some forms but not others. The brain regions damaged in amnesia, while necessary for many or most kinds of learning and memory, may not be required for certain other kinds. Consistent with this view is the recent work that has identified a domain of information processing that is spared in amnesia (Cohen & Squire 1980, Cohen 1981). This work also sheds light on the important observation that patients can demonstrate certain memory capabilities without having explicit access to their memory via verbal report (Weiskrantz 1978).

Declarative and Procedural Knowledge

Normal subjects can learn to read geometrically inverted or otherwise transformed text and can retain such pattern-analyzing skills for months (Kolers 1979). Amnesic patients (case N. A., patients with Korsakoff syndrome, and patients receiving ECT) and control subjects were asked to read sets of words that were reversed by a mirror. The amnesic patients improved their skill at this task at a normal rate over a period of three days and retained the skill at a normal level three months later. This occurred despite amnesia for aspects of the testing situation and despite profound amnesia for the specific words that had been read (Cohen & Squire 1980).

This finding suggests a distinction between information that is based on rules or procedures and information that is based on specific items or data. Thus, amnesic patients can learn the procedures needed for acquisition and retention of mirror-reading skills, but cannot remember the specific data, i.e. the words, that result from applying these procedures. Procedural learning, in contemporary psychology (Rumelhart 1981, Cohen 1981), is considered to result in modification or tuning of existing schemata (Bartlett 1932) (processes specialized for the interpretation of environmental events and for operating in the world). In the case of mirror-reading, schemata involved in reading can apparently be modified in long-lasting ways.

This distinction was developed in the artificial intelligence literature (Winograd 1975) and is similar to previous distinctions concerning the representation of knowledge: knowing how and knowing that (Ryle 1949); habit memory and pure memory (Bergson 1910); memory without record and memory with record (Bruner 1969). The evidence now suggests that such a distinction is honored by the nervous system, and that these two domains of information depend on fundamentally different kinds of neurological organization. Although this distinction may not permit all tasks to be neatly dichotomized, it should be useful in predicting what is affected or spared in amnesia.

The possible usefulness of the declarative-procedural distinction to questions about the neurological organization of memory has recently been developed in some detail (Cohen 1981). Thus, the ability to develop and store declarative memory (all the bits of specific-item information that are the subject of conventional memory experiments like faces, words, and shapes) depends on the integrity of the particular bitemporal and diencephalic brain structures affected in amnesia. By contrast, motor skills and mirror-reading are taken as examples of procedural learning, which can occur in a normal way in the absence of these particular brain structures. Moreover, procedural knowledge is considered by its nature to be implicit —accessible only by engaging in or applying the procedures in which the knowledge is contained. Many skills, like playing golf or tennis, proceed despite poor access to "the specific instances that led to the perfection of the [skill]." These instances "change the rules by which [one] operates, but are virtually inaccessible in memory as specific encounters" (Bruner 1969, p. 254). The critical feature that is procedural here is that there can develop in memory a representation based on experience that changes the way an organism responds to the environment, without affording access to the specific instances that led to this change. Accordingly, procedural learning applies to more than just the acquisition of motor skills.

The neural systems specialized for interpreting and operating in the environment that subserve procedual learning may be phylogentically more primitive than those subserving declarative learning. The acquisition of declarative knowledge is dependent on specific structures in the bitemporal and diencephalic regions. Procedural learning appears to be independent of these brain structures and, since it is considered to involve changes in existing schemata, may be intrinsic to the neural systems in which it occurs. Habituation and sensitization, which can be demonstrated in invertebrates possessing relatively simple nervous systems (Kandel 1976), would appear to be examples of what has been termed procedural learning. Modification of synaptic efficacy can occur in existing networks, resulting in a specific change in how the organism operates in the environment. However, memory for the specific instances that cumulated in this change, i.e. declarative knowledge, does not seem to be necessary in these cases.

Indeed, it is an intriguing possibility that the capacity for declarative learning emerged relatively late in evolution, perhaps as greater mobility and longer life spans made it important to maintain a (declarative) record of time and place information. Some forms of classical conditioning, as recently described in *Aplysia* (Walters et al 1979), might also reflect procedural learning. It is interesting that two amnesic patients (one post-encephalitic and one with Korsakoff syndrome) were able to acquire aversive (eye-blink) conditioning despite being unable, while still seated in front of the apparatus, to describe the learning experience or recount that they had received air puffs to the eye (Weiskrantz & Warrington 1979). Further work is needed to know if this or other forms of classical conditioning can in fact be acquired at a normal rate in amnesia and whether it should therefore be regarded as a preserved capacity.

Semantic vs Episodic Memory and Reference vs Working Memory

It is helpful to contrast the declarative-procedural distinction to two distinctions that have been advanced recently in the context of amnesia and brain function: *semantic* vs *episodic memory* (Kinsbourne & Wood 1975) and *reference* vs *working memory* (Olton et al 1979). It seems unlikely that the deficit in amnesic patients involves episodic autobiographical memory that has a specific time and place, but not semantic memory (context-free knowledge of facts, language, or concepts). Amnesic patients have seemingly equal difficulty acquiring both new episodic information (e.g. the events of a visit to Sacramento) and semantic information (e.g. the capital of California is Sacramento), and their retrograde amnesia applies to both spheres as well (Cohen & Squire 1981).

Likewise, the working memory-reference memory distinction developed from studies with experimental animals does not seem to fit the facts of human bitemporal amnesia. Reference memory, postulated to be spared in the case of hippocampal damage, is that aspect of a task that is constant from trial to trial. Working memory refers to aspects of a task that are useful for only one trial and not for subsequent trials. Thus, the constant features of a task are considered to be protected from the effects of hippocampal damage. In amnesia, however, simple repetition is not sufficient to insure learning (Drachman & Arbit 1966). Moreover, procedural learning is thought to be possible in amnesia because of the special nature of this information, not because such information is repeatedly available.

The notion of spatial maps and hippocampal function, developed also from study of experimental animals (O'Keefe & Nadel 1978), may have relevance to the ideas presented here. Whereas a literal version of this proposal does not fit well the findings from human amnesia (Squire 1979, Squire & Zola-Morgan 1982), a more abstract notion of spatial mapping might be made consistent with the idea that the brain regions damaged in amnesia are involved in establishing a specific kind of representation in memory.

Summary

The neuropsychological data, together with ideas founded in the cognitive sciences, suggest a distinction between procedural and declarative knowledge as a way of understanding what is spared and what is not spared in human amnesia. Procedural learning includes motor skills and certain cognitive skills and can proceed normally in amnesia. Declarative learning includes specific facts and data that are the subject of most contemporary memory studies and is impaired in amnesia.

ANATOMY OF AMNESIA

The preceding sections review neuropsychological aspects of diencephalic and bitemporal amnesia. It has also been of considerable interest to identify the specific brain structures which when damaged cause amnesia. Since this topic is the subject of several recent reviews (Brierley 1977, Horel 1978, Mair et al 1979, Squire 1980b), it is considered only briefly here. Most neuroanatomical discussions begin with the simplifying assumption that amnesia results when damage occurs to one of a group of functionally interrelated structures. In this sense the idea that damage in the hippocampal region (Scoville & Milner 1957) or in the mammillary bodies (Brierley 1977) can cause amnesia has been easy to accept because of their close anatomical relationship. Indeed, a functional link between hippocampus, fornix, and mammillary bodies was proposed decades ago (Papez 1937).

Yet the suggestion that diencephalic and bitemporal amnesia might be distinct entities raises the possibility that this functional link might not be so obligatory as was previously thought. Recent anatomical and neuropsychological data support this idea. In the rhesus monkey a substantial projection from the subiculum of the hippocampal formation is directed not only through the fornix but also to a variety of cortical and subcortical structures including amygdala, cingulate gyrus, entorhinal cortex, perirhinal cortex, and medial frontal cortex (Rosene & Van Hoesen 1977). Since the fornix projection to the mammillary bodies need not be viewed as the only significant output of the hippocampal formation, these structures need not be viewed as a unitary anatomical system that either is or is not involved in amnesia. That is, as the connectivity between these regions is relaxed, it becomes easier to understand how lesions in the two regions could lead to different neuropsychological findings. In any case, it is well known that separate lesions in two brain regions, even if they are connected to some degree, produce markedly different patterns of brain disorganization and reorganization (Lynch 1976).

These neuroanatomical considerations notwithstanding, if both these regions were part of a unitary functional system, then damage to the fornix should also cause amnesia. However, in 50 cases that could be identified as having fornix damage, only three involved memory loss (Squire & Moore 1979). Though these data can often be faulted for lack of neuropathological confirmation or neuropsychological testing, it is worth mentioning that the best known case of bilateral fornix damage and memory loss (Sweet et al 1959) had a relatively mild amnesia (a difference between IQ and Wechsler Memory Quotient of only 13 points), compared to the amnesias described in virtually all contemporary studies. Fornix transection in the monkey has been reported to produce a memory deficit (Gaffan 1974), but also has been reported to be without effect on tasks sensitive to amnesia (Moss et al 1980). The discussion that follows accepts as a working hypothesis that diencephalic and bitemporal amnesia are distinct entities and considers what specific brain structures in each region have been linked to amnesia.

Diencephalic Amnesia

Our understanding of this topic comes largely from clinico-pathological studies of Korsakoff syndrome. Recent reviews (Brierley 1977, Mair et al 1979) agree that damage to the mammillary bodies correlates invariably with this syndrome, but uncertainty remains as to which lesions are correlated most strongly with the memory disorder itself. In their influential monograph on the Wernicke-Korsakoff syndrome, Victor et al (1971) identified five cases in their series where brain damage was apparently limited to the mammillary bodies and where memory loss was not observed. They suggested that the dorsomedial thalamic nucleus was the critical structure, since it was damaged in all 38 of their cases who exhibited amnesia. Since the mammillary bodies were also damaged in all these cases, their data are just as consistent with the hypothesis that lesions in both structures are required to cause amnesia.

Two thoroughly studied cases recently available for autopsy (Mair et al 1979) bear on these conclusions. Both these patients had bilateral lesions in the medial nuclei of the mammillary bodies and also a band of gliosis lying bilaterally between the third ventricle and the dorsomedial thalamic nucleus. Uncertainty about nomenclature, together with the indistinct boundaries of the medial thalamic nuclei, were considered to complicate comparisons between these lesions and those described by others. Mair et al (1979) suggested that lesions in both regions might be needed to cause amnesia, or that lesions in mammillary bodies alone might be sufficient if they were large enough.

Unfortunately, the number of cases with adequate neuropsychological and pathological information is not sufficient to permit further resolution of these issues in a definitive way. In one series of 11 cases of Korsakoff syndrome for which neuropathological material was available, each patient was reported on the basis of clinical examination to have amnesia, and each had damage to the mammillary bodies. Thalamic nuclei were involved in some cases and not in others. The dorsomedial thalamic nucleus was damaged in only seven of the 11 cases (Brion & Mikol 1978). Whereas this report appears to favor the view that damage in the mammillary bodies is preeminent in diencephalic amnesia, at least in the case of Korsakoff syndrome, it also appears that damage limited to the region of the dorsomedial thalamic nucleus can be sufficient to cause amnesia. Memory loss, in association with vertical gaze apraxia and decreased alertness, has been reported in patients with infarctions affecting this region (Mills & Swanson 1978). Further, stereotaxic lesions of dorsomedial nucleus caused severe memory disturbances that were reported to subside within one year (Orchinik 1960). However, memory testing here depended on the Wechsler Memory Scale, which can be an insensitive measure of amnesia (Mair et al 1979), and no pathological information was provided. Finally, case N. A. has been shown by CAT scan to have damage in the left dorsal thalamus, in a region corresponding to the position of the dorsomedial nucleus (Squire & Moore 1979).

The available literature does not lead to easy conclusions about the relative roles of the mammillary bodies and dorsomedial nucleus in human amnesia. Perhaps lesions in either region can cause some degree of amnesia, and lesions in both regions cause more severe amnesia. In any case, the traditional view that damage to the fornix-mammillary system causes amnesia seems less secure than it once was, while damage to the dorsal thalamus seems able to cause amnesia.

The effects of dorsomedial nucleus or mammillary body lesions in experimental animals seem consistent in a general way with the findings in human patients, but many questions remain. Lesions of the mammillary bodies seems not to affect learning and memory in a global way (see Woody & Ervin 1966 and references therein), though the learning of an alternation task was impaired in rats (Rosenstock et al 1977). Dorsomedial nucleus lesions have been reported to affect memory in cats (Pectel et al 1955) and monkeys (Schulman 1964). These findings may need to be reevaluated in the light of the demonstration that certain kinds of learning can proceed normally in human amnesia (Cohen & Squire 1980). That is, before concluding on the basis of negative evidence that a specific brain structure is not involved in learning and memory, it must be clear that the behavioral tasks used to assess memory are the kind at which human amnesics do not succeed.

Bitemporal Amnesia

Information about bitemporal amnesia comes largely from surgical cases in which portions of the temporal lobes have been removed bilaterally in an effort to relieve intractable epilepsy. As is now well known, the medial temporal region became clearly linked to memory after 1953 when it was discovered that bilateral resection of this region resulted in profound and lasting amnesia (Scoville & Milner 1957). The removals extended posteriorly along the medial surface of the temporal lobes for a distance of approximately 8 cm from the temporal poles and included uncus, amygdala, hippocampal gyrus, and the anterior two-thirds of the hippocampus. The noted case, H. M., was one of two patients to undergo this procedure. Severe amnesia was also observed in two other well-studied cases (P. B. and F. C.), who sustained left unilateral resections, and who had preexisting pathology of the right temporal lobe (Penfield & Milner 1958).

Several lines of evidence have suggested that damage to the hippocampal formation may be responsible for the amnesia in these cases. In five of six cases where bilateral resections extended posteriorly 4.5 to 6 cm so as to include uncus, amygdala, but only anterior hippocampus, the memory loss was not so severe as in case H. M. (Scoville & Milner 1957). [The remaining case, a paranoid schizophrenic (Case D. C.), was considered to have as severe amnesia as H. M. despite this limited removal.] In another case in the same series, where the resection was limited to the uncus and the amygdala, no amnesia resulted (Case I. S.). In addition, in the case of unilateral temporal lobe resections, which are associated with verbal or nonverbal (material-specific) memory deficits, the severity of the deficit was correlated with the extent of involvement of the hippocampal zone (Milner 1974). Finally, the notion that hippocampal damage is critical to the amnesic effects of medial temporal resections seems supported also by neuropsychological study of patients sustaining left or right amygdalotomy (Andersen 1978). These patients exhibited some behavioral deficits but were not amnesic and scored normally on tests of delayed recall.

The interpretation of these findings is complicated to some extent by the lack of a reported case of well-documented amnesia with bilateral damage limited to the hippocampus. All cases with hippocampal damage have damage to other structures; and this includes the surgical cases just reviewed as well as post-encephalitic cases, where lesions also occur in cingulate gyrus and frontal lobes (Hierons et al 1978), as well as cases of vascular disease, anoxia, and degenerative disease (see review by Horel 1978). Accordingly, although one could suppose from the surgical cases that hippocampus is the critical structure, the data are also consistent with the view that amnesia depends on conjoint damage to the hippocampus and the more anterior structures included in the resections, i.e. amygdala and uncus. Recently, these two hypotheses were tested with monkeys who had received separate or combined lateral damage to hippocampus and amygdala (Mishkin 1978). Only monkeys with the combined lesion were severely impaired on delayed nonmatching to sample, a test of the sort at which human amnesics fail. In this task, the monkey first displaces a single stimulus object to find food. Then, after a delay, in this case up to two minutes, the monkey is given a choice between a novel object and the original object. Food is hidden under the novel object. Whether the hippocampus and amygdala contribute in different ways to the ability to perform this task, and whether hippocampal damage alone might be sufficient to cause impairment on some tasks, is not yet clear.

Recently, traditional views of bitemporal amnesia were challenged with the suggestion that amnesia might depend not on damage to hippocampus but on damage to the albal stalk or temporal stem, a band of white matter lying above and in close proximity to the hippocampus (Horel 1978). Horel contended that the temporal stem was necessarily damaged in the anterior approach used in operating on case H. M. Moreover, because of the position of the temporal stem relative to the hippocampus, the more posteriorly a hippocampal lesion is extended, the greater the damage to the temporal stem. This interpretation of bitemporal amnesia has now been directly tested.

Monkeys with lesions of the temporal stem (TS) and monkeys with hippocampal-amygdala (HA) lesions were tested on a delayed nonmatching to sample task, involving delays up to 10 min, and also on a two-choice pattern discrimination (Zola-Morgan et al 1981). Although histological confirmation of these lesions is not yet available, the TS lesions were done under visual guidance. The intended lesion of the temporal stem included 10–15 mm of its anterior-posterior extent using the wall of the lateral ventricle as a visual guide. The HA lesions were done by a combined anterior and lateral approach, designed to avoid damage to the temporal stem. The results were that HA lesions severely disrupted the ability to retain information across a delay, whereas TS lesions had no effect. By contrast, TS lesions severely disrupted pattern discrimination learning, which is consistent with the effects of temporal neocortical lesions (Mishkin

1954, Mishkin & Pribram 1954), whereas HA lesions produced only a mild impairment. These results do not support the recent hypothesis that the temporal stem has a critical role in memory functions. Retarded acquisition of visual discrimination habits in monkeys with temporal stem lesions, together with their normal ability to retain information across a 10-minute delay, suggests that the temporal stem, perhaps by virtue of its connections with temporal neocortex (See Horel 1978, for review), contributes to the ability to process visual information.

The effects of medial temporal lesions in monkeys have often seemed difficult to reconcile with data from studies of human bitemporal amnesia (Weiskrantz & Warrington 1975, Iversen 1976). The demonstration of a domain of learning and memory that is spared in amnesia (Cohen & Squire 1980) may help to bring these two areas of research into agreement. Thus, those tasks in which monkeys with hippocampal-amygdala lesions are not impaired may be those kinds of tasks that are spared in human amnesia. Conversely, deficits do seem to appear in those tasks that involve declarative memory and that are sensitive to human amnesia (Squire & Zola-Morgan 1982).

Summary

In agreement with the neuropsychological findings, available anatomical data from patients with diencephalic or bitemporal amnesia suggest that these amnesias need not result from damage to a single functional system. In the case of diencephalic amnesia, the mammillary bodies and the dorsomedial thalamic nucleus have been implicated, but it is not yet clear which structure deserves the greater emphasis. In the case of bitemporal amnesia, the evidence suggests that the hippocampal formation plays a crucial role in memory functions; the possibility also needs to be considered that the hippocampal formation and amygdala may function conjointly in this regard. An alternative possibility that temporal stem damage is responsible for amnesia now seems quite unlikely.

PERSPECTIVE

Not long ago it was reasonable to think of amnesia as a unitary disorder reflecting damage to a specific, tightly connected neuroanatomical system. Accordingly, memory tended to be viewed rather monolithically as information that could be stored and retrieved so long as this system were intact. The neuropsychological and anatomical facts now tell a richer story in which amnesia may reflect either of two disorders corresponding to damage in the diencephalic or the medial temporal regions; and in which memory normally depends on the separate contributions of both these regions. Of course, in considering the data from any area of inquiry, one always seeks to discover patterns of similarity and difference. Here certain differences have been emphasized between the disorders associated with diencephalic and medial temporal dysfunction with the thought that these differences may point the way to more detailed understanding of how the brain accomplishes memory storage. Appreciation of these differences, however, should not obscure the many similarities that can also be found among the amnesias; nor the fact that to infer the existence of two distinct entities from these similarities and differences is to some extent arbitrary.

The importance to memory of the diencephalic and bitemporal structures affected in amnesia is believed to lie in their role in the establishment of memory at the time of learning and in the consolidation or elaboration of memory for a time after learning so as to permit effective retrieval. It also seems clear that this role is narrower than it once appeared to be, in the sense that it applies to particular domains of learning and memory and not to all domains. Thus, motor skills and certain cognitive skills have been proposed to belong to a class of learning that is termed "procedural." This kind of learning is spared in amnesia and therefore is independent of the diencephalic and medial temporal structures that are affected in amnesia.

Our understanding of memory and its neural substrates is still rudimentary. Yet the experimental work suggests a framework for thinking about these matters that should prove useful in neuropsychological studies of memory as well as in cellular studies of simpler forms of behavioral plasticity. While technological advances in the neurosciences make feasible ever more detailed analysis of the nervous system and its parts, it is useful to remember that to know about the function of the nervous system entails understanding the behavior that it subserves. If these various levels of analysis can all be applied to the problem of behavioral plasticity, i.e. memory, then we can expect to achieve eventually a good understanding of how neural activity can give rise to behavior. And that is an exciting prospect.

ACKNOWLEDGMENTS

My work is supported by the Medical Research Service of the Veterans Administration and by NIMH Grant MH24600-08. I thank Drs. Lynn Nadel, Neal Cohen, and Stuart Zola-Morgan for numerous stimulating discussions of the ideas and data presented here; and Elizabeth Statzer for preparing the manuscript.

Literature cited

Albert, M. S., Butters, N., Levin, J. 1979. Temporal gradients in the retrograde amnesia of patients with alcoholic Korsakoff's disease. *Arch. Neurol.* 36:211–16

Albert, M. S., Butters, N., Levin, J. 1980. Memory for remote events in chronic alcoholics and alcoholic Korsakoff patients. *Adv. Exp. Med. Biol.* 126:719–30

Andersen, R. 1978. Cognitive changes after amygdalotomy. *Neuropsychologia* 16:439–51

Bartlett, F. C. 1932. *Remembering*. Cambridge: Cambridge Univ. Press. 317 pp.

Bennett, E. L., Rosenzweig, M. R., Morimoto, H., Hebert, M. 1979. Maze training alters brain weights and cortical RNA/DNA ratios. *Behav. Neurol. Biol.* 26:1–22

Bergson, H. L. 1910. *Matter and Memory*. Authorized transl. N. M. Paul, W. S. Palmer. London: Allen

Brierley, J. B. 1977. Neuropathology of amnesic states. In *Amnesia*, ed. C. W. M. Whitty, O. L. Zangwill, pp. 199–223, London: Butterworths

Brion, S., Mikol, J. 1978. Atteinte du noyau lateral dorsal du thalamus et syndrome de Korsakoff alcoolique. *J. Neurol. Sci.* 38:249–61

Brooks, D. N., Baddeley, A. 1976. What can amnesic patients learn? *Neuropsychologia* 14:111–22

Bruner, J. S. 1969. Modalities of memory. In *The Pathology of Memory*, ed. G. A. Talland, N. C. Waugh, pp. 253–59. New York: Academic

Buell, S. J., Coleman, P. D. 1979. Dendritic growth in the aged human brain and failure of growth in senile dementia. *Science* 206:854–56

Butters, N., Cermak, L. S. 1980. *Alcoholic Korsakoff's Syndrome: An Information Processing Approach to Amnesia*. New York: Academic. 188 pp.

Cermak, L. S., Butters, N., Moreines, J. 1974. Some analyses of the verbal encoding deficit of alcoholic Korsakoff patients. *Brain Lang.* 1:141–50

Cermak, L. S., Lewis, R., Butters, N., Goodglass, H. 1973. Role of verbal mediation in performance of motor tasks by Korsakoff patients. *Percept. Mot. Skills* 37:259–62

Cermak, L. S., Reale, L. 1978. Depth of processing and retention of words by alcoholic Korsakoff patients. *J. Exp. Psychol.* 4:165–74

Cermak, L. S., Reale, L., Baker, E. 1978. Alcoholic Korsakoff patients' retrieval from semantic memory. *Brain Lang.* 5:215–26

Cermak, L. S., Uhly, B., Reale, L. 1980. Encoding specificity in the alcoholic Korsakoff patient. *Brain Lang.* 11:119–27

Cohen, N. J. 1981. *Neuropsychological evidence for a distinction between procedural and declarative knowledge in human memory and amnesia.* PhD thesis. Univ. Calif., San Diego. 175 pp.

Cohen, N. J., Squire, L. R. 1980. Preserved learning and retention of pattern analyzing skill in amnesia: Dissociation of knowing how and knowing that. *Science* 210:207–9

Cohen, N. J., Squire, L. R. 1981. Retrograde amnesia and remote memory impairment. *Neuropsychologia.* 19:337–56

Coons, E. E., Miller, N. E. 1960. Conflict versus consolidation of memory traces to explain retrograde amnesia produced by ECS. *J. Comp. Physiol. Psychol.* 53:524–31

Corkin, S. 1968. Acquisition of motor skill after bilateral medial temporal lobe excision. *Neuropsychologia* 6:225–65

Craik, F. I. M., Tulving, E. 1975. Depth of processing and the retention of words in episodic memory. *J. Exp. Psychol. Gen.* 104:268–94

Drachman, D. A., Arbit, J. 1966. Memory and the hippocampal complex. *Arch. Neurol.* 15:52–61

Freud, S. S. 1901. *The Psychopathology of Everyday Life*. Stand. Ed. 6, p. 134. London: Hogarth (1960)

Freud, S. S. 1930. *Civilization and Its Discontents.* Stand. Ed. 21, p. 69. London: Hogarth

Gaffan, D. 1974. Recognition impaired and association intact in the memory of monkeys after transection of the fornix. *J. Comp. Physiol. Psychol.* 86:1100–9

Glosser, G., Butters, N., Samuels, I. 1976. Failures in information processing in patients with Korsakoff's syndrome. *Neuropsychologia* 14:327–34

Gold, P. E., King, R. A. 1974. Retrograde amnesia: Storage failure vs retrieval failure. *Psychol. Rev.* 81:465–69

Greenough, W. T., Juraska, J. M., Volkmar, F. R. 1979. Maze training effects on dendritic branching in occipital cortex of adult rats. *Behav. Neurol. Biol.* 26:287–97

Hierons, R., Janota, I., Corsellis, J. A. N. 1978. The late effects of necrotizing encephalitis of the temporal lobes and limbic areas: A clinico-pathological study of 10 cases. *Psychol. Med.* 8:21–42

Horel, J. A. 1978. The neuroanatomy of amnesia: A critique of the hippocampal memory hypothesis. *Brian* 101:403–45

Huppert, F. A., Piercy, M. 1978. Dissociation between learning and remembering in organic amnesia. *Nature* 275:317–18

Huppert, F. A., Piercy, M. 1979. Normal and abnormal forgetting in organic amnesia: Effect of locus of lesion. *Cortex* 15:385–90

Inglis, J. 1970. Shock, surgery, and cerebral symmetry. *Br. J. Psychiatr.* 117:143–48

Iversen, S. D. 1976. Do hippocampal lesions produce amnesia in animals? *Int. Rev. Nuerobiol.* 19:1–49

Juraska, J. M., Greenough, W. T., Elliott, C., Mack, K. J., Berkowitz, R. 1980. Plasticity in adult rat visual cortex: An examination of several cell populations after differential rearing. *Behav. Neurol. Biol.* 29:157–67

Kandel, E. R. 1976. *Cellular Basis of Behavior.* New York: Freeman. 727 pp.

Kaushall, P. J., Zetin, M., Squire, L. R. 1981. Amnesia: Detailed report of a noted case. *J. Nerv. Ment. Dis.* 169:383–89

Kinsbourne, M., Wood, F. 1975. Short-term memory processes and the amnesic syndrome. In *Short-Term Memory,* ed. D. Deutsch, J. A. Deutsch, pp. 258-91. New York: Academic

Kinsbourne, M., Winocur, G. 1980. Response competition and interference effects in paired-associate learning by Korsakoff amnesics. *Neuropsychologia* 18:541–48

Kolers, P. A. 1979. A pattern-analyzing basis of recognition. In *Levels of Processing in Human Memory,* ed. L. S. Cermak, F. I. M. Craik, pp. 363–84. Hillsdale, NJ: Erlbaum Assoc.

Korsakoff, S. S. 1887. Disturbance of psychic function in alcoholic paralysis and its relation to the disturbance of the psychic sphere in multiple neuritis of nonalcoholic origin. *Vestnik Psichiatrii.* 4:2

Loftus, E. F., Loftus, G. R. 1980. On the permanence of stored information in the human brain. *Am. Psychol.* 35:409–20

Luria, A. R. 1960. Frontal lobe syndromes. In *Handbook of Clinical Neurology,* ed. P. J. Vinken, G. W. Bruyn, 2:725–57. New York: Wiley

Lynch, G. S. 1976. Some difficulties associated with the use of lesion techniques in the study of memory. In *Neural Mechanisms of Learning and Memory,* ed. M. Rosenzweig, E. Bennett, pp. 544–46. Cambridge, Mass: MIT Press

Mair, W. G. P., Warrington, E. K., Weiskrantz, L. 1979. Memory disorder in Korsakoff's psychosis: A neuropathological and neuropsychological investigation of two cases. *Brain* 102:749–83

Marslen-Wilson, W. D., Teuber, H. L. 1975. Memory for remote events in anterograde amnesia: Recognition of public figures from news photographs. *Neuropsychologia* 13:353–64

Mayes, A. R., Meudell, P. R., Neary, D. 1978. Must amnesia be caused by either encoding or retrieval disorders? In *Practical Aspects of Memory,* ed. M. M. Gruneberg, P. E. Morris, R. N. Sykes, pp. 712–19. London: Academic

Mayes, A. R., Meudell, P. R., Neary, D. 1980. Do amnesics adopt inefficient encoding strategies with faces and random shapes? *Neuropsychologia* 18:527–40

McGaugh, J. L., Herz, M. M. 1972. *Memory Consolidation.* San Francisco: Albion. 204 pp.

Meudell, P. R., Northern, B., Snowden, J. S., Neary, D. 1980. Long-term memory for famous voices in amnesic and normal subjects. *Neuropsychologia* 18:133–39

Miller, R. R., Springer, A. D. 1973. Amnesia, consolidation and retrieval. *Psychol. Rev.* 80:69–73

Mills, R. P., Swanson, P. D. 1978. Vertical oculomotor apraxia and memory loss. *Ann. Neurol.* 4:149–53

Milner, B. 1962. Les troubles de la memoire accompagnant des lesions hippocampiques bilaterales. In *Physiologie de l'Hippocampe.* Paris: Cent. Natl. Rechesche Scientifique

Milner, B. 1963. Effects of different brain lesions on card sorting. *Arch. Neurol.* 9:100–10

Milner, B. 1966. Amnesia following operation on the temporal lobes. In *Amnesia,* ed. C. W. M. Whitty, O. L. Zangwill, pp. 109–33. London: Buttersworths

Milner, B. 1968. Disorders of memory after brain lesions in man. Preface: Material-specific and generalized memory loss. *Neuropsychologia* 6:175–79

Milner, B. 1974. Hemispheric specialization: Scope and limits. In *The Neurosciences: Third Study Program,* ed. F. O. Schmitt, F. G. Worden, pp. 75–89. Cambridge, Mass. MIT Press

Milner, B., Corkin S., Teuber, H.-L. 1968. Further analysis of the hippocampal amnesic syndrome: 14-year follow-up study of H. M. *Neuropsychologia* 6:215–34

Mishkin, M. 1954. Visual discrimination performance following partial ablations of the temporal lobe. II. Ventral surface vs hippocampus. *J. Comp. Physiol. Psychol.* 147:187–93

Mishkin, M. 1978. Memory in monkeys severely impaired by combined but not by separate removel of amygdala and hippocampus. *Nature* 273:297–98

Mishkin, M., Pribram, K. H. 1954. Visual discrimination performance following partial ablations of the temporal lobe: Ventral vs lateral. *J. Comp. Physiol. Psychol.* 47:14–20

Mishkin, M., Spiegler, B. J., Saunders, R. C., Malamut, B. J. 1981. An animal model of global amnesia. In *Toward a Treatment of Alzheimer's Disease,* ed. S. Corkin, K. L. Davis, J. H. Growdon, E. Usdin, R. J. Wurtman. New York: Raven. In press

Mortensen, E. L. 1980. The effects of partial information in amnesic and normal subjects. *Scand. J. Psychol.* 21:75–82

Moscovitch, M. 1981. Multiple dissociations of function in the amnesic syndrome. In *Human Memory and Amnesia,* ed. L. S. Cermak. Hillsdale, NJ: Erlbaum Assoc. In press

Moss, M., Mahut, H., Zola-Morgan, S. 1980. Associative and recognition memory impairments in monkeys after hippocampal resections. *Soc. Neurosci. Abstr.* 6:192

Norman, D. A., Rumelhart, D. E. 1975. *Explorations in Cognition.* San Francisco: Freeman. 430 pp.

O'Keefe, J., Nadel, L. 1978. *The Hippocampus as a Cognitive Map.* London: Oxford Univ. Press

Olton, D. S., Becker, J. T., Handelmann, G. E. 1979. Hippocampus, space, and memory. *Behav. Brain Sci.* 2:313–65

Orchinik, C. W. 1960. Some psychological aspects of circumscribed lesions of the diencephalon. *Confin. Neurol.* 20:292–310

Oscar-Berman, M. 1973. Hypothesis testing and focusing behavior during concept formation by amnesic Korsakoff patients. *Neuropsychologia* 11:191–98

Papez, J. W. 1937. A proposed mechanism of emotion. *Arch. Neurol. Psychiatr.* 38:725–43

Parker, E. S., Noble, E. 1977. Alcoholic consumption and cognitive functioning in social drinkers *J. Studies Alcohol* 38:1224–32

Pectel, C., Masserman, J. H., Schreiner, L., Levitt, M. 1955. Differential effects of lesions in the mediodorsal nuclei of the thalamus on normal and neurotic behavior in the cat. *J. Nerv. Ment. Dis.* 121:26–33

Penfield, W., Milner, B. 1958. Memory deficit produced by bilateral lesions in the hippocampal zone. *AMA Arch. Neurol. Psychiatr.* 79:475–97

Piercy, M. F. 1977. Experimental studies of the organic amnesic syndrome. In *Amnesia,* ed. C. W. M. Whitty, O. L. Zangwill. London: Buttersworth. 2nd ed.

Purves, D., Lichtman, J. W. 1980. Elimination of synapses in the developing nervous system. *Science* 210:153–57

Ribot, T. 1882. *Diseases of Memory.* New York: Appleton. 127 pp.

Rosene, D. L., Van Hoesen, G. 1977. Hippocampal efferents reach widespread areas of cerebral cortex and amygdala in the Rhesus monkey. *Science* 198:315–17

Rosenstock, J., Fields, T. D., Greene, E. 1977. The role of mammillary bodies in spatial memory. *Exp. Neurol.* 55:340–52

Rosenzweig, M. R. 1979. Responsiveness of brain size to individual experience. Behavioral and evolutionary implication. In *Development and Evolution of Brain Size: Behavioral Implications,* ed. M. E. Hahn, C. Jensen, B. Dudek, pp. 263–94. New York: Academic

Rumelhart, D. E. 1981. Schemata: The building blocks of cognition. In *Theoretical Issues in Reading Comprehension,* ed. R. Spiro, B. Bruce, W. Brewer. Hillsdale, NJ: Erlbaum Assoc. In press

Rumelhart, D. E., Norman, D. A. 1978. Accretion, tuning and restructuring: Three modes of learning. In *Semantic Factors in Cognition,* ed. J. W. Cotton, R. Klatzky, pp. 37–53. Hillsdale, NJ: Erlbaum Assoc.

Russell, W. R., Nathan, P. W. 1946. Traumatic amnesia. *Brain* 69:280–300

Ryan, C., Butters, N., Montgomery, L. 1980. See Albert et al 1980, pp. 701–18

Ryle, G. 1949. *The Concept of Mind.* London: Hutchinson. 334 pp.

Sanders, H. I., Warrington, E. K. 1971. Memory for remote events in amnesic patients. *Brain* 94:661–68

Schulman, S. 1964. Impaired delayed response from thalamic lesions. Studies in monkeys. *Arch. Neurol.* 11:477–99

Scoville, W. B., Milner, B. 1957. Loss of recent memory after bilateral hippocampal lesions. *J. Neurol. Neurosurg. Psychiatr.* 20:11–21

Seltzer, B., Benson, D. F. 1974. The temporal pattern of retrograde amnesia in Korsakoff's disease. *Neurology* 24:527–30

Squire, L. R. 1975. See Kinsbourne & Wood 1975, pp. 1–40

Squire, L. R. 1979. The hippocampus, space

and human amnesia. *Behav. Brain Sci.* 2:514–15

Squire, L. R. 1980a. Specifying the defect in human amnesia: Storage, retrieval, and semantics. *Neuropsychologia* 18:368–72

Squire, L. R. 1980b. The anatomy of amnesia. *Trends Neurosci.* 3:52–54

Squire, L. R. 1981a. Neuropsychology of ECT. In *Electroconvulsive Therapies: Biological Foundations and Clinical Application,* ed. W. B. Essman, R. Abrams. Jamaica, NY: Spectrum. In press

Squire, L. R. 1981b. Two forms of human amnesia: An analysis of forgetting. *J. Neurosci.* 1:635–40

Squire, L. R., Chace, P. M., Slater, P. C. 1976. Retrograde amnesia following electroconvulsive therapy. *Nature* 260:775–77

Squire, L. R., Cohen, N. J. 1979. Memory and amnesia: Resistance to disruption develops for years after learning. *Behav. Neurol. Biol.* 25:115–25

Squire, L. R., Cohen, N. J. 1981. Remote memory, retrograde amnesia, and the neuropsychology of memory. See Moscovitch 1981, in press

Squire, L. R., Cohen, N. J. 1982. Human memory and amnesia. In *Handbook of Behavioral Neurobiology,* ed. J. McGaugh, R. Thompson, Vol. 10. New York: Plenum. In press

Squire, L. R., Fox, M. M. 1981. Assessment of remote memory: Validation of the television test by repeated testing during a seven-year period. *Behav. Res. Methods Instrum.* 12:583–86

Squire, L. R., Moore, R. Y. 1979. Dorsal thalamic lesion in a noted case of chronic memory dysfunction. *Ann. Neurol.* 6:503–6

Squire, L. R., Nadel, L., Slater, P. C. 1981. Anterograde amnesia and memory for temporal order. *Neuropsychologia.* 19: 141–45

Squire, L. R., Schlapfer, W. T. 1981a. Memory and memory disorders: A biological and neurologic perspective. In *Handbook of Biological Psychiatry,* Pt. 4, ed. H. M. van Praag, M. H. Lader, O. J. Rafaelsen, E. J. Sachar, pp. 309–41. New York: Dekker

Squire, L. R., Slater, P. C. 1975. Forgetting in very long-term memory as assessed by an improved questionnaire technique. *J. Exp. Psych.* 104:50–54

Squire, L. R., Slater, P. C. 1977. Remote memory in chronic anterograde amnesia. *Behav. Biol.* 20:398–403

Squire, L. R., Slater, P. C. 1978. Anterograde and retrograde memory impairment in chronic amnesia. *Neuropsychologia* 16: 313–22

Squire, L. R., Slater, P. C., Chace, P. M. 1975. Retrograde amnesia: Temporal gradient in very long-term memory following electroconvulsive therapy. *Science* 187:77–79

Squire, L. R., Slater, P. C., Miller, P. L. 1981b. Retrograde amnesia following ECT: Long-term follow-up studies. *Arch. Gen. Psychiatr.* 38:89–95

Squire, L. R., Zola-Morgan, S. 1982. The neurology of memory: The case for correspondence between the findings for man and non-human primate. In *The Physiological Basis of Memory,* ed. J. A. Deutsch. New York: Academic Press, 2nd ed. In press

Squire, L. R., Wetzel, C. D., Slater, P. C. 1978. Anterograde amnesia following ECT: An analysis of the beneficial effect of partial information. *Neuropsychologia* 16:339–47

Sweet, W. H., Talland, G. A., Ervin, F. R. 1959. Loss of recent memory following section of fornix. *Trans. Am. Neurol. Assoc.* 84:76–82

Talland, G. A. 1965. *Deranged Memory.* New York: Academic. 356 pp.

Teuber, H.-L. 1964. The riddle of frontal lobe function in man. In *The Frontal Granular Cortex and Behavior,* ed. J. M. Warren, K. Akert. New York: McGraw-Hill

Teuber, H.-L., Milner, B., Vaughan, H. G. 1968. Persistent anterograde amnesia after stab wound of the basal brain. *Neuropsychologia* 6:267–82

Thompson, R. 1981. Rapid forgetting of a spatial habit in rats with hippocampal lesions. *Science* 212:959–60

Victor, M., Adams, R. D., Collins, G. H. 1971. In *The Wernicke-Korsakoff Syndrome,* ed. F. Plum, F. H. McDowell. Philadelphia: Davis. 206 pp.

Walters, E. T., Carew, T. J., Kandel, E. R. 1979. Classical conditioning in *Aplysia California. Proc. Natl. Acad. Sci.* 76: 6675–79

Warrington, E. K., Weiskrantz, L. 1968. A new method of testing long-term retention with special reference to amnesic patients. *Nature* 217:972–74

Warrington, E. K., Weiskrantz, L. 1970. The amnesic syndrome: Consolidation or retrieval? *Nature* 228:628–30

Warrington, E. K., Weiskrantz, L. 1974. The effect of prior learning on subsequent retention in amnesic patients. *Neuropsychologia* 12:419–28

Warrington, E. K., Weiskrantz, L. 1978. Further analysis of the prior learning effect in amnesic patients. *Neuropsychologia* 16:169–77

Weiskrantz, L. 1978. A comparison of hippocampal pathology in man and other animals. In *Functions of the Septo-hippocampal System*, CIBA Found. Symp. 58. Oxford: Elsevier

Weiskrantz, L., Warrington, E. K. 1970. Verbal learning and retention by amnesic patients using partial information. *Psychonom. Sci.* 20:210–11

Weiskrantz, L., Warrington, E. K. 1975. The problem of the amnesic syndrome in man and animals. In *The Hippocampus*, ed. R. L. Isaacson, K. H. Pribram, 2:411–28. New York: Plenum

Weiskrantz, L., Warrington, E. K. 1979. Conditioning in amnesic patients. *Neuropsychologia* 17:187–94

Wetzel, C. D., Squire, L. R. 1980. Encoding in anterograde amnesia. *Neuropsychologia* 18:177–84

Wetzel, C. D., Squire, L. R. 1981. Cued recall in anterograde amnesia. *Brain Lang.* In press

Reference added in proof:

Lhermitte, F., Signoret, J.-L. 1972. Analyse neuropsychologique et differenciation des syndromes amnesiques. *Rev. Neurol. Paris* 126:161–78

Wicklegren, W. A. 1979 Chunking and consolidation: A theoretical synthesis of semantic networks, configuring in conditioning, S-R v. cognitive learning, normal forgetting, the amnesic syndrome and the hippocampal arousal system. *Psychol. Rev.* 86:44–60

Winograd, R. 1975. Frame representations and the declarative-procedural controversy. In *Representation and Understanding*, ed. D. Bobrow, A. Collins. New York: Academic

Woody, C. D., Ervin, F. R. 1966. Memory function in cats with lesions of the fornix and mammillary bodies. *Physiol. Behav.* 1:273–80

Zangwill, O. L. 1977. The amnesic syndrome. In *Amnesia*, ed. C. W. M. Whitty, O. L. Zangwill, pp. 104–117. London: Buttersworths. 2nd ed.

Zola-Morgan, S., Mishkin, M., Squire, L. R. 1981. The anatomy of amnesia: Hippocampus and amygdala vs. temporal stem. *Soc. Neuro. Abstr.* 7:236

Ann. Rev. Neurosci. 1982. 5:275–96
Copyright © 1982 by Annual Reviews Inc. All rights reserved

CEREBELLAR CONTROL OF THE VESTIBULO-OCULAR REFLEX— AROUND THE FLOCCULUS HYPOTHESIS

Masao Ito

Department of Physiology, Faculty of Medicine, University of Tokyo, Bunkyoku, Tokyo 113, Japan

INTRODUCTION

The vestibulo-ocular reflex (VOR) is an elementary reflex, which attracted the attention of a number of classical neurophysiologists who studied, in particular, the reflexive compensatory eye movements induced by head movements (cf Magnus 1924, Lorente de Nó 1931). The major pathway for this reflex is the well-defined trineuronal arc, composed of primary vestibular neurons, secondary vestibular neurons, and oculomotor neurons (Ramon y Cajal 1909, Lorente de Nó 1931, 1933, Szentágothai 1943, 1950, 1964, Brodal & Pompeiano 1957). The pioneering electrophysiological work by Lorente de Nó (1939) and by subsequent investigators recognized that the vestibulo-ocular reflex involves both excitation and inhibition (Szentágothai 1950, Cohen et al 1964). More recent data about the VOR have been reviewed by Brodal (1974), Cohen (1974), Precht (1978), and Wilson & Melvill Jones (1979).

In recent years, the vestibulo-ocular reflex has been reexamined in detail because of its close relationship with the cerebellum. Supported by remarkable advances in cerebellar physiology and eye movement studies, the VOR has become an interesting and important subject of neurophysiology, especially for investigation of cerebellar control mechanisms. This article reviews the efforts made in the past decade toward understanding the

275

0147-006X/82/0301-0275$02.00

cerebellum through relationships between the VOR and the cerebellum; however, it is not intended to extensively review all relevant literature, but to focus on the narrower but critical examination of the "flocculus hypothesis of the vestibulo-ocular reflex," proposed a decade ago (Ito 1970, 1972, 1974). Various lines of evidence supporting the hypothesis are presented, and problems remaining are discussed.

DEVELOPMENT OF BASIC IDEAS

Around 1970

A decade ago, neurophysiology reached a turning point. Although elementary neuronal processes had been revealed and neuronal network structures had been determined in many parts of the central nervous system, there was a great deal of uncertainty about how to extrapolate from the cellular level to "higher" nervous system functions such as perception, control of complex movement, learning, and symbolic thought in speech and language. Investigation on the cerebellum was at the forefront of this change in approach. The structure of the cerebellar neuronal network had been analyzed in detail (cf Eccles et al 1967), and an ingenious constructive model of the function of the cerebellum was proposed (Marr 1969, Albus 1971). Similarly, a vast pool of data concerning functional disorders arising from destruction of the cerebellum was available (Dow & Moruzzi 1958). Yet, there was no clear, concrete idea of what the cerebellum does. Therefore, I decided to apply this general knowledge to a detailed examination of a specific cerebellar system.

Vestibulocerebellum

The vestibulocerebellum (including the flocculus, ventral paraflocculus, nodulus, and ventral folia of the uvula) is an attractive system for examining specific aspects of cerebellar function. It has been suggested that this phylogenetically old part of the cerebellum may contain a prototype of cerebellar systems (cf Herrick 1924, Dow 1938). The vestibulocerebellum receives primary vestibular afferents directly as a mossy fiber input (Brodal & Høivik 1964, Precht & Llinás 1969). Purkinje cells in the vestibulocerebellum project directly to vestibular nuclei (Dow 1938, Voogd 1964, Angaut & Brodal 1967, Van Rossum 1969). Thus, it offers an opportunity to investigate cerebellar mechanisms in connection with well-defined vestibular functions. An electrophysiological investigation in rabbits revealed a specific connection from Purkinje cells of the flocculus to relay cells of the vestibulo-ocular reflex in vestibular nuclei (Ito et al 1970, Fukuda et al 1972). A similar connection was also reported in cats (Baker et al 1972).

Feedforward Control

Although the reflex arcs of the vestibulo-ocular reflex and the vestibulospinal reflex are similar, relay cells of the latter could not be shown to have an inhibitory connection with the vestibulocerebellum (Fukuda et al 1972). This peculiar difference between the vestibulo-ocular and vestibulospinal reflexes was correlated to their difference as control systems (Ito 1970); the VOR is a feedforward control (a class of open loop controls), while the vestibulospinal reflex is a typical feedback control. In the vestibulospinal reflex, serving to maintain constant head position, the final output (head position) readily influences the vestibular organ, the input to the reflex arc. By contrast, the final output of the vestibulo-ocular reflex is the constancy of retinal images, monitored by vision. However, there is no straightforward visual feedback to the vestibular organ.

Because of lack of feedback, performance of a feedforward control system is inevitably susceptible both to external disturbances and to changes in internal parameters. Any misperformance of such a system would persist or worsen unless corrected by some device that replaces the feedback. Collewijn (1980) points out that this functional deficiency of the VOR, requiring frequent recalibration, was recognized a long time ago by Rönne (1923). This problem led me to suggest that the cerebellar flocculus is a recalibration device for the VOR (Ito 1970). However, this hypothesis does not imply that the cerebellum itself performs a feedforward control; the cerebellum only assists the feedforward control performed by the vestibulo-ocular reflex.

This idea inevitably leads to a suggestion that the flocculus receives visual signals monitoring the constancy of retinal images (Ito 1970). Maekawa & Simpson (1973) revealed the existence of a visual pathway to the flocculus mediated by climbing fiber afferents. Later, another visual pathway to the flocculus via mossy fiber afferents was found (Maekawa & Takeda 1975, 1976).

Lack of a vestibulocerebellar contribution to the vestibulospinal reflex may indicate that the vestibulospinal reflex is a closed loop control system that does not require a recalibration device. However, not all reflexes with a feedback are free of cerebellar contribution. For example, the optokinetic response, an eye movement driven by movement of the whole visual field, is essentially a closed loop control. Its output, eye movement, will readily influence its input, vision. Yet, the optokinetic response is closely connected to the flocculus, as is discussed below. Apparently, the optokinetic response has a low efficacy at relatively fast velocity ranges because of the limited capacity of the visual system, and this low efficacy is compensated for by the aid of the flocculus. Another example is the pupillary light reflex, which is also facilitated by the cerebellum at relatively fast frequencies of light intensity change (Tsukahara et al 1973). Thus, the cerebellum appears to intervene even in a closed loop control under conditions in which the control does not operate by itself effectively.

Flocculus Hypothesis

Neuronal organization of the flocculo-vestibulo-ocular system suggests three possible roles of the flocculus in controlling the vestibulo-ocular reflex (Ito 1970, 1972, 1974): (a) as a sidepath to the major VOR arc, the flocculus contributes to dynamic characteristics of the VOR; (b) through visual feedback, the flocculus rapidly corrects performance of the VOR so as to maintain constancy of retinal images (rapid readjustment of the VOR by vision); (c) when correction is repeated, there occurs a progressive change in the internal parameters of the flocculus ("learning") so that performance of the VOR will be improved (adaptive modification of the VOR by vision). Here, the term "learning" is used in a wide sense, including simple forms of adaptation (for further discussion, see Ito 1976b).

The third role of the flocculus is in good agreement with Marr's model of the cerebellum, which relates the unique dual input system of the cerebellum (mossy fiber and climbing fiber afferents) to possible learning capabilities (Marr 1969). The model postulates that signal transfer across the cerebellar cortex, from mossy fiber afferents through granule cells and their axons (parallel fibers) to Purkinje cells, can be reorganized by "instruction" signals of climbing fiber afferents. Thus, the visual climbing fiber afferents to the flocculus may reorganize neuronal networks of the flocculus according to retinal error signals, so as to improve performance of the VOR (Ito 1972).

Simultaneously to my proposal of this "flocculus hypothesis of the vestibulo-ocular reflex control," Melvill Jones reported striking results of an experiment on human subjects wearing dove prism goggles (Gonshor & Melvill Jones 1976a,b); the gain of the horizontal VOR can be reduced adaptively and even reversed in polarity during continuous reversal of the visual field on the horizontal plane. Robinson (1976) then proceeded to test the flocculus hypothesis by applying dove prism goggles to cats, while I also started to investigate the vestibulo-ocular reflex of rabbits using Amanomori & Takahira's (1969) combined vestibular-visual stimulation techniques (Ito et al 1974a,b). Takemori & Cohen (1974a,b) also investigated "visual suppression of the VOR" in monkeys; they suggested that the visual pathway to the flocculus plays a key role in this phenomenon.

Microzones

A complicating feature of the flocculo-vestibulo-ocular system is the presence of multiple parallel component pathways arising from different endorgans of the labyrinth and terminating in different extraocular muscles. An electrophysiological reflex testing study (Ito et al 1977) revealed that the flocculus of rabbits contains subgroups of Purkinje cells that independently control component pathways subserving vertical, horizontal, and rotatory canal-ocular reflexes. Differential localization of these Purkinje cell subgroups in the flocculus was visualized by horseradish peroxidase techniques in rabbits (Yamamoto & Shimoyama 1977, Yamamoto 1978) and cats (Voogd & Bigaré 1980).

In rabbit flocculus, five zones (I–V) have been distinguished in anatomical and physiological studies (Figure 1). Only one zone (II) has been related to the horizontal VOR. Purkinje cells in this zone project specifically to the rostral area of the medial vestibular nucleus that contains relay cells of the horizontal VOR. Zone II can be specified by applying repetitive, stimulating pulses through a microelectrode, producing a horizontal abduction of the ipsilateral eye, often conjugated with adduction of the contralateral eye (Dufossé et al 1977). Zone II is also specified by climbing fiber inputs from

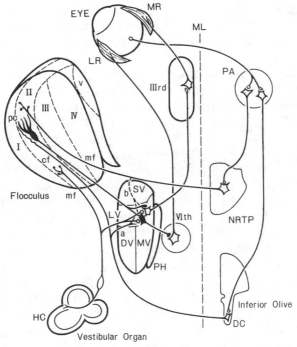

Figure 1 Neuronal circuitry for flocculus control of the horizontal VOR. ML, mid line. IIIrd, VIth, cranial nuclei. mf, mossy fiber. cf, climbing fiber. pc, Purkinje cell. I-V, zones in the flocculus. SV, MV, LV, DV, superior, medial, lateral, descending vestibular nuclei. PH, nucleus prepositus hypoglossi. NRTP, nucleus reticularis tegmenti pontis. DC, dorsal cap. HC, horizontal canal. PA, pretectal area. a, collateral of climbing fiber afferents. b, postulated collateral of visual mossy fiber afferents (see text).

the ipsilateral retina (Ito et al 1977, Yamamoto 1979). This functional localization within the flocculus supports the idea that the cerebellum consists of a number of longitudinal, narrow bands (Groenewegen & Voogd 1977) and that each band can further be divided into functionally distinct microzones (Andersson & Oscarsson 1978).

Action of Purkinje Cells

In extracellular recording from a Purkinje cell within a microzone, *complex* spikes represent activation through a climbing fiber afferent, while *simple* spikes reflect an integrated effect of excitatory inputs from granule cells, of inhibitory inputs from basket and superficial stellate cells, and also of secondary influence from inhibition of granule cells by Golgi cells (Eccles et al 1967, Thach 1968). *Simple* spikes, occurring at relatively high rates (around 50/sec), are the major output signals of Purkinje cells, while *complex* spikes, discharging only at a low (around 1/sec), irregular rate, make only a subsidiary contribution to the whole output of a Purkinje cell.

By recording from flocculus Purkinje cells of monkeys, Lisberger & Fuchs (1974, 1978) found that the discharge frequency of *simple* spikes is modulated in phase with head velocity, during visual suppression of the VOR induced by sinusoidal head rotation. Since discharges in primary vestibular afferents are modulated in phase with head velocity (Fernandez & Goldberg 1971), it was postulated that visual suppression of the VOR is effected through cancellation of excitatory vestibular inputs by inhibitory Purkinje cell signals, both modulating in phase with head velocity, on vestibulo-ocular relay cells. However, in recording from the rabbit's flocculus, both in-phase and 180° out-phase modulation of *simple* spike discharge was observed (Ghelarducci et al 1975). Out-phase modulation of inhibitory Purkinje cell signals, waning and waxing opposite to excitatory vestibular signals, should enhance the vestibulo-ocular reflex. The flocculus can therefore either depress the vestibulo-ocular reflex through inphase-modulating Purkinje cells, or enhance it through outphase-modulating Purkinje cells; the vestibulo-ocular reflex can be modified by altering in-phase and out-phase Purkinje cells (Ito 1976a). This view was substantiated by examination of *simple* spike modulation during adaptive modification of the vestibulo-ocular reflex in rabbits (Dufossé et al 1978a). The amplitude of out-phase modulation exhibited a significant increase or decrease in accordance with increase or decrease of the vestibulo-ocular reflex gain.

Reconstruction of the System

Since neuronal circuitry of the flocculo-vestibulo-ocular system has been dissected and impulse activities occurring within the system are known, the next step is a quantitative evaluation of characteristics at each step of signal transfer in the system. This should permit construction of a model that faithfully reproduces the operations of the flocculus in controlling the vestibulo-ocular reflex.

As a first step, electrical pulse signals may be used in place of natural vestibular and visual stimuli, because these signals can be better controlled and give better temporal resolution between stimulus and response. Very recently, it was shown that application of sinusoidally frequency-modulated train pulses to a vestibular nerve induces either in-phase or out-phase modulation of *simple* spike discharges in flocculus Purkinje cells, similar to responses to sinusoidal head rotation (Ito et al 1981a). In addition, conjunctive stimulation of climbing fiber afferents and vestibular mossy fiber afferents causes a drastic depression of the responsiveness of flocculus Purkinje cells to vestibular mossy fiber afferents (Ito et al 1981a,b). This observation supports the hypothesis that the cerebellar cortex is a modifiable neuronal network (see below).

Knowledge of neurotransmitter substances involved in the system has been useful in evaluating the system. Several lines of evidence suggest that L-glutamic acid is the neurotransmitter liberated from granule cells (Young et al 1974, McBride et al 1976, Hudson et al 1976, Herndon & Coyle 1977, Sandoval & Cotman 1978, Hacket et al 1979). Application of climbing fiber signals to a Purkinje cell leads to a simultaneous depression of sensitivity of that Purkinje cell to L-glutamic acid (Ito et al 1981a,b). This suggests that the modifiability of the cerebellar neuronal networks involves a change in the chemosensitivity of Purkinje cells.

SUPPORT FOR THE FLOCCULUS HYPOTHESIS DERIVED FROM EYE MOVEMENT STUDIES

The three roles postulated for the flocculus in controlling the vestibulo-ocular reflex (see above) have been substantiated by recent eye movement studies. Dynamic characteristics of the VOR have been evaluated in terms of the gain and phase. Rapid readjustment of the VOR by vision (second role) has been defined explicitly in terms of interaction of the VOR with visually evoked eye movements. In rabbits, rapid readjustment of the VOR by vision can fully be accounted for by a linear interaction of the VOR and optokinetic response (Baarsma & Collewijn 1974, Batini et al 1979). Impairment of the rapid readjustment of the VOR by vision after flocculectomy (Ito et al 1974a, Hassul et al 1976) is explicable by reduction of the optokinetic response gain (Ito et al 1981c). Hence, the flocculus facilitates the optokinetic response and thereby contributes to rapid readjustment of the VOR by vision. Visual suppression of the VOR in monkeys may also represent this second role of the flocculus. Monkeys, however, exhibit well-developed smooth pursuit eye movements, which are virtually absent in rabbits. Therefore, it is possible that visual suppression of the VOR reflects an interaction of the VOR with both the optokinetic response and smooth pursuit eye movements. The optokinetic response is a pure reflex, while the smooth pursuit eye movement represents a voluntary motor control. It is possible, therefore, that this second role of the flocculus involves different neuronal mechanisms in rabbits and monkeys. Adaptation of the VOR (third role) has been demonstrated as a progressive increase or decrease of the VOR gain (Gonshor & Melvill Jones 1976a,b, Ito et al 1974b, 1979b, Robinson 1976, Miles & Fuller 1974). Deficiencies in these eye movement functions arising from destruction of the flocculus and related structures provide evidence supporting the flocculus hypothesis of the VOR control.

Flocculectomy

Effects of flocculectomy have been investigated in various animal species. Manni (1950) demonstrated that ablation of guinea pig flocculus causes abnormal tonic eye movements under head tilting. Carpenter (1972) examined the horizontal VOR by applying sinusoidal head rotation to decerebrate cats, and found that when the cerebellum was ablated or cooled the gain was reduced and the phase was advanced. However, Robinson (1974) obtained a much milder effect in chronically cerebellectomized cats, and cast doubt upon the assumption that a decerebrate cat maintains normal cerebellar functions. With chronic ablation of cat vestibulocerebellum, including the flocculus, Robinson (1976) obtained an abnormally high gain of the horizontal vestibulo-ocular reflex; however, flocculectomy in rabbits regularly caused an abnormally low gain (Ito et al 1981c). Flocculectomy also impairs rapid readjustment of the VOR by vision in rabbits (Ito et al 1974a, 1981c), monkeys (Takemori & Cohen 1974a,b), and chinchillas (Hassul et al 1976). Adaptive modification of the VOR was shown to be abolished by flocculectomy in cats (Robinson 1976) and rabbits (Ito et al 1974b, 1981c).

Two complications, however, were involved in these experiments. First, lesions were usually bilateral and frequently exceeded the borders of the flocculus. This raises the question of whether effects of such an extensive lesion can be related specifically to the unilateral projection from the flocculus to vestibular nuclei (Dow 1938, Angaut & Brodal 1967, Ito et al 1977). Second, ablation of the flocculus induces prominent retrograde degeneration in olivocerebellar neurons in the dorsal cap of the inferior olive, the source of climbing fiber afferents to the flocculus (Barmack & Simpson 1980, Ito et al 1980). This raises the possibility that flocculectomy has side effects due to degeneration of olivocerebellar neurons, which may arise through collaterals of climbing fiber afferents innervating vestibular nuclei

(Balaban et al 1981). Effects of flocculectomy in rabbits were therefore reinvestigated using unilateral surgical or chemical lesions (Ito et al 1981c). Chemical flocculectomy was achieved by injecting small amounts of kainic acid into the flocculus; this destroys Purkinje cells without causing loss of olivocerebellar neurons (Ito et al 1980).

Functional deficiencies arising from unilateral flocculectomy in rabbits correspond to the postulated three roles of the flocculus: (a) lowering of the vestibulo-ocular reflex gain accompanied by a slight phase delay in the eye on the lesioned side; (b) gain reduction and phase delay in the optokinetic response, and the corresponding impairment of rapid readjustment of the VOR by vision, on the lesioned side (monocular stimulation); and (c) loss of adaptive modification of the VOR during sustained, combined vestibular-visual stimulation. These deficiencies were absent in the eye contralateral to lesions, except for transient depression of the VOR in an early postoperative period. This is in excellent agreement with the specific connectivity of the flocculus with VOR pathways (Figure 1). Flocculus Purkinje cells have inhibitory connections only with relay cells mediating signals from the ipsilateral horizontal canal to the medial and lateral rectus muscles of the ipsilateral eye. Other relay cells mediating signals from the contralateral horizontal canal to the ipsilateral eye, or from either horizontal canal to the contralateral eye, are free of such inhibitory connections (Ito et al 1977).

Interruption of the Visual Climbing Fiber Pathway

Lesions placed in the dorsal cap of the inferior olive, which relays the visual climbing fiber pathway (Alley et al 1975, Maekawa & Takeda 1977), abolished adaptive modification of the vestibulo-ocular reflex (Ito & Miyashita 1975, Haddad et al 1980). However, the VOR gain was also reduced; this suggests a complicating issue in the cerebellum accompanying death of olivocerebellar neurons (Ito & Miyashita 1975). This suggestion was substantiated by a later observation in rabbits with lesions in the dorsal cap; electrical stimulation of the flocculus no longer induced eye movement, thus implying that the action of the Purkinje cells on vestibulo-ocular relay neurons is impaired (Dufossé et al 1978b). In addition, in rabbits, the inhibitory action of vermal Purkinje cells on Deiters neurons is attenuated soon after lesion of the olivary neurons (Ito et al 1979a). In another test of this phenomenon rats were poisoned with 3-Acetylpyridine, which fairly specifically destroys olivocerebellar neurons (Desclin & Escubi 1974, Llinás et al 1975). In these rats, fewer Deiters neurons were inhibited by vermal Purkinje cells than in controls, and the latency of the inhibition was delayed (Ito et al 1978b). However, in a recent experiment by Montarolo et al (1981) reduction in the rate of occurrence of the inhibition could not be confirmed. Furthermore, in organized cultures of mouse cerebellum, Purkinje cells form inhibitory synapses with nuclear neurons in the absence of climbing fiber afferents (Marshall et al 1980). Therefore, the reduction of Purkinje cell inhibition after deprivation of climbing fiber afferents may not be all-or-none, but graded, thus requiring a quantitative evaluation of the magnitude and time course of inhibition in control and experimental animals. Nevertheless, caution is required in testing functional roles of climbing fiber afferents with destruction of olivocerebellar neurons, because the effects may also extend to attenuation of Purkinje cell inhibition.

The effect of olivary destruction upon Purkinje cell output was avoided by placing lesions rostral to the dorsal cap (Ito & Miyashita 1975); this abolished adaptive modification of the vestibulo-ocular reflex without influencing the reflex gain. Rapid readjustment of the VOR was not affected, indicating that the visual climbing fiber pathway is not essential for the optokinetic response.

Interruption of the Visual Mossy Fiber Pathway

The visual mossy fiber pathway to the flocculus was selectively interrupted by placing lesions in the nucleus reticularis tegmenti pontis of Bechterew, which relays this input to the flocculus (Maekawa & Takeda 1978). Unilateral lesions of the nucleus resulted in reduction of the optokinetic response gain in the contralateral, but not ipsilateral, eye during monocular stimulation (Miyashita et al 1980). By contrast, adaptive modification of the vestibulo-ocular reflex was not impaired by the lesions. These observations demonstrate differential roles of the visual mossy fiber and climbing fiber pathways to the flocculus; the mossy fiber pathway serves the optokinetic response and rapid readjustment of the VOR by vision (second role of the flocculus), while the climbing fiber pathway is involved specifically in adaptive modification of the VOR (third role of the flocculus).

Impairment of the optokinetic response after bilateral destruction of the nucleus reticularis tegmenti pontis was reported in cats by Precht & Strata (1980), who, however, considered this nucleus as a relay for a visual pathway to vestibular nuclei, rather than as a relay for the visual mossy pontis fiber pathway to the flocculus. A projection from nucleus reticularis tegmenti to vestibular nuclei has been demonstrated in cats (Hoddevik 1978) and rabbits (C. D. Balaban, unpublished observation). Thus, it is possible that this nucleus is a relay site of visual pathways both to vestibular nuclei and the flocculus. This raises the possibility that the flocculus forms a sidepath to the brainstem visual pathway to vestibular nuclei, just as the flocculus forms a sidepath to the primary vestibular afferents to vestibular nuclei (Figure 1).

Although Keller & Precht (1979) stressed that cerebellectomy does not have a major effect on the optokinetic response in cats, their Figure 9 shows a clear effect of cerebellectomy in a relatively fast velocity range of optokinetic stimuli. In rabbits, as well, flocculectomy influences the optokinetic response more prominently in a relatively fast velocity range (Ito et al 1981c). Since the optokinetic response is essentially a closed loop control, it may not require assistance of the cerebellum, provided that the open loop gain of the system is sufficiently high. It appears that the cerebellum facilitates the optokinetic response in a fast velocity range, where its open loop gain declines because of limited capacity of the visual system.

SUPPORT DERIVED BY RECORDING FROM THE PURKINJE CELLS OR THE FLOCCULUS

In rabbits, responses of flocculus Purkinje cells have been recorded during natural vestibular and visual stimulation in connection with the vestibulo-ocular reflex, optokinetic response, rapid readjustment of the VOR by vision, and adaptive modification of the VOR (Ghelarducci et al 1975, Ito 1977, Dufossé et al 1978a, Neverov et al 1980). In monkeys, Purkinje cell responses have been investigated in relationship to the VOR, smooth pursuit eye movement, saccadic eye movement, visual suppression of the VOR, and adaptive modification of the VOR (Lisberger & Fuchs 1974, 1978, Miles & Fuller 1975, Miles et al 1980, Noda & Suzuki 1979a,b,c).

Vestibular Stimulation

Sinusoidal rotation of a rabbit on a horizontal turntable in darkness produces modulation of *simple* spike discharges of flocculus Purkinje cells, with the same period as the rotation. In zone II of rabbit flocculus, which is specifically related to the horizontal vestibulo-ocular reflex (see above), the majority of Purkinje cells exhibit modulation 180° out-phase with head velocity; a minority show an in-phase or intermediate type of modulation (Dufossé et al 1978a). It is peculiar that Purkinje cells in other zones also respond to horizontal head rotation, frequently in phase with head velocity, in spite of their having no direct relevance to the horizontal vestibulo-ocular reflex. This observation offers a warning against merely relying upon response types without specifying their microzonal relations when sampling Purkinje cells. Dominance of outphase-modulating Purkinje cells in zone II is in good agreement with the fact that flocculectomy causes a decrease of the VOR gain, as an out-phase modulation in Purkinje cells should have an action of facilitating the horizontal VOR (see above).

Out-phase modulation in rabbit flocculus Purkinje cells may represent vestibular mossy fiber inputs to the flocculus. The response may be induced by inphase-modulating vestibular signals from the ipsilateral labyrinth, if they are converted to out-phase through cortical inhibitory neurons, and

also by outphase-modulating vestibular signals from the contralateral labyrinth. In monkeys, however, the dominance of eye velocity inputs in determining responsiveness of flocculus Purkinje cells to head rotation has been
emphasized (Lisberger & Fuchs 1978). Since the majority of Purkinje cells
in monkey flocculus exhibit little modulation during rotation in darkness,
it has been assumed that vestibular and eye velocity signals cancel each
other on flocculus Purkinje cells; i.e. vestibular inputs preferentially induce
in-phase modulation, while eye velocity inputs cause out-phase modulation,
which cancels out the former when the eye moves. If this interpretation is
applied to rabbit flocculus, the out-phase modulating during head rotation
in darkness might represent eye velocity inputs overwhelming vestibular
inputs. However, this possibility is unlikely because, as is discussed below,
there is no evidence indicating a powerful eye velocity input to zone II of
rabbit flocculus.

Horizontal rotation in darkness induces a modulation of *complex* spike
discharge in some flocculus Purkinje cells; this suggests that climbing fiber
inputs to the flocculus monitor head rotation (Ghelarducci et al 1975). This
is consistent with an earlier report that caloric stimulation of the labyrinth
induces *complex* spike discharges in Purkinje cells of the cat flocculus
(Ferin et al 1971). However, zone II Purkinje cells of rabbit flocculus do
not respond to head rotation with *complex* spike modulation, which apparently is an event in another zone.

Visual Stimulation

Sinusoidal movement of a vertical narrow slit of light induces modulation
of *simple* spike discharges accompanying the optokinetic response. In zone
II, the modulation usually occurs such that discharges are facilitated by
backward movement and depressed by forward movement of the visual
stimulus (Ito 1977). This modulation was absent in rabbits whose visual
mossy fiber pathway was interrupted at the nucleus reticularis tegmenti
pontis (Miyashita & Nagao 1981). Since repetitive stimulation of zone II
Purkinje cells causes horizontal abduction of the ipsilateral eye, enhanced
discharges during backward movement of optokinetic stimuli and depressed
discharges during forward movement should facilitate the optokinetic response. This view is consistent with the fact that flocculectomy induces a
lowering of the gain of the optokinetic response (see above).

In monkeys, modulation of *simple* spikes induced by a moving visual
target is correlated with the velocity of eye movement, rather than the
velocity of the visual target (Lisberger & Fuchs 1978). It has been claimed
that monkey flocculus receives little visual information, and that Purkinje
cell responses during smooth pursuit eye movement chiefly reflect eye
velocity inputs. The following two experiments indicate that this is not true

in rabbit flocculus. First, Neverov et al (1980) observed that in rabbits, eye movements during reversed afternystagmus in darkness were accompanied by *simple* spike discharges in the Purkinje cells of the flocculus. In contrast to the predominant changes in *simple* spike discharge rates during the optokinetic response (31 out of 89 cells tested), only a small portion of Purkinje cells respond to eye movement in darkness (13 of 89 cells). In Neverov et al's (1980) experiment, unidirectional movement of an optokinetic drum induced nystagmus. Only 8% of the Purkinje cells showing increased discharges during backward slow phase of the afternystagmus (similar to out-phase modulation in sinusoidal head rotation) responded to eye movements in darkness (7 of 89 cells). Second, Miyashita & Nagao (1981) recorded from flocculus Purkinje cells in rabbits whose eye movements were heavily depressed by lesions in vestibular nuclei. In spite of the very low gain of the optokinetic response in these rabbits, the amplitude and phase of the modulation of *simple* spikes in zone II Purkinje cells of lesioned animals was identical with control rabbits. These observations suggest that eye velocity does not provide a major input to rabbit flocculus, at least to zone II.

In zone II of rabbit flocculus, *complex* spike discharges are also modulated by optokinetic stimuli (Ghelarducci et al 1975). Modulation of *complex* spikes occurs in the direction opposite to that of *simple* spikes; *complex* spikes are facilitated by forward, and depressed by backward, movement of visual stimuli. As would be expected, this *complex* spike modulation remains intact after interruption of the visual mossy fiber pathway to the flocculus, but it is abolished after interruption of the climbing fiber pathway to the flocculus (Miyashita & Nagao 1981).

Rapid Readjustment of the VOR by Vision

During rapid, visually guided readjustment of the vestibulo-ocular reflex in rabbits, flocculus Purkinje cells exhibit modulation of *simple* spike discharges, which is apparently due to summation of the modulation accompanying the VOR and optokinetic response separately (Ghelarducci et al 1975). Thus, it appears that flocculus Purkinje cell responses are determined by converging vestibular and visual mossy fiber inputs. During visual suppression of the VOR in monkeys, the Purkinje cells of the flocculus exhibit *simple* spike modulation in phase with head velocity (Lisberger & Fuchs 1978). Lisberger & Fuchs (1978) postulated that this in-phase modulation represents vestibular inputs to the flocculus that are isolated from eye velocity inputs, because visual suppression eliminates eye velocity.

Complex spike discharges are also modulated markedly during rapid readjustment of the VOR by vision in rabbit flocculus (Ghelarducci et al

1975). This modulation, however, simply reflects an optokinetic effect involved in the combined vestibular-visual stimulation.

Adaptive Modification of the VOR

In rabbits, an adaptive increase of the vestibulo-ocular reflex gain is accompanied by an increase in the out-phase modulation of *simple* spike discharges from zone II Purkinje cells. Similarly, an adaptive decrease of the VOR gain is accompanied by a decrease in out-phase modulation. Because of technical difficulties, it has not been possible to follow the time course of these changes in single Purkinje cells, but a population study revealed statistically significant differences in the amplitude of out-phase modulation before and after adaptation (Dufossé et al 1978a). Since out-phase modulation should facilitate the VOR (see above), it is probable that increased or decreased *simple* spike modulation is the cause of the adaptive increase or decrease of the VOR gain; however, an entirely different explanation is possible. If out-phase modulation of *simple* spike discharges reflects mainly eye velocity inputs, changes in the out-phase modulation may merely result from changes in eye movements due to adaptation. The latter explanation is unlikely, though, because eye velocity inputs to zone II of rabbit flocculus do not seem to make a significant contribution to *simple* spike modulation (see above).

In the monkey flocculus, *simple* spike responsiveness of Purkinje cells has been investigated in control and adapted states of the vestibulo-ocular reflex (Miles et al 1980). Gain of the horizontal VOR was either increased by adaptation to X 2 spectacles or decreased by dove prism goggles. A significant increase and decrease of head velocity sensitivity occurred in connection with the increase and decrease of the VOR gain. In monkeys, however, head velocity sensitivity is represented by *simple* spike modulation obtained during smooth pursuit tracking of a visual target moving with the head. The VOR is completely suppressed under these conditions. This modulation is in phase with head velocity, and, therefore, if the modulation is forwarded to relay cells of the horizontal VOR, it should produce an effect opposite to what actually occurs. Mills et al (1980) have argued that such changes in Purkinje cell responsiveness could not underlie the adaptive changes in the VOR and are more probably a secondary consequence of adaptive changes.

These studies on monkeys have not, however, considered the microzonal structure of the flocculus. No attempt was made to apply local stimuli to the flocculus of monkeys for revealing relationships of sampled Purkinje cells to the vestibulo-ocular reflex as done in rabbits (Dufossé et al 1977). One may question whether Purkinje cells examined in monkey flocculus are all related to the horizontal vestibulo-ocular reflex. It must be emphasized that even in rabbit flocculus the population study failed to detect any meaningful change in simple spike modulation correlated with VOR adaptation, unless zone II Purkinje cells were selected with the local stimulation method (Dufossé et al 1978a).

The changes in *simple* spike modulation associated with vestibulo-ocular reflex adaptation in rabbits can be explained in terms of the modifiable neuronal network hypothesis of the cerebellar cortex. Out-phase modulation in zone II Purkinje cells during rotation in darkness may arise from a combination of two factors:

1. inhibition from the ipsilateral horizontal canal,
2. facilitation from the contralateral horizontal canal.

These may be antagonized by another two factors:

3. facilitation from the ipsilateral horizontal canal,
4. inhibition from the contralateral horizontal canal.

Under the conditions of combined vestibular-visual stimulation, where amplitude of out-phase modulation increases adaptively, visual climbing fiber afferents are activated in phase with head velocity (Ghelarducci et al 1975) so that there is a high probability that visual climbing fiber and ipsilateral horizontal canal signals arrive simultaneously in the flocculus. According to the Marr-Albus hypothesis, this pattern of input should selectively influence inhibition from the ipsilateral horizontal canal and/or facilitation from the ipsilateral canal. To increase out-phase modulation, inhibition should be enhanced, or facilitation should be depressed. Similarly, under conditions in which out-phase modulation is decreased adaptively, visual climbing fiber afferents are activated out-phase with head velocity (Ghelarducci et al 1975) so that there is an increased probability that visual climbing fiber and contralateral horizontal canal signals simultaneously arrive in the flocculus, which should influence facilitation from the contralateral horizontal canal and/or inhibition from the contralateral canal. To decrease out-phase modulation, facilitation should be depressed or inhibition should be enhanced. In both cases, facilitatory canal influences should be depressed, while inhibitory canal influences should be enhanced, through simultaneous arrival of climbing fiber and mossy fiber signals. Under the assumption that the major site of plasticity in the cerebellar cortex is at excitatory synapses between granule cells (parallel fibers) and Purkinje cell dendrites (Marr 1969, Albus 1971), it can be expected that adaptive changes in out-phase modulation are produced by depression at parallel fiber-Purkinje cell synapses involved in facilitation from the ipsilateral or contralateral canal.

Depression at parallel fiber-Purkinje cell synapses was indeed postulated by Albus (1971), in place of Marr's (1969) original suggestion that these synapses may be facilitated plastically. Reciprocal changes seen in *complex* and *simple* spike activities in monkey cerebellum during adaptation of hand movement control has also been interpreted as supporting Albus' assumption (Gilbert & Thach 1977). The very recent finding that conjunctive electrical stimulation of a vestibular nerve and the inferior olive causes a drastic depression of the facilitatory effect of the vestibular nerve on flocculus Purkinje cells (Ito et al 1981a,b) is in accordance with the above speculation and supports Albus' assumption.

PROBLEMS TO BE SOLVED

There are still a number of problems with the flocculus hypothesis of the vestibulo-ocular reflex control that need to be solved. These problems include (*a*) neuronal connections within the flocculo-vestibulo-ocular system, which have links still missing in spite of their functional importance, and (*b*) animal species differences in structure and presumably also in functional involvement of the flocculus.

Neuronal Connections of Zones I–III

While Purkinje cells in zone II of rabbit flocculus project to medial vestibular nucleus neurons relaying the horizontal vestibulo-ocular reflex, Purkinje cells in zones I and III project to superior vestibular nucleus neurons relaying vertical and rotatory VORs. Therefore, zones I–III form the vestibulo-ocular part of rabbit flocculus. Afferent pathways to zones I–III have not yet been fully elaborated. For example, even though three components of the visual climbing fiber pathway to zones I–III have successfully been described (cf Ito et al 1977, 1978a, Ito 1980), other components are left unidentified. Here, the principle to be established is that each subgroup of flocculus Purkinje cells, connected with a certain component(s) of the VOR arc, is equipped with a climbing fiber pathway to monitor the performance of that component of the vestibulo-ocular reflex pathway. So far, the principle has been confirmed to hold for only a part of the vestibulo-ocular reflex pathway (Ito et al 1977).

The possible functional significance of a microzone in the flocculus and its relationship to other structures also raises a number of problems. The collateral innervation of vestibular nuclei by visual climbing fiber afferents, which has been suggested for consistency with cerebellar nuclei and Deiters' nucleus (Ito et al 1977, cf also Barmack & Simpson 1980) is supported by a recent horseradish peroxidase study (Balaban et al 1981). However, the implication of this collateral innervation in the control of the VOR is not clear. Recently, mossy fiber inputs to the flocculus have been found to arise from vestibular nuclei (Alley 1977, Yamamoto 1979) and from oculomotor nuclei (Kotchabhakdi & Walberg 1978, Yamamoto 1979). These projections may serve as internal feedback from the VOR arc to the flocculus.

In addition to vestibular and visual mossy fiber afferents, the flocculus receives neck afferents (Wilson et al 1975, 1976) and eye muscle proprioceptive afferents (Maekawa & Kimura 1980). Since neck afferent signals induce a cervico-ocular reflex, it is probable that the neck afferents incorporate the flocculus as a sidepath to the cervico-ocular reflex arc. Extraocular proprioceptive afferents may also incorporate the flocculus in the stretch reflex of extraocular muscles (Mitsui et al 1979). Thus, the flocculus may provide sidepaths for a number of reflexes, serving to stabilize retinal images. Transfer characteristics across these pathways may also be adaptively adjusted so as to minimize retinal error signals conveyed by the visual climbing fiber pathway. In this manner, the flocculus may coordinate a number of ocular reflexes toward stable vision during movements (Figure 2). This idea expands the "flocculus hypothesis of the vestibulo-ocular reflex control" to the "flocculus hypothesis of the coordinated ocular reflex control." This expanded hypothesis may be useful in understanding complex features in neuronal organization found in and around the flocculus.

Conflict Between the Results Derived from Studies of Rabbits and Monkeys

To solve conflicts in applying the flocculus hypothesis of the VOR control to rabbits and monkeys, it is necessary to investigate the mosaic, zonal structure of the monkey flocculus. There are major differences both in types of eye movements made by rabbits and monkeys and in the number of folia in the flocculus. Eye movements are predominantly reflexive in rabbits, consisting of the vestibulo-ocular reflex and optokinetic response. The flocculus contains three to four folia. In monkeys, by contrast, eye movements are predominantly controlled voluntarily, and feature well-developed saccades and smooth pursuit eye movements. The flocculus contains as many

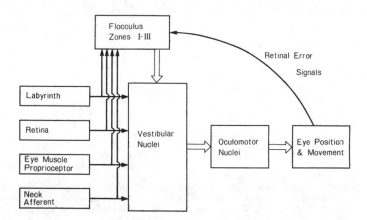

Figure 2   Block diagram for the coordinated ocular reflex control by the flocculus.

as ten folia. Thus, one may suspect that functional involvement of the flocculus is substantially different between the two animal species.

One possible scheme for resolving differences between rabbits and monkeys is that cerebellar control of reflexive eye movements (vestibulo-ocular reflex and optokinetic response) is executed by sidepath connections of zones I–III of the flocculus to the major vestibulo-ocular reflex arc, whereas cerebellar control of voluntary eye movements (saccade and smooth pursuit eye movement) is executed by the cerebellocerebral communication loop involving both the cerebellum and cerebrum. In rabbits, zone IV seems to be incorporated in the cerebellocerebral communication loop. It projects to the caudal portion of the lateral cerebellar nucleus, which sends axons to the pulvinar in cats (Itoh & Mizuno 1979). Zone IV receives mossy fiber inputs from the pontine nucleus, which is the major relay site for cerebral descending signals to the cerebellum. Zone IV does not receive inputs from the nucleus reticularis tegmenti pontis, which relays visual inputs to zones I–III. Zone IV receives climbing fiber afferents from rostral areas of the principal and medial accessory olive, as opposed to that from the dorsal cap and ventrolateral outgrowth of the principal olive to zones I–III (Yamamoto 1979). The function of zone IV in rabbits is not yet clear. Nevertheless, its presence in rabbits raises the possibility that monkey flocculus contains an equivalent zone, and that this zone, as a part of the cerebellocerebral communication loop, contributes to voluntary eye movement control. The possibilities to be tested experimentally are whether horizontal gaze velocity Purkinje cells, as investigated by Miles et al (1980) in connection with smooth pursuit eye movements, are involved in a zone equivalent to zone IV of rabbits that has no direct relevance with the vestibulo-ocular reflex, and whether monkey flocculus contains another zone equivalent to zone II of rabbits that is directly related to the horizontal vestibulo-ocular reflex and exhibits adaptive modification of Purkinje cell responsiveness to head velocity, as predicted by the flocculus hypothesis of the vestibulo-ocular reflex control.

Literature Cited

Albus, J. S. 1971. A theory of cerebellar function. *Math. Biosci.* 10:25–61

Alley, K. A. 1977. Anatomical basis for interaction between cerebellar flocculus and brainstem. In *Control of Gaze by Brain Stem Neurons,* ed. R. G. Baker, A. Berthoz, pp. 109–17. Amsterdam: Elsevier. 514 pp.

Alley, K. A., Baker, R. G., Simpson, J. I. 1975. Afferents to the vestibulocerebellum and the origin of the visual climbing fibers in the rabbit. *Brain Res.* 98:585–89

Amanomori, Y., Takahira, K. 1969. *The physiology of nystagmus. Kansai Ika Daigaku Zasshi* 21:394–99. (In Japanese)

Andersson, G., Oscarsson, O. 1978. Climbing fiber microzones in cerebellar vermis and their projection to different groups of cells in the lateral vestibular nucleus. *Exp. Brain Res.* 32:565–79

Angaut, P., Brodal, A. 1967. The projection of the vestibulocerebellum onto the vestibular nuclei in the cat. *Arch. Ital. Biol.* 105:441–79

Baarsma, E. A., Collewijn, H. 1974. Vestibulo-ocular and optokinetic reactions to rotation and their interaction in the rabbit. *J. Physiol.* 238:603–25

Baker, R. G., Precht, W., Llinás, R. 1972. Cerebellar modulatory action on the vestibulo-trochlear pathway in the cat. *Exp. Brain Res.* 15:364–85

Balaban, C. D., Kawaguchi, Y., Watanabe, E. 1981. Evidence of a collateralized climbing fiber projection from the inferior olive to the flocculus and vestibular nuclei in rabbits. *Neurosci. Lett.* 22:23–29

Barmack, N. J., Simpson, J. I. 1980. Effects of microlesions of dorsal cap of inferior olive of rabbits on optokinetic and vestibulo-ocular reflexes. *J. Neurophysiol.* 43:182–206

Batini, C., Ito, M., Kado, R. T., Jastreboff, P. J., Miyashita, Y. 1979. Interaction between the horizontal vestibulo-ocular reflex and optokinetic response in rabbits. *Exp. Brain Res.* 37:1–15

Brodal, A. 1974. Anatomy of the vestibular nuclei and their connections. In *Handbook of Sensory Physiology*, Vol. 6/Pt.1: *Vestibular System*, ed. H. H. Kornhuber, pp. 239–352. New York: Springer. 676 pp.

Brodal, A., Høivik, B. 1964. Site and mode of termination of primary vestibulocerebellar fibres in the cat. *Arch. Ital. Biol.* 102:1–21

Brodal, A., Pompeiano, O. 1957. The vestibular nuclei in the cat. *J. Anat.* 91:438–54

Carpenter, R. H. S. 1972. Cerebellectomy and the transfer function of the vestibulo-ocular reflex in the decerebrate cat. *Proc. R. Soc. London Ser. B* 181:353–74

Cohen, B. 1974. The vestibulo-ocular reflex arc. See Brodal 1974, pp. 477–540

Cohen, B., Suzuki, J. I., Bender, M. B. 1964. Eye movements from semicircular canal nerve stimulation in the cat. *Ann. Otol.* 73:153–69

Collewijn, H. 1980. The modifiability of the adult vestibulo-ocular reflex. *Trends Neurosci.* 3:98–101

Desclin, J. C., Escubi, J. 1974. Effects of 3-Acetylpyridine on the central nervous system of the rat, as demonstrated by silver methods. *Brain Res.* 77:349–64

Dow, R. S. 1938. Efferent connections of the flocculo-nodular lobe in macca mulatta. *J. Comp. Neurol.* 68:297–305

Dow, R. S., Moruzzi, G. 1958. *The Physiology and Pathology of the Cerebellum.* Minneapolis: Univ. Minnesota Press. 675 pp.

Dufossé, M., Ito, M., Miyashita, Y. 1977. Functional localization in the rabbit's cerebellar flocculus determined in relationship with eye movements. *Neurosci. Lett.* 5:273–77

Dufossé, M., Ito, M., Jastreboff, P. J., Miyashita, Y. 1978a. A neuronal correlate in rabbit's cerebellum to adaptive modification of the vestibulo-ocular reflex. *Brain Res.* 150:511–616

Dufossé, M., Ito, M., Miyashita, Y. 1978b. Diminution and reversal of eye movements induced by local stimulation of rabbit cerebellar flocculus after partial destruction of the inferior olive. *Exp. Brain Res.* 33:139–41

Eccles, J. C., Ito, M., Szentágothai, J. 1967. *The Cerebellum as a Neuronal Machine.* New York/Berlin/Heidelberg: Springer-Verlag. 335 pp.

Ferin, M., Grigorian, R. A., Strata, P. 1971. Mossy and climbing fiber activation in the cat cerebellum by stimulation of the labyrinth. *Exp. Brain Res.* 12:1–17

Fernandez, C., Goldberg, J. M. 1971. Physiology of peripheral neurons innervating semicircular canals of the squirrel monkey. II. Response to sinusoidal stimulation and dynamics of peripheral vestibular system. *J. Neurophysiol.* 34:661–75

Fukuda, J., Highstein, S. M., Ito, M. 1972. Cerebellar inhibitory control of the vestibulo-ocular reflex investigated in rabbit IIIrd nucleus. *Exp. Brain Res.* 14:511–26

Ghelarducci, B., Ito, M., Yagi, N. 1975. Impulse discharges from flocculus Purkinje cells of alert rabbits during visual stimulation combined with horizontal head rotation. *Brain Res.* 87:66–72

Gilbert, P. F. C., Thach, W. T. 1977. Purkinje cell activity during motor learning. *Brain Res.* 128:309–28

Gonshor, A., Melvill Jones, G. 1976a. Short-term adaptive changes in the human vestibulo-ocular reflex arc. *J. Physiol.* 256:361–79

Gonshor, A., Melvill Jones, G. 1976b. Extreme vestibulo-ocular adaptation induced by prolonged optical reversal of vision. *J. Physiol.* 256:381–414

Groenewegen, H. J., Voogd, J. 1977. The parasagittal zonation within the olivocerebellar projection. I. Climbing fiber distribution in the vermis of the cerebellum. *J. Comp. Neurol.* 174:417–88

Hacket, J. T., Hou, S.-M., Cochran, S. L. 1979. Glutamate and synaptic depolarization of Purkinje cells evoked by parallel fibers and by climbing fibers. *Brain Res.* 170:377–80

Haddad, G. M., Demer, J. L., Robinson, D. A. 1980. The effect of lesions of the dorsal cap of the inferior olive on the vestibulo-ocular and optokinetic systems of the cats. *Brain Res.* 185:265–75

Hassul, M., Daniels, P. D., Kimm, J. 1976. Effects of bilateral flocculectomy on the vestibulo-ocular reflex in the chinchilla. *Brain Res.* 118:339–43

Herndon, R. M., Coyle, J. T. 1977. Selective destruction of neurons by a transmitter agonist. *Science* 198:71–72

Herrick, C. P. 1924. Origin and evolution of the cerebellum. *Arch. Neurol. Psychiatr. Chicago* 11:621–52

Hoddevik, G. H. 1978. The projection from nucleus reticularis tegmenti pontis onto the cerebellum in the cat. *Anat. Embryol.* 153:227–42

Hudson, D. B., Valcana, T., Bean, G., Timiras, P. S. 1976. Glutamic acid: A strong candidate as the neurotransmitter of the cerebellar granule cell. *Neurochem. Res.* 1:73–81

Ito, M. 1970. Neurophysiological aspects of the cerebellar motor control system. *Int. J. Neurol.* 7:162–76

Ito, M. 1972. Neural design of the cerebellar motor control system. *Brain Res.* 40: 81–84

Ito, M. 1974. The control mechanisms of cerebellar motor system. In *The Neurosciences, Third Study Program,* ed. F. O. Schmitt, F. G. Worden, pp. 293–303. Boston: MIT Press. 1107 pp.

Ito, M. 1976a. Cerebellar learning control of vestibulo-ocular mechanisms. In *Mechanisms in Transmission of Signals for Conscious Behavior,* ed. T. Desiraju, pp. 1–22. Amsterdam: Elsevier. 376 pp.

Ito, M. 1976b. Adaptive control of reflexes by the cerebellum. In *Understanding the Stretch Reflexes,* ed. R. Granit, S. Homma, pp. 435–44. Amsterdam: Elsevier. 507 pp.

Ito, M. 1977. Neuronal events in the cerebellar flocculus associated with an adaptive modification of the vestibulo-ocular reflex of the rabbit. In *Control of Gaze by Brain Stem Neurons,* ed. R. G. Baker, A. Berthoz, pp. 391–98. Amsterdam: Elsevier. 514 pp.

Ito, M. 1980. Roles of the inferior olive in the cerebellar control of vestibular functions. In *The Inferior Olivary Neurons,* ed. J. Courville, C. de Montigny, Y. Lamarre, pp. 367–77. New York: Raven. 396 pp.

Ito, M., Miyashita, Y. 1975. The effects of chronic destruction of inferior olive upon visual modification of the horizon-tal vestibulo-ocular reflex of rabbits. *Proc. Jpn. Acad.* 51:716–60

Ito, M., Highstein, S. M., Fukuda, J. 1970. Cerebellar inhibition of the vestibulo-ocular reflex in rabbit and cat and its blockage by picrotoxin. *Brain Res.* 17:524–26

Ito, M., Shiida, T., Yagi, N., Yamamoto, M. 1974a. Visual influence on rabbit horizontal vestibulo-ocular reflex presumably effected via the cerebellar flocculus. *Brain Res.* 65:170–74

Ito, M., Shiida, T., Yagi, N., Yamamoto, M. 1974b. The cerebellar modification of rabbit's horizontal vestibulo-ocular reflex induced by sustained head rotation combined with visual stimulation. *Proc. Jpn. Acad.* 50:85–89

Ito, M., Nisimaru, N., Yamamoto, M. 1977. Specific patterns of neuronal connexions involved in the control of the rabbit's vestibulo-ocular reflexes by the cerebellar flocculus. *J. Physiol.* 265: 833–54

Ito, M., Miyashita, Y., Ueki, A. 1978a. Functional localization in the rabbit's inferior olive determined in connection with the vestibulo-ocular reflex. *Neurosci. Lett.* 8:283–87

Ito, M., Orlov, I., Shimoyama, I. 1978b. Reduction of the cerebellar stimulus effect on rat Deiters neurons after chemical destruction of the inferior olive. *Exp. Brain Res.* 33:143–45

Ito, M., Nisimaru, N., Shibuki, K. 1979a. Destruction of inferior olive induces rapid depression in synaptic action of cerebellar Purkinje cells. *Nature* 277:568–69

Ito, M., Jastreboff, P. J., Miyashita, Y. 1979b. Adaptive modification of the rabbit's horizontal vestibulo-ocular reflex during sustained vestibular and optokinetic stimulation. *Exp. Brain Res.* 37:17–30

Ito, M., Jastreboff, P. J., Miyashita, Y. 1980. Retrograde influence of surgical and chemical flocculectomy upon dorsal cap neurons of the inferior olive. *Neurosci. Lett.* 20:45–48

Ito, M., Sakurai, M., Togroach, P. 1981a. Evidence for modifiability of parallel fiber—Purkinje cell synpases. In *Advances in Physiological Sciences,* 2:97–105. Oxford: Pergamon

Ito, M., Sakurai, M., Tongroach, P. 1981b. Climbing fibre induced depression of both mossy fibre responsiveness and glutamate sensitivity of cerebellar Purkinje cells. *J. Physiol. London.* In press

Ito, M., Jastreboff, P. J., Miyashita, Y. 1981c. Specific effects of unilateral lesions in the flocculus upon eye movements of rabbits. *Exp. Brain Res.* In press

Itoh, K., Mizuno, N. 1979. A cerebello-pulvinar projection in the cat as visualized by the use of anterograde transport of horseradish peroxidase. *Brain Res.* 171:131–34

Keller, E. L., Precht, W. 1979. Visual-vestibular responses in vestibular nuclear neurons in the intact and cerebellectomized, alert cat. *Neuroscience* 4:1599–1613

Kotchabhakdi, N., Walberg, F. 1978. Cerebellar afferents from neurons in motor nuclei of cranial nerves demonstrated by retrograde axonal transport of horseradish peroxidase. *Exp. Brain Res.* 31:13–29

Lisberger, S. G., Fuchs, A. F. 1974. Responses of flocculus Purkinje cells to adequate vestibular stimulation in the alert monkey: Fixation vs. compensatory eye movements. *Brain Res.* 69:347–53

Lisberger, S. G., Fuchs, A. F. 1978. Role of primate flocculus during rapid behavioral modification of vestibulo-ocular reflex. I. Purkinje cell activity during visually guided horizontal smooth-pursuit eye movements and passive head rotation. *J. Neurophysiol.* 41:733–63

Llinás, R., Walton, K., Hillman, D. E., Sotelo, C. 1975. Inferior olive: Its role in motor learning. *Science* 190:1230–31

Lorento de Nó, R. 1931. Ausgewalte Kapitel aus der vergleichenden Physiologie des Labyrinthes. Die Augenmuskel-reflexe beim Kaninchen und ihre Grundlagen. *Ergeb. Physiol. Biol. Chem. Exp. Pharmakol.* 32:73–242

Lorente de Nó, R. 1933. Vestibulo-ocular reflex arc. *Arch. Neurol. Psychiatr. Chicago* 30:245–91

Lorente de Nó, R. 1939. Transmission of impulses through cranial motor nuclei. *J. Neurophysiol.* 2:402–64

Maekawa, K., Kimura, M. 1980. Mossy fiber projections to the cerebellar flocculus from the extraocular muscle afferents. *Brain Res.* 191:313–25

Maekawa, K., Simpson, J. I. 1973. Climbing fiber responses evoked in vestibulocerebellum of rabbit from visual system. *J. Neurophysiol.* 36:649–66

Maekawa, K., Takeda, T. 1975. Mossy fiber responses evoked in the cerebellar flocculus of rabbits by stimulation of the optic pathway. *Brain Res.* 98:590–95

Maekawa, K., Takeda, T. 1976. Electrophysiological identification of the climbing and mossy fiber pathways from the rabbit's retina to the contralateral cerebellar flocculus. *Brain Res.* 109:169–74

Maekawa, K., Takeda, T. 1977. Afferent pathways from the visual system to the cerebellar flocculus of the rabbit. See Ito 1977, pp. 187–95

Maekawa, K., Takeda, T. 1978. Origin of the mossy fiber projection. In *Integrative Control Functions of the Brain*, ed. M. Ito, N. Tsukahara, K. Kubota, K. Yagi, 1:93–95. Tokyo: Kodansha; Amsterdam: Elsevier. 457 pp.

Magnus, R. 1924. *Körperstellung.* Berlin: Springer. 740 pp.

Manni, D. E. 1950. Localizzazioni cerebellari corticali nella cavia nota 29: Effetti di lesioni delle "parti vestibolari" del cerevelletto. *Arch. Fisiol.* 50:110–23

Marr, D. 1969. A theory of cerebellar cortex. *J. Physiol.* 202:437–70

Marshall, K. C., Wojtowicz, J. M., Hendelman, W. J. 1980. Patterns of functional synaptic connections in organized cultures of cerebellum. *Neuroscience* 5:1847–57

McBride, W. J., Nadi, N. S., Altman, J., Aprison, M. H. 1976. Effects of selective doses of X irradiation on the levels of several amino acids in the cerebellum of the rat. *Neurochem. Res.* 1:141–52

Miles, F. A., Fuller, J. H. 1974. Adaptive plasticity in the vestibulo-ocular responses of the rhesus monkey. *Brain Res.* 80:512–16

Miles, F. A., Fuller, J. H. 1975. Visual tracking and the primate flocculus. *Science* 189:1000–2

Miles, F. A., Braitman, D. J., Dow, B. M. 1980. Long-term adaptive changes in primate vestibuloocular reflex. IV. Electrophysiological observations in flocculus of adapted monkeys. *J. Neurophysiol.* 43:1477–93

Mitsui, Y., Hirai, K., Akazawa, K., Matsuda, K. 1979. The sensorimotor reflex and strabisms. *Jpn. J. Ophthal.* 23:227–56

Miyashita, Y., Ito, M., Jastreboff, P. J., Maekawa, K., Nagao, S. 1980. Effect upon eye movements of rabbits induced by severance of mossy fiber visual pathway to the cerebellar flocculus. *Brain Res.* 198:210–15

Miyashita, Y., Nagao, S. 1981. Signal contents of Purkinje cell responses in rabbit flocculus to optokinetic stimuli. *J. Jpn. Physiol. Soc.* 43:317

Montarolo, P. G., Raschi, F., Strata, P. 1981. Are the climbing fibres essential for the Purkinje cell inhibitory action? *Exp. Brain Res.* 42:215–18

Neverov, V. P., Šterc, J., Bureš, J. 1980. Electrophysiological correlates of the reversed postoptokinetic nystagmus in the rabbit: Activity of vestibular and floccular neurons. *Brain Res.* 189:355–67

Noda, H., Suzuki, D. A. 1979a. The role of the flocculus of the monkey in saccadic eye movements. *J. Physiol* 294:317–34

Noda, H., Suzuki, D. A. 1979b. The role of the flocculus of the monkey in fixation and smooth pursuit eye movements. *J. Physiol.* 294:335–48

Noda, H., Suzuki, D. A. 1979c. Processing of eye movement signals in the flocculus of the monkey. *J. Physiol* 294:349–64

Precht, W. 1978. *Neuronal Operations in the Vestibular System.* Berlin/Heidelberg/New York: Springer-Verlag. 223 pp.

Precht, W., Llinás, R. 1969. Functional organization of the vestibular afferents to the cerebellar cortex of frog and cat. *Exp. Brain Res.* 9:30–52

Precht, W., Strata, P. 1980. On the pathway mediating optokinetic responses in vestibular nuclear neurons. *Neuroscience* 5:777–87

Ramon y Cajal, S. 1909. *Histologie du Système Nerveux de l'Homme et des Vertébrés.* 1:754–73. Paris: Maloine. 986 pp.

Robinson, D. A. 1974. The effect of cerebellectomy on the cat's vestibulo-ocular integrator. *Brain Res.* 71:195–207

Robinson, D. A. 1976. Adaptive gain control of vestibulo-ocular reflex by the cerebellum. *J. Neurophysiol.* 39:954–69

Rönne, H. 1923. Movements apparents, produits a la vision par verres de lunettes, et la correction de ces movements par les cannaux semicirculaires. *Acta Otolaryngol.* 5:108–10

Sandoval, M. E., Cotman, C. W. 1978. Evaluation of glutamate as a neurotransmitter of cerebellar parallel fibres. *Neuroscience* 3:199–206

Szentágothai, J. 1943. Die zentrale Innervation der Augenbewegungen. *Arch. Psychiatr. Nervenkr.* 116:721–60

Szentágothai, J. 1950. The elementary vestibulo-ocular reflex arc. *J. Neurophysiol.* 13:395–407

Szentágothai, J. 1964. Pathways and synaptic articulation patterns connecting vestibular receptors and oculomotor nuclei. In *Oculomotor System,* ed. M. B. Bender, pp. 205–23. New York: Hoeber

Takemori, S., Cohen, B. 1974a. Visual suppression of vestibular nystagmus in rhesus monkey. *Brain Res.* 72:203–12

Takemori, S., Cohen, B. 1974b. Loss of visual suppression of vestibular nystagmus after flocculus lesions. *Brain Res.* 72:213–24

Thach, W. T. 1968. Discharge of Purkinje and cerebellar nuclear neurons during rapidly alternating arm movements in the monkey. *J. Neurophysiol.* 26:785–97

Tsukahara, N., Kiyohara, T., Ijichi, Y. 1973. The mode of cerebellar control of pupillary light reflex. *Brain Res.* 60:244–48

Van Rossum, J. 1969. *Corticonuclear and Corticovestibular Projections of the Cerebellum.* Assen, Neth.: Van Gorcum & Comp. 169 pp.

Voogd, J. 1964. *The Cerebellum of the Cat. Structure and Fibre Connections.* Assen, Neth.: Van Gorcum & Comp. 215 pp.

Voogd, J., Bigaré, F. 1980. Topographical distribution of olivary and corticonuclear fibers in the cerebellum. A Review. See Ito 1980, pp. 207–34

Wilson, V. J., Melvill Jones, G. 1979. *Mammalian Vestibular Physiology.* New York/London: Plenum. 365 pp.

Wilson, V. J., Maeda, M., Franck, J. I. 1975. Input from neck afferents to the cat flocculus. *Brain Res.* 89:133–38

Wilson, V. J., Maeda, M., Franck, J. I., Shimazu, H. 1976. Mossy fiber neck and second-order labyrinthine projections to cat flocculus. *J. Neurophysiol.* 39: 301–10

Yamamoto, M. 1978. Localization of rabbit's flocculus Purkinje cells projecting to the cerebellar lateral nucleus and the nucleus prepositus hypoglossi, investigated by means of the horseradish peroxidase retrograde axonal transport. *Neurosci. Lett.* 7:197–202

Yamamoto, M. 1979. Topographical representation in rabbit cerebellar flocculus for various inputs from the brainstem investigated by means of retrograde axonal transport of horseradish peroxidase. *Neurosci. Lett.* 12:29–34

Yamamoto, M., Shimoyama, I. 1977. Differential localization of rabbit's flocculus Purkinje cells projecting to the medial and superior vestibular nuclei, investigated by means of the horseradish peroxidase retrograde axonal transport. *Neurosci. Lett.* 5:279–83

Young, A. B., Oster-Granite, M. L., Herndon, R. M., Snyder, S. H. 1974. Glutamic acid: Selective depletion by viral induced granule cell loss in hamster cerebellum. *Brain Res.* 73:1–13

Society for Neuroscience Presidential Symposium

October 1981, Los Angeles, California

Preface: The Origins of Modern Neuroscience,
Eric Kandel

Squid Axon Membrane: Impedance Decrease to
Voltage Clamp, *Kenneth S. Cole*

The Synapse: From Electrical to Chemical
Transmission, *John C. Eccles*

Developmental Neurobiology and the Natural
History of Nerve Growth Factor,
Rita Levi-Montalcini

Cortical Neurobiology: A Slanted Historical
Perspective, *David H. Hubel*

THE ORIGINS OF MODERN NEUROSCIENCE

The task of modern neuroscience is as simple as it is formidable. Stripped of detail, its main aim is to provide an intellectually satisfying set of explanations in cellular and molecular terms of normal mentation: of perception, motor coordination, feeling, thought, and memory. In addition, neuroscientists would ultimately also like to account for the disorders of functions produced by neurological and psychiatric disease.

This task is not new; it was assumed at the turn of the century by the leaders of several subdisciplines that now make up neuroscience: by the neuroanatomist, Ramon y Cajal; by the physiologists, Sherrington and Adrian; by the biochemical pharmacologist, Langley; and by the students of behavior, Pavlov and Freud. There is, however, a fundamental difference between the attitudes of these men and the attitude we now hold. Whereas Pavlov, Freud, and Sherrington, for example, did not influence each other's thinking, contemporary neuroscience is distinguished by a sense of common purpose. As recently as 30 years ago, neuroanatomy, neurophysiology, neurochemistry, and behavior were distinct and isolated disciplines; they were isolated from each other but, equally important, because of their own complexities and parochialisms, neurophysiology and neuroanatomy were isolated from other areas of biology as well. As currently structured, neuroscience has woven into one cloth these previously independent scientific strands. This new arrangement is meaningful both historically and scientifically.

From a historical point of view, the emergence of modern neuroscience is but one of several examples, molecular genetics being another, in which inspired rearrangements of scientific disciplines provided opportunities for novel interactions—interactions which substantially changed the perspective, the technical power, and the excitement of a field.

This, the fifth anniversary of the founding of the *Annual Review of Neuroscience,* and the tenth anniversary of the Society for Neuroscience, therefore seemed an appropriate time to examine the scientific origins and current directions of our field. Indeed, neuroscience is at an interesting juncture, a point of scientific transition. Still actively exploring the gains initiated by a cell-biological approach that had its origins in the mid 1930s, neuroscience is already benefitting from the recent application of new biochemical, immunological, and genetic techniques that go beyond the individual nerve cell to the specific molecules within it. Moreover, many of the major pioneers who brought cell biology to the nervous system are still active, and we can profit from their perspective and the direct accounting of their experience.

The Presidential Symposium of the eleventh Annual Meeting of the Society for Neuroscience, to which the accompanying chapters are contributions, offered the opportunity to bring together leading investigators representing the various strands of neural science in order for them to describe the major influences on their work. I have selected four topics for consideration: (a) excitable membranes, (b) synaptic transmission, (c) developmental neurobiology, and (d) the cerebral cortex and the investigation of higher functions. These are representative selections, intended to highlight key developments rather than to cover systematically the broad panorama of neuroscience. Each illustrates how the emergence of modern neuroscience required the gradual fusion of the component subdisciplines.

It is fair to say that modern study of the biophysical properties of the neuron began with the discovery by Cole and Curtis, in 1938, that the action potential of the squid giant axon is accompanied by a dramatic increase in the resting conductance of the membrane. By achieving a creative merger of physics and cell biology, Cole and Curtis provided the initial evidence for what was later, in its more complete form, termed the "ionic hypothesis." In one dramatic experiment, Cole and Curtis demonstrated that the membrane becomes very leaky to ions during the peak of the action potential. That this leakiness occurs without significant change in the membrane capacity suggested, for the first time, that the increased conductance to ions does not require a fundamental change in the structure of the membrane. This advance was soon followed by the almost simultaneous demonstration by Hodgkin and Huxley and by Curtis and Cole that the action potential does not simply short circuit the resting potential to zero, as predicted by Bernstein, but overshoots the resting potential and reverses it by 50 mV. What was responsible for this overshoot? To answer this question Hodgkin and Katz substituted various ions and found that the overshoot is due to a sudden inrush of Na^+ down its concentration gradient. Cole explored the problem from another point of view, by measuring the constituent currents of the action potential. To do this Cole modified the space clamp, developed for current clamping by his post-doctoral fellow George Marmont, so that membrane voltage was the controlled variable. Thus the voltage clamp was born; a powerful new tool was added to the armamentarium of cellular biophysics. The voltage clamp allowed Cole to hold the voltage of the membrane at any preset potential and to determine the currents that flow through the membrane at that potential as a function of time. Using the voltage clamp, Cole was the first to show that depolarization of the membrane gives rise to an inward current followed by an outward current. In a subsequent detailed analysis, Hodgkin, Huxley, and Katz applied and modified Cole's voltage-clamp techniques and found that the inward current is due to Na^+ and the final outward current to K^+. The experiments by Cole and by Hodgkin, Huxley, and Katz laid the foundation for most of our knowledge of the excitable properties of the axon membrane and the mechanisms of impulse initiation. Cole's paper summarizes his recollections, with emphasis on the development of impedance measuring techniques and the origins of the voltage clamp.

A second advance in the modern era is the great understanding we have obtained of the mechanisms of neuromuscular and central synaptic transmission. This achievement has resulted from combining the tools of biochemical pharmacology developed by Loewi, Dale, and Feldberg with those of modern cellular neurophysiology. In the 1930s the analysis of synaptic transmission focused on the question: Is synaptic transmission at the nerve-muscle synapse and in the central nervous system electrical or chemical? There was strong evidence for chemical transmission at the nerve-muscle synapse, both from pharmacological studies and from the elegant localized extracellular recordings made by Stephen Kuffler on single skeletal muscle fibers. Nonetheless, investigators could begin to answer this question directly only in 1950, following the development of techniques for recording intracellularly from single nerve cells and the establishment of the ionic hypothesis of the resting and action potential. As is often the case in persistent debates, both positions proved correct: depending on the particular synapse, transmission can be either chemical or electrical. Thus, Katz and Fatt analyzed the synapse between motor neurons and skeletal muscles of the frog, and John Eccles and his colleagues analyzed the central connections made by the stretch receptor (1A) afferent fibers onto the motor neurons of the spinal cord. Their combined results clearly demonstrated that transmission at both of these classes of synapses is chemical. The transmitters exert their excitatory or inhibitory effects by acting on specific receptors on the membrane of the postsynaptic cell to open up specific, chemically gated, ion channels. Other synapses were found to operate by electrical means. In his contribution John Eccles describes the exciting "Soup-Sparks" controversy of transmission and recounts how this conflict was experimentally resolved. He then details the subsequent analysis of synaptic transmission at both chemical and electrical synapses.

A complete understanding of the nervous system requires not only an appreciation of how nerve cells function and how they interconnect, but also of how nerve cells differentiate to produce their characteristic actions and interconnections. To understand neuronal differentiation, we will have to identify the signals that influence gene transcription and the post-transcriptional regulatory processes that act within the cell during various phases of a neuron's developmental program. Few neuron-specific regulatory signals have so far been uncovered and analyzed. The first, most interesting, and best understood example is nerve growth factor (NGF), a substance discovered by Rita Levi-Montalcini and Viktor Hamburger in the early 1950s and subsequently brilliantly investigated by Levi-Montalcini and by others. NGF is essential for the survival of sympathetic neurons and certain sensory neurons, and stimulates the outgrowth of their processes. Antibodies to NGF produce an immunosympathectomy—a selective destruction of sympathetic neurons. Historians of science are likely to regard the discovery of NGF and the first isolations of chemical transmitter substances as the beginning of a molecular exploration of the nervous system (predating by a decade the analysis of the acetylcholine receptor). In the case of NGF, its analysis reflects a new partnership between neuroembryology and modern biochemistry, particularly protein chemistry. NGF has provided a model of the level of understanding that needs to be attained for the actions of other factors that govern the development and functioning of the nervous system. In her discussion, Rita Levi-Montalcini recounts the discovery and exploration of NGF.

The ultimate goal of neuroscience is to understand mentation: how the brain perceives and initiates action, how it learns and remembers. Of the four areas considered here, these higher functions have traditionally been and remain the most difficult to analyze. Nevertheless, one of the major accomplishments of the last 15 years, an accomplishment recognized this year by the Nobel Prize in Medicine and Physiology, has been the remarkable progress in the investigation of higher functions. Particularly surprising has been how much we have learned by applying cellular techniques to the study of cerebral cortex. By exploiting the recent revolution in neuroanatomical methodology, investigators of cortical function, such as David Hubel and Torsten Wiesel, have united modern morphological and electrophysiological approaches with those of sensory psychophysics to give us insights of surpassing importance into how structure determines function and how single cells interconnect to produce alterations in the coding of visual information in the cerebral cortex. These experiments have given rise to a family of concepts that are now intrinsic to our understanding of cortical function: the transformation of receptive fields at different relay points, binocular interaction on the level of individual cells, the existence of columns for axes of orientation as well as for ocular dominance, and the plasticity that the cells in these columns show in response to experimentally produced visual deprivation or eye deviation (strabismus). In a relatively short period of 25 years, our view of the cerebral cortex has changed dramatically. In his contribution David Hubel details aspects of this explosion in our understanding of the visual cortex.

Eric Kandel

President, Society for Neuroscience

Ann. Rev. Neurosci. 1982. 5:305–23

SQUID AXON MEMBRANE:
Impedance Decrease to Voltage Clamp

Kenneth S. Cole

Department of Neurosciences, School of Medicine, University of California, San Diego, La Jolla, California 92093

Introduction

Galvani started the serious investigations of electrical nerve excitability with his frog leg experiment and, in his controversy with Volta, helped the start of electrochemistry. Both fields progressed with the newest apparatus and best concepts—I've heard that for a while a nerve-muscle preparation was considerably the most sensitive electrical measurer. Among the 20 good experimenters and theorists, Nernst led the way with applications of his thermodynamic concepts of excitability. Following him was Bernstein, who produced a most durable and reasonable theory of nerve structure and function.

My first efforts on nerve followed the St. Louis School—Erlanger, Blair, and Bishop—in which excitability was measured in terms of itself. In this I found that a number of phenomena could be related by the principle of superposition and I published this in the first 1933 Cold Spring Harbor Symposium on Quantitative Biology (Cole 1933) to help replace drop-outs who didn't think our symposium would fly. I was barely ahead of Rashevsky, Monnier, and A. V. Hill with their two-factor, excitation and accommodation, formulations. This era was nicely wrapped up by Katz with his book, "Electric Excitation of Nerve," in which he was just able to include our squid axon membrane impedance change as I was about to give it at the Physiological Congress in 1938.

I'd heard of L. W. Williams's 1912 squid monograph in 1927, but it meant nothing until John Zed Young rediscovered the giant axons that he told us about in 1936 at Cold Spring Harbor. This was a most dramatic passing on of the baton, from an anatomist to biophysicists. Young was clearly the pivotal character who turned axon research from the poorly

305

0147-006X/82/0301-0305$02.00

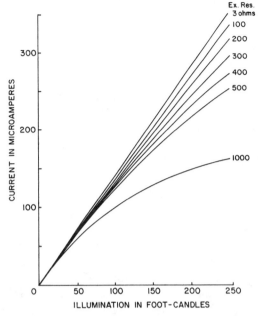

Figure 1 Photronic cell characteristics. Illumination vs current for external resistances as marked. After (Romain 1933).

specified multifiber terms to the clearly defined physical and electrical axon properties—such as μF cm^{-2}, phase angle, Ω cm, and cycles sec^{-1}—that have dominated my career.

Early Years

Some fifty years ago, I first heard of the selenium photovoltaic "Photronic Cell" from Weston Instruments. The published data of output current as a function of illumination (Figure 1) showed that the lower external resistances not only gave larger currents, but they became linear in light intensity when extrapolated to a zero external resistance (Romain 1933). The Poggendorff potentiometer was a favorite in those days, measuring potentials without current flow, but how was a current flowing without an external potential difference to be measured? I designed, but didn't build, a circuit similar to that published by Wood (1936). It was also about the same as the one Ussing & Zehran (1951) used to prove that frog skin was a sodium pumper. In retrospect, this was my first encounter with a voltage clamp, and the power of interchanging currents and potentials made a deep impression on me.

Kenneth S. Cole in 1952.

During graduate work in physics at Cornell, I picked up some bits of information that were to come in handy. I remember being impressed by the nonlinear work of the engineer, van der Pol, on relaxation oscillators (1926) as well as Barkhausen's negative resistances (1926). I heard later about van der Pol's (1934) oscillation theory, Thévenin's theorem (Guillemin 1935) (which was to reinforce my Photronic experience), and the complex biquadratic equation treating a second reactance. Somewhere along the line, perhaps from Steimel (1930), I learned about stability conditions—as apparently evey physicist does.

My introduction to biology, as I've told so often (Cole 1979), had been in 1923 in Fricke's laboratory where I checked the calibrations of his fabulously complicated Wheatstone bridge for measuring impedances. I managed to avoid a similar arrangement until 1935, when I finally had to spend the year designing and building the beautiful instrument that is still operating at Woods Hole. In it (Cole & Curtis 1937) we balanced the unknown against a similar arrangement on the opposite side and then, for high accuracy, replaced the unknown by a carefully calibrated variable electrolytic cell and capacitors. (An example of its performance is Figure 3.) Then H. J. Curtis joined me and we rushed to get data for a paper at the 1936 Cold Spring Harbor Symposium.

Squid Axon Impedance

After Young (1936) sold us on the squid giant axon at tbe Cold Spring Harbor Symposium, Curtis and I used *Nitella* (1937) during the winter and went to Woods Hole for squid (1938) to do our first external electrode work on single cells. The transverse impedances both gave about 1 μF cm^{-2} membrane capacities, with dielectric loss. However, in 1938, with better equipment, and with *Nitella* again leading the way, we got impedance decreases of usually a few percent during the passage of an impulse (Figure 2) (Cole & Curtis 1939). Alan Hodgkin paid us a surprise visit when we had the squid axon impedance change on the oscilloscope—he literally jumped up and down. Ralph Gerard was the first to reproduce this figure, which became a classic in his "Unresting Cells." Then, with an assist from Sten Knudsen, it appeared 15 years later in the Danish "La Femme" as a wall decoration in a newlywed's apartment, to be "most modern art greatly admired by all." When I showed the slide in Seattle, Walt Woodbury said they were furnishing their house in Danish modern—wouldn't I send him a print? Yes, but only if he'd give me a picture, which he did in 1965. We also replaced the time axis by the impedance change on the scope. This I called an owl, which Bill Adelman painted for my birthday two summers ago.

The impedance decrease did not begin until the inflection in the action potential occurred and, in squid, went to about 30 X the resting conductance of 1 mmho cm^{-2} that Hodgkin and I found later (Cole & Hodgkin 1939). A neat analysis (Figure 3) showed the membrane as an almost fixed 1 μF cm^{-2} capacitor, shunted only by a resistor decreasing with activity. These measurements showed for the first time that this action potential involved only an increased conductance to ions. They further made it evident that there could be no more than a slight

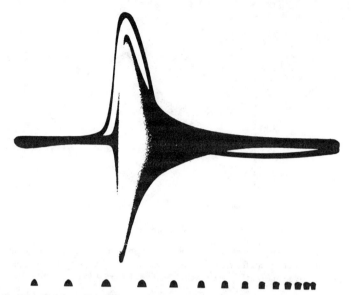

Figure 2   Time in msec vs action potential, line, and impedance decrease band, during passage of impulse in giant axon (Cole 1968, Figure 2:12).

change of membrane structure, either general or localized. This work has been described as "the beginning of a new era of axonology."

Baker and I did the same sort of a high frequency experiment (Cole & Baker 1941a), with a similar result for applied currents—a steady state increased conductance, independent of frequency above 5 kHz. We then knew how many ions got through the membrane and when during an action potential, but I had no idea how the two were related. Baker and I found the inductive reactance, which Hodgkin and I had found at low frequency, to be in the membrane at about 0.1 henry cm^{-2} (Cole & Baker 1941b) and it was subsequently explained by potassium ion nonlinearity or "delayed rectification." This and a similar anomalous capacity could be explained with a potassium permeable membrane between high and low potassium concentrations. A sudden change from a steady state field would give, first, an ohmic or linear change of current. Then as the ions began to move, the conductance would increase or decrease respectively, as the number of ions in the membrane was increased or decreased (Cole 1968, Figures 2:38 and 2:50).

We then (Cole & Curtis 1941) turned to early oscilloscope records of internal potential changes for currents applied externally and—ignoring the initial oscillations and impulses—we had perfectly smooth V,I curves. I could convert these into uniform membrane potentials and current densities, by what Hodgkin called Cole's theorem, with a high rectification ratio of about 100 to 1.

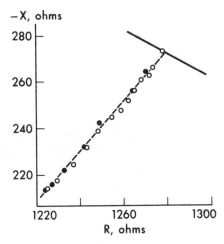

Figure 3 Complex plane impedance, real, R, vs imaginary part, X. *Solid points* for rising phases of action, *open circles* for recovery. *Solid arc* for rest, *dashed arc* for pure membrane conductance increase at 10 kHz, best data (Cole 1968, Figure 2:15).

So we had confusing membrane properties: high frequency conductance increases, independent of frequency for a passing impulse and for current flow; an inductive reactance; and highly nonlinear V,I steady state curves, showing almost no trace of the impulse threshold. Although I didn't think that clearly at the time, around 1940, all of the excitation and propagation effects must have been happening between about 5 kHz and direct current.

It had long been known that the foot of an action potential was usually an exponential function of time as the impulse bore down on the recording point. Before the start of the conductance increase, the Kelvin cable equation (Figure 4) could be juggled (Cole 1968, p. 142) to give the conductance current V,I within the impulse (Figure 5). The negative resistance of –45 Ω cm^2 in the rising potential portion had to be considered very seriously as the driving force for the impulse. But what kind of a negative resistance was it? We have usually made a potential difference the independent variable and the current flow dependent so a V,I plot corresponds to the usual x,y plot. Then the usual resistances or conductances go northeast and southwest. But if a device has a northwest and southeast region, such as AB of Figure 6, we consider that region as a negative resistance or conductance segment, which produces energy instead of consuming it. And it is simple to specify the two types of devices shown as N or S according to the Roman capital letter they resemble on Figure 6. Bell Laboratories had shown me their thermistor about 1940—a uranium bead between two platinum wires —which gave a steady state S characteristic as shown on a V,I diagram, along with many other systems. But there were also N characteristics, and

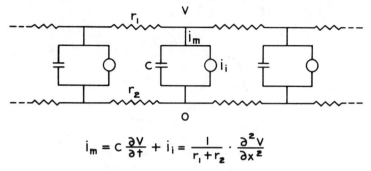

$$i_m = C \frac{\partial V}{\partial t} + i_i = \frac{1}{r_1 + r_2} \cdot \frac{\partial^2 V}{\partial x^2}$$

Figure 4   Kelvin cable schematic, above; equation, below; C, membrane capacity; r_2, r_1 internal, external resistance; i_m, membrane current; i_i conduction current (Cole 1968, Figure 2:21).

how could we choose between them when we'd seen only the region AB in an impulse?

Iron Wire

James Bartlett, a theoretical physics graduate student I'd known at Harvard and Leipzig, came in 1940 to Cornell Medical School in New York for his sabbatical. He proposed to work on the Ostwald-Lillie passive iron wire nerve analogue of iron in concentrated nitric acid. In this, a bit of zinc can activate a region that will grow, propagate, and recover. A potential applied to a small region produces a current on the left-hand passive limb of Figure 7a until the dashed line is reached, which the current and potential follow

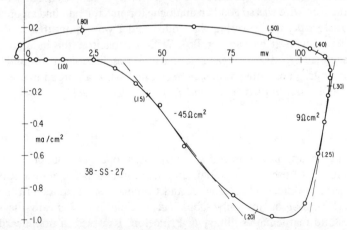

Figure 5   Membrane potential vs conduction current during impulse, after passive foot. Time in msec in parentheses (Cole 1968, Figure 2:23).

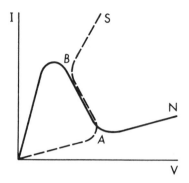

Figure 6   Schematics of negative resistances, potential vs current. *Solid line* for N characteristics, *dashed line* for S (Cole 1968, Figure 2:62).

to the right-hand solid, active, region. But recovery can only take place along the lower dashed line, as indicated. I was thoroughly puzzled by these absolutely straight parallel transient lines that he'd found, in old literature, for the hysteresis loop between the passive and active states. Finally, the light dawned: they were the load lines for the external equipment, and the iron-acid surface had an N characteristic. I insisted, and Jim agreed, that the only way to investigate the transition region was with as low resistance external circuitry as possible to give only one solution, B (Figure 7b), instead of three, as for a higher external resistance, A. Although he could use a small uniform surface, his results were far from simple (Bartlett 1945). Since the iron wire was so good an analogue for an axon and had an obvious N characteristic, my wavering was ended and I was committed to the N axon behavior used by van der Pol. With a capacitor, the inductor and three-segment N resistor, I was able to explain oscillations, threshold, and a spike by graphical integrations, as recounted—I'm afraid ad nauseum—in my book, *Membranes, Ions, and Impulses* (Cole 1968).

War

The battle of Britain was on and as our overseas colleagues fought it, George Marmont and I used longitudinal squid axon impedance measurements to investigate the effects of external K and Ca ions. I spent much of the year of Pearl Harbor on my only sabbatical at Fine Hall, beneath the joint physics and mathematics library at Princeton. I reveled in nonlinearity—mostly what I could glean from Russian and Japanese equations and illustrations, although I could not see how to use any of their techniques for a squid axon.

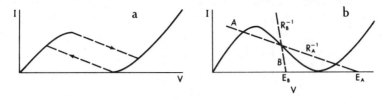

Figure 7   Sketches of iron wire analogue behaviors, potential vs current. Left, typical hysteresis loop. Right, dashed load lines; A, high resistance unstable in negative region; B, low resistance stable through negative region (Cole 1968, Figure 2:66).

Marmont and I did war chores for four years but, near the end, I painfully slugged through a few membrane admittance (reciprocal impedance) loci, such as Figure 8 for 2 × K. The low frequency conductance and inductive limb were almost certainly potassium. The excursion into the left half plane was a negative conductance, but it and the higher frequency capacity limb were complete mysteries. I also started work on the first two parts of Minorsky's *Introduction to Non-linear Mechanics* (1947), which had been leaked to me through an unclassified channel. I put most of my prewar work in order (Cole 1968, p. 218), and I found I could propagate an impulse graphically.

Post War

Marmont and I shifted our base to the University of Chicago just in time to return to Woods Hole for squid in 1946. It was a disastrous summer, as we tried to get effects of current flow on longitudinal impedance. Near the end of the summer, Jimmie Savage, a mathematician, showed up. He'd had a Rockefeller Fellowship foisted on him so he could work with me. He asked, why didn't we just put an electrode in the axoplasm and measure across the membrane to an outside electrode? In spite of my objections that an unruly electrode polarization impedance would swamp any membrane effect, Marmont put Will Rall to work drawing 100 μm glass tubes while Marmont worked on a photographically developed silver chloride electrode.

Space and Current Control

After we returned to Chicago, Marmont presented me with a well-thought out experimental plan. He would line up 500 μm diameter holes in plastic discs spaced 3 mm apart and place a concentric spiral electrode in each of the three outside chambers (Figure 9). After pulling the axon through the holes, he would run a 100 μm glass tube, coated at the end for 12 mm with his treated silver electrode, down the center of the axon. The end outside electrodes served as guards to prevent things happening at the ends from

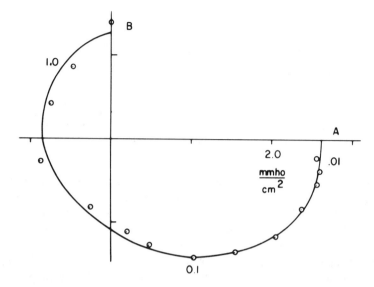

Figure 8 Membrane ionic admittance locus; A, real, vs B, imaginary, parts as derived from longitudinal impedance data at 2 X external K (Cole 1968, Figure 1:51).

reaching the center measuring electrode. All potentials were measured between the axial electrode and the center external electrode. The Kelvin cable equation (Figure 4) was a partial differential equation with distance and time as independent variables. But with inside and outside electrodes, r_2 and r_1 were negligible, spatial variations were eliminated, and the remaining terms were just an ordinary differential equation in time—the axon was "space clamped." Marmont would then control the center chamber current by electronic feedback of the difference between it and a command.

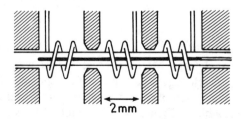

Figure 9 Sketch of space clamp for axon, with axial electrode inside, outside measuring electrode between guard electrodes and separated by insulating spacers (Cole 1968, Figure 3:2).

In this control system a signal proportional to the membrane current and the desired, command, current were fed into the inputs of a differential amplifier. The amplifier output was then applied to the axial electrode and the potential recorded. Marmont probably got the idea for this system from H. W. Bode of Bell Labs and used it in his war work. Thus, no matter how the membrane potential changed, the current density would be under control. As far as I know it was the first application of such an electronic control for a biological system. I added that, conversely, it would be possible to control the membrane potential and to measure the current necessary to hold it; Marmont agreed, but without any particular enthusiasm.

At Woods Hole in 1947, after we got the bugs out of the new equipment, we made quite a series of tests, many of them near and above the excitation threshold, and measured directly many numbers that we'd had to calculate or guess at before (Marmont 1949). I was more interested in linear behaviors near rest than in excitation thresholds, and the results for a short, small current step were useful (Figure 10). There were jumps of potential from the resistance in series with the membrane with a nearly linear rise between, as the 1 μF cm^{-2} membrane capacity was charged. But the jumps were surprisingly large. For external sea water, the calculated axoplasm resistivity averaged about 100 Ω cm as compared with Hodgkin's and my less direct determination of 1.4 X sea water or about 30 Ω cm. I had suspected from the large and erratic internal resistivity calculations on other cells that there might be a series resistance directly associated with the membrane capacity. On this basis, the squid membrane had a resistivity of 5 to 3 Ω cm^2 depending on whether the axoplasm resistivity was once or twice that of sea water.

As I saw more of the space-current clamped action potentials after short stimuli, I realized that—with zero externally applied current—the membrane ion conduction current must be supplied by the membrane capacity charge—and similar to Figure 5, which had no propagating factor. I became

23 (14)

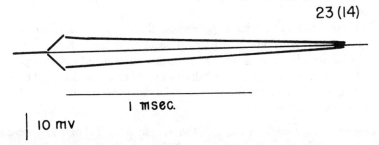

I msec.

10 mv

Figure 10   Current control: Time vs potential. Showing series resistance, membrane capacity, and conductance (Cole 1968, Figure 3:7).

convinced that the capacity had to be kept from functioning and the only way to do this was to keep a constant potential across it. Then we would have only the conduction current between the inner and outer electrodes. These currents could be measured as a function of time and there should be no threshold of current or potential for excitation if the membrane did indeed have the N characteristic that seemed almost obvious. Then everything fell into place.

Potential Control

We had to measure the membrane current through an external short circuit —as for the Photronic cell. But all of this was only bits and pieces of theory, approximations, and assumptions and, as I've said often, theories come at about a dime a dozen. Nonetheless, it could now be tried with a real, live axon. Further, a constant potential across a membrane had never been used before.

Marmont, however, was firmly opposed to wasting time and effort on any such silly and unphysiological work and he tried to divert me into intriguing threshold experiments. But he had a spare terminal block to connect the axon to the electronics and I wired it for potential control (Figure 11). Eventually he gave me permission to use the equipment for my experiment and finally even agreed to run the electronics—making it completely clear that he would take no responsibility for the design of such an experiment or for whatever the results might be. With this cooperation, I managed to do only about four experiments. But for the only time in my long memory, even the first experiment worked and all agreed in general (Figure 12). I had made excitation stand still in space, if not in time.

1. There were the short initial capacity transients to be expected as the command potential was changed from rest to less negative constant potentials.
2. There was not even a slight indication of threshold response as the potential was changed and probably the N characteristic was quasistable.
3. For moderate depolarizing potentials, there was a transient inward "wrong way" current—which could account for excitation.
4. At high depolarizations, the early inward current did not appear.
5. All of the early currents eventually turned later to the outward current to be expected for potassium—except for the droop from the axial electrode polarization.

I wanted to see such behavior on the V,I plane but I didn't know where to put the early current. Probably only because it was easy, I took the maxima as an approximation. This, however, brought me face to face with

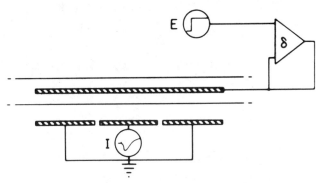

Figure 11   Potential control: Current, I, through membrane and center, guarded external electrode to give axial electrode potential command, E (Cole 1968, Figure 3:15).

the resistance in series with the capacity. I knew it reasonably well from current clamp experiments and I put it into the calculation to give Figure 13. The near horizontal dashed line has the slope of the resting membrane conductance and the voltage step aims through the capacity transient to one of the points shown by an X. About as it arrives there, the early inward current begins and the point goes down the series conductance line to a solid point for the inward maximum. Here it turns and goes back up the line for the open circle outward current maximum.

Here was a clear indication of the negative resistance—but I was horrified to see how close the series resistance line came to it and instability. If the series conductance had been only a little less and/or the membrane negative conductance only a little larger, the two lines would have had three intersections and would have been unstable between the two outer ones.

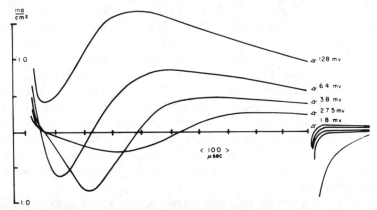

Figure 12   Potential control: Time vs membrane current after indicated depolarizations from rest potentials (Cole 1949; Cole 1968, Figure 3:16).

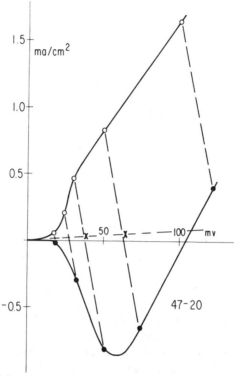

Figure 13   Potential vs current under potential control. Passive capacity transients head to x points, along series resistance *dashed lines* to early current peaks, *solid points,* and then to steady state, *open circles* (Cole 1968, Figure 3:19).

These curves I integrated graphically for a space-clamped response (Figure 14). Up to depolarizations of 18 mV there was a slight current outward from the axon. But between 18 mV and 27.5 mV the inflow became large enough to produce a net inflow and a threshold. This current then increased to its largest value of about 0.8 mA cm^{-2}, while charging the capacity at 800 V sec^{-1}, which was found at 40–50 mV or about half way up the action potential. The spike height was given by the potential at which the early current vanished—somewhere between 64 mV and 128 mV above the resting potential. It was harder to estimate whether or not such an action would propagate. Then at later times, as the clamp membrane current changed its direction, the action potential went down through its point of inflection and returned to rest potential, ready for another impulse.

I was ecstatic that my assumptions and approximations all along the line had not been so bad as to keep me from proving my major goals experimentally. The conductance increase and the threshold and all-or-none charac-

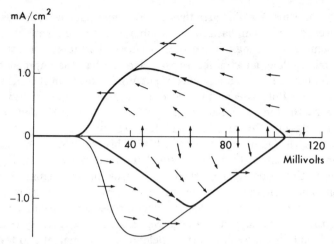

Figure 14 V vs I. Graphical action potential path, *heavy lines;* potential control data, *light lines* (Cole 1968, Figure 3:21).

teristics of the spike could be explained by the negative slope characteristic and capacity of the membrane as seen in Figures 5 and 14. But I persisted in trying to explain my data on the basis of a single independent variable —such as potassium. Hodgkin had written, near the end of the war when there were no new or major radar problems, that he had convinced himself the action potential overshoot could and must be caused by an increase of sodium permeability. However, I did not appreciate this—I was addicted to calcium, which had such dramatic effects (Cole 1968, Figure 1:49).

Hodgkin's Visit

As Hodgkin related in his Physiological Society Centennial address (Hodgkin 1976), I broke the news of what we'd been up to in the Fall of 1947. This he wanted to see and we made arrangements for him to visit Chicago —probably in early April, 1948. In order to take his Rockefeller money out of England for the trip, he had to have a foreign invitation, so I arranged that. But he was disgusted to find that I'd also arranged for him to give the annual Biology Division lecture. He had slides and gave his usual compelling and convincing dissertation to a full house on his postwar work with Katz. This essentially proved that external sodium ions, invoked for the first time since Overton did so in 1902, were the cause of the action potential overshoot beyond the resting potential (Hodgkin & Katz 1949).

I tried telling Hodgkin my story much as I have here, but he seemed more interested in my experiments rather than how I got there. I have no recollection of his commenting on my voltage clamp, my N characteristic or the stability problems. We did agree that a second internal electrode was obviously needed, and I emphasized my trouble with the series resistance.

I remember his telling me that he and Huxley had talked of a string of micropipettes along an axon for a space clamp. I had the impression that neither he nor Huxley had any idea of potential control until I mentioned it in my letter. But he seemed to accept the concept without question and was already thinking ahead about using it to test his and Huxley's carrier theory (Hodgkin 1976) that they had been working on—especially when frozen out of the lab by lack of heat in 1947. This required a near instantaneous maximum inward current for small and moderate depolarizing voltage clamps. Consequently, he argued, my records, such as Figure 12, would be wrong and their relatively slow rise of inward current must be a fault of the equipment. Although he had no specific objections, he could only say that anything could happen in a negative resistance feedback system such as ours. I had only a weak theory to defend and no analogue to test, but Hodgkin granted that, if my measurements were valid, they "shot down" the carrier theory and we left it at that. Hodgkin and Huxley borrowed all of the various pieces of apparatus to repeat my experiments and, by late 1948, Hodgkin wrote that they had confirmed my results!

Paris Symposium

I moved to the Naval Medical Research Institute in early 1949 and went to the Paris Symposium on "Electrophysiologie des Transmissions et Facteurs Ioniques" in honor of Louis Lapicque in April. I gave only a short introductory paper, telling of my earlier work and Marmont's current clamp development, and closed with my Figure 12 and a short paragraph on my voltage clamp results (Cole 1949). But I was surprised the next day when Hodgkin gave the paper with Huxley and Katz (1949), using both the terms "potential control" and "voltage clamp," confirming most of our results with details of their modification of my circuit, the effects of changing external sodium, and giving the Hodgkin and Huxley carrier model.

It seemed obvious to me that he had thought, at Chicago, of my early inward currents as sodium, although I remember no mention of it. I'm still curious as to when he thought of feedback compensation for the resistance in series with the membrane capacity—particularly after I. I. Rabi's glowing report to the Radiation Laboratory of Hodgkin's—a physiologist's—war achievements in the Royal Air Force radar work.

In the summer of 1949, they had given up the carrier model, added external guard electrodes and, going to low temperatures, got most of the data for their 1952 series (Hodgkin 1976).

1952 Series

In the late Spring of 1952, Hodgkin sent me the manuscripts of their five papers to appear in the *Journal of Physiology* and at about the same time I was invited to attend the upcoming Cold Spring Harbor Symposium. I was profoundly impressed by the magnitude of the difficulties they had overcome and by the breadth and depth of detail of the work. As a whole, the 91 pages (Hodgkin et al 1952, Hodgkin & Huxley, 1952a–d) were a truly magnificent advance in axonology. Hodgkin gave an excellent summary at Cold Spring Harbor (Hodgkin & Huxley 1952e) of the entire work and, as he said in discussion, they had confirmed my 1947 results. Most remarkable of all, however, was that the entire achievement was based on a measurement concept that had been recognized and proven only five years before.

In spite of my uncompensated series resistance and polarizing axial electrode, they had proven that my early inward currents were indeed sodium while the late outward currents were potassium as I'd suspected. There was, however, one startling difference. My time scales were roughly 10 X faster and so the capacity transients appeared much longer. But these they could have explained by their finding (Hodgkin et al 1952) of the high temperature coefficient for the ion conduction processes. At Woods Hole we had no air conditioning, a lot of electronics, and no local cooling, so our axon temperatures were 25°C to 30°C—corresponding to their Q_{10} of 3!

They might also have pointed out that their first clamp currents (Figures 11 and 12, Hodgkin et al 1952) and their final experimental patterns, shown to test their analytical analyses (Figure 11, Hodgkin & Huxley 1952d), were very similar to those Hodgkin had seen in Chicago in 1948 (Figure 14).

Conclusion

A long, torturous path, leading to a test of the voltage clamp concept in 1947, established it as the valid basis for the Hodgkin and Huxley contribution in 1952. Following their spectacular success, similar approaches have been used in so many and such diverse fields as to justify the statement, "The voltage clamp revolutionized electrobiology." It is difficult to fault those who attribute the voltage clamp to Hodgkin and Huxley. I finally realized that I was probably the only one who could reorganize and present the pertinent material—mostly from my book (Cole 1968), and my memory —as I have now done.

ACKNOWLEDGMENTS

A condensed version of this article was given at the Presidential Symposium of the Annual Meeting of the Society for Neuroscience, October 19, 1981 in Los Angeles, California. President Eric Kandel has been most helpful and Dr. Max Cowan has also been very patient.

Literature Cited

Barkhausen, H. 1926. Warum kehren die für den Lichtbogen gültigen Stabilitätes-bedingungen bei Elektronenröhren um? *Phys. Z.* 27:43–46

Bartlett, J. H. Jr. 1945. Transient anode phenomena. *Trans. Electrochem. Soc.* 87: 521–45

Cole, K. S. 1933. Electric excitation in nerve. *Cold Spring Harbor Symp. Quant. Biol.* 1:131–37

Cole, K. S. 1949. Dynamic electrical characteristics of the squid axon membrane. *Arch. Sci. Physiol.* 3:253–58

Cole, K. S. 1968. *Membranes, Ions and Impulses.* Berkeley: Univ. Calif. Press, 569 pp. (Reprinted 1972)

Cole, K. S. 1979. Mostly membranes. *Ann. Rev. Physiol.* 41:1–24

Cole, K. S., Baker, R. F. 1941a. Transverse impedance of the squid giant axon during current flow. *J. Gen. Physiol.* 24:535–49

Cole, K. S., Baker, R. F. 1941b. Longitudinal impedance of the squid giant axon. *J. Gen. Physiol.* 24:771–88

Cole, K. S., Curtis, H. J. 1937. Wheatstone bridge and electrolytic resistor for measurements over a wide frequency range. *Rev. Sci. Inst.* 8:333–39

Cole, K. S., Curtis, H. J. 1938. Electric impedance of *Nitella* during activity. *J. Gen. Physiol.* 22:37–64

Cole, K. S., Curtis, H. J. 1939. Electric impedance of squid giant axon during activity. *J. Gen. Physiol.* 22:649–70

Cole, K. S., Curtis, H. J. 1941. Membrane potential of the squid giant axon during current flow. *J. Gen. Physiol.* 24:551–63

Cole, K. S., Hodgkin, A. L. 1939. Membrane and protoplasm resistance in the squid giant axon. *J. Gen. Physiol.* 22:671–87

Curtis, H. J., Cole, K. S. 1937. Transverse electric impedance of *Nitella. J. Gen. Physiol.* 21:189–201

Curtis, H. J., Cole, K. S. 1938. Transverse electric impedance of the squid giant axon. *J. Gen. Physiol.* 21:757–65

Guillemin, E. A. 1935. *Communication Networks,* 2:181. New York: Wiley. Vol. 2, 587 pp.

Hodgkin, A. L. 1976. Chance and design in electrophysiology: An informal account of certain experiments on nerve carried out between 1934 and 1952. *J. Physiol.* 263:1–21

Hodgkin, A. L., Huxley, A. F. 1952a. Currents carried by sodium and potassium ions through the membrane of the giant axon of *Loligo. J. Physiol.* 116:449–72

Hodgkin, A. L., Huxley, A. F. 1952b. The components of membrane conductance in the giant axon of *Loligo. J. Physiol.* 116:473–96

Hodgkin, A. L., Huxley, A. F. 1952c. The dual effect of membrane potential on sodium conductance in the giant axon of *Loligo. J. Physiol.* 116:497–506

Hodgkin, A. L., Huxley, A. F. 1952d. A quantitative description of membrane current and its application to conduction and excitation in nerve. *J. Physiol.* 117:500–44

Hodgkin, A. L., Huxley, A. F. 1952e. Movements of sodium and potassium ions during nervous activity. *Cold Spring Harbor Symp. Quant. Biol.* 17:43–52

Hodgkin, A. L., Huxley, A. F., Katz, B. 1949. Ionic currents underlying activity in the giant axon of the squid. *Arch. Sci. Physiol.* 3:129–50

Hodgkin, A. L., Huxley, A. F., Katz, B. 1952. Measurement of current-voltage relations of the giant axon of *Loligo. J. Physiol.* 116:424–48

Hodgkin, A. L., Katz, B. 1949. The effect of sodium ions on the electrical activity of the giant axon of the squid. *J. Physiol.* 108:37–77

Marmont, G. 1949. Studies on the axon membrane. I. A. new method. *J. Cell. Comp. Physiol.* 34:351–82

Minorsky, N. 1947. *Introduction to Non-linear Mechanics.* Ann Arbor, MI: Edwards. 447 pp.

Overton, E. 1902. Beitrage zur allgemeinen Muskel—und Nervenphysiologie. *Pflüg. Arch. Ges. Physiol.* 92:346–86

Romain, B. P. 1933. Notes on the Western Photronic photoelectric cell. *Rev. Sci. Inst.* 4:83–85

Steimel, K. 1930. Die Stabilität und die Selbserregung elektrischer Kreise mit organen fallender Charakteristik. *Z. Hochfrequenz* 36:161–72

Ussing, H. H., Zehran, K. 1951. Active transport of sodium as the source of electric current in the short-circuited isolated frog skin. *Acta Physiol. Scand.* 23: 110–27

van der Pol, B. 1926. On relaxation oscillations. *Philos. Mag.* 7th Ser., 2:978–92

van der Pol, B. 1934. The nonlinear theory of electric oscillations. *Proc. Inst. Radio Eng.* 22:1051–86

Wood, L. A. 1936. Zero-potential circuit for blocking-layer photo-cells. *Rev. Sci. Inst.* 7:157

Young, J. Z. 1936. Structure of nerve fibers synapses in some invertebrates. *Cold Spring Harbor Symp. Quant. Biol.* 4:1–6

Ann. Rev. Neurosci. 1982. 5:325–39

THE SYNAPSE:
From Electrical to Chemical Transmission

John C. Eccles

Max-Planck-Institut für biophysikalische Chemie, Göttingen, Federal Republic of Germany

The controversy as to whether synaptic transmission in the central nervous system is electrical or chemical played a key role in the origin of modern neuroscience, and I shall attempt to give it a historical perspective.

Almost 100 years ago (1886–1890) it was recognized by several neuroanatomists, including His, Forel, and Ramón y Cajal, that nerve cells or neurons were independent biological units and were not mere nodes in the complex net-like structure that was proposed in the reticular theory of Gerlach and Golgi. Implicit in the neuron theory was the assumption that neurons must enter into functional connection by contiguity, not continuity. Neurohistologists soon demonstrated this postulated contiguity, various kinds of microstructures, such as knobs, baskets, or boutons being described from 1890 onwards. Meanwhile the mechanism of action at these contact regions was being considered by Sherrington, who in 1897 derived the word *synapse* from the Greek, and this name has been universally adopted, with such derivatives as synaptic knob, synaptic cleft, synaptic vesicles, presynaptic, postsynaptic, etc. In that early period Sherrington was particularly interested in the convergence of many pathways onto a particular neuron and in the recognition of two antagonistic types of synaptic action, excitation and inhibition. The modes of action across the synaptic clefts were left undefined. There had been suggestions of chemical transmission across the neuro-muscular synapse by Dubois-Reymond in 1877, and for the synapses made by autonomic nerves onto effector cells by Elliott, Dixon, and Dale. The first suggestion of electrical transmission was by Hermann (1879), with later developments by Lapicque and Forbes. It was an obvious conjecture

325

that transmission between two electrically generating and responsive structures could be electrical. It is important to recognize the long lineages of the two hypotheses.

The first definitive experiments were made in 1921 by Loewi, who demonstrated that vagal inhibition of the heart was mediated by a chemical substance that was later identified as acetylcholine (ACh), and that the action of the nervi accelerantes was due to an adrenalin-like substance, now identified as noradrenalin (Loewi 1921). The slow action of these indubitable chemical transmissions was revealed in 1932 by a systematic study (Brown & Eccles 1934) on the action of a single vagal volley on the rhythm of the heart beat. There was a latent period of about 100 msec, even when the vagal fibers were stimulated in the region of the sino-auricular node. The slowing reached a maximum at about 0.4 sec and gradually passed off over many seconds. The ACh mediation was demonstrated by the depressant action of atropine on the inhibitory slowing, and by its enhancement and prolongation by the anticholinesterase drug, eserine. I continued for many years to regard this slow time course of an indubitable chemical mediation by ACh as a paradigm of all chemical transmissions, as was recently pointed out by MacIntosh & Paton in their biographical memoir on G. L. Brown (1974):

> But the analysis (of the vagal inhibition) may have helped by its very elegance to perpetuate the erroneous view that chemical transmitters are in general, or even of necessity, transmitters of long duration, acting for at least some tenths of a second after being released. A corollary opinion was that any briefer transmission process could not be chemical; thus it was supposed that transmission at neuro-muscular and neuronal synapses, which in many cases is accomplished within a few milliseconds, must be mediated by the electrical currents associated with the nerve impulse.

Until the advent of electronmicroscopy in the 1950s there was no clear evidence that the vagal innervation of heart muscle gave a much more remote liberation of ACh than occurred in the extremely close contact of the neuro-muscular synapse on striated muscle. A further factor unknown at that time was that the muscarinic and nicotinic actions of ACh were effected by quite different postsynaptic mechanisms that were sharply distinguished by their speeds of action—muscarinic being slow, nicotinic fast.

The concept that synaptic transmission was effected by ACh acting in its nicotine-like manner was first suggested for sympathetic ganglia in 1933 and 1934 by Feldberg & Gaddum, who showed that when a sympathetic ganglion was stimulated, ACh was liberated into the eserinized perfusate. This hypothesis was further developed by Feldberg & Vartiainen in 1934 on the basis of experiments on the liberation of ACh and on the stimulating action of ACh when it was given by close intra-arterial injection and registered by the contraction of the nictitating membrane. A full account of the intensive

Feldberg and Eccles at a Ciba meeting in 1957.

investigations of the Dale school in the memorable years of 1933 to 1936 is given in my review of 1936, which was written with much help from G. L. Brown. It was a triumph of neuropharmacology.

As I remember the cumulative impacts of these research reports, I recall that I was sceptical, not about the role of ACh as a transmitter, *but about its exclusive role.* I had been investigating synaptic transmission in the superior cervical ganglion, by electrically recording, and I was impressed by the short synaptic delay (3 msec) and by the one-to-one relationship of a preganglionic volley and the postganglionic discharge. Such fast transmission was so different from the well-established chemical transmissions referred to above. There were other difficulties:

1. Large doses of eserine given intravenously had no appreciable effect on the transmission of a single preganglionic volley. Neither the ganglionic action potential nor the negative and positive after-potentials showed any change.
2. Almost all the cholinesterase had been shown to be in, or on, the preganglionic fibers and not on the postsynaptic membrane where it should be located for most effective inactivation of the liberated ACh.

The climax to this controversy came in May 1935 when there was a very tense encounter. I presented to the British Physiological Society the results of repetitively stimulating the pathway to the nictitating membrane either presynaptically or postsynaptically. With maximal stimuli the contractions recorded with a very sensitive optical myograph were identical (within 1%) except for a slight delay in decline for presynaptic stimulation in some experiments. With eserinization of the superior cervical ganglion (by local or intravenous application), the delayed decline became a prominent feature and lasted for about 15 sec. It was argued that ACh accumulation was responsible for the after-discharge and that the slow contraction was matched by that produced by close intra-arterial injection of ACh into the carotid artery and so to the ganglion (Eccles 1935).

Two explanations were proposed for the initial fast response:

1. The fast response was also due to ACh.
2. The fast response was due to electrical excitation.

The first explanation was criticized because the rapid decline of the initial action was not slowed by eserine inactivation of the acetylcholinesterase. For example, the facilitation curve for submaximal preganglionic stimulation showed that synaptic activation declined in an exponential manner for about 150 msec and this facilitation curve was not slowed by eserine.

Comparable examples of fast and slow transmitter actions, with the fast not giving the expected pharmacological response, were known at this time, and were used to support a double transmitter theory.

1. Henderson & Roepke (1934) reported that stimulation of the parasympathetic innervation to the bladder results first in a quick, then in slow contraction of the bladder wall, only the slow being paralyzed by atropine and mimicked by ACh.
2. Monnier & Bacq (1935) and Eccles & Magladery (1936) observed that with the adrenergic contraction of the nictitating membrane, the anti-adrenalin drug 933F blocked the slow electrical response and the associated contraction but not the fast response.

Dale (1934) recognized the difficulty raised by these selective blockages but made the valuable suggestion that the fast responses are due to very close apposition of the releasing presynaptic terminals and the slow responses to a remote release. This explanation has turned out to be correct. It illustrates the penetrating insight with which he was gifted.

Nevertheless it can be recognized in retrospect that there were grounds on which to build the alternative electrical transmitter story. At that time, and for almost two decades later, there was an inadequate knowledge of the microstructure of the nerve endings. The best histological pictures of synaptic boutons or loops showed a synaptic gap of about 1 μm, which, for chemically mediated synaptic transmission, would result in a delay of many milliseconds. Only in 1961 did Gray & Guillery demonstrate that the commonly used silver strains were not displaying the entire synaptic knob, but only the neurofibrillar loops within the knob. Not until the early 1950s did we have the wonderful new insights provided by electron microscopy which showed the close apposition (200 Å) of presynaptic and postsynaptic elements across the synaptic cleft, and led to the discovery of synaptic vesicles and all the other marvelous machinery concerned in chemical transmission (Palay 1958, de Robertis 1958).

Sydney, 1942. Kuffler, Eccles and Katz, walking to the University.

The transmission from nerve impulse to skeletal muscle had long been recognized as being very fast and unitary. A single nerve impulse evokes the discharge of a single muscle impulse with a synaptic delay of only 1 msec. It contrasted so sharply with the transmissions that were clearly chemically mediated that Loewi stated in a 1933 lecture, "Personally I do not believe in a humoral mechanism existing in the case of striated muscle." I had a similar belief at that time. I can remember being quite outraged when Dale & Feldberg suggested in 1934 that their discovery of ACh release on nerve stimulation of muscle indicated that ACh was the transmitter. As mentioned above, we had recently shown that the action of a single vagal volley on the heart had a latency of at least 0.1 sec and a duration of several seconds. However, the evidence rapidly accumulated, thanks particularly to the elegant experiments of Brown (1937) on muscle stimulation by close intra-arterial injection of ACh and its block by curare and on the analysis of the action of eserine. It was certainly impressive that the single response was transformed by eserine into a brief waning tetanus. There were fewer anomalies than with ganglionic transmission. In my 1936 review I concluded that, although there was much evidence supporting the hypothesis of ACh as the transmitter, further experiments were necessary. However, it was there admitted that the alternative electrical transmission hypothesis was so vaguely formulated as to be virtually useless in guiding further experiments. When later it was formulated in a model (Figure 1; Eccles 1949) it was found to be untenable because of the extreme mismatch between the surface areas of a nerve fiber and the many muscle fibers that it innervates. In the early 1940s, Katz, Kuffler, and I made a detailed electrical study of the endplate potential of a curarized muscle and the effect of anticholinesterase drugs upon it. This work showed that the initial fast transmission was due to ACh, as well as the quite distinct slow response, so we went on record in complete support of ACh transmission (Eccles, Katz & Kuffler 1942).

I have an entertaining exchange of correspondence with Dale that occurred at that time.

J. C. E. to H. H. D.—28 July 1943
My dear Sir Henry,
    I heard recently from some source—which at present escapes me—that you and Brown were glad that we had at last admitted the full significance of ACh at the neuro-muscular junction! And now I have the unkindness to reopen the argument by a paper on eserine action on the sympathetic ganglion. It seems that the picture is more complicated than we yet realize . . .

H. H. D. to J. C. E.—22 August 1943
My dear Eccles,
    I have just got your Airgraph letter. I do not know from what source you may have heard of our reaction to your paper in the *Journal of Neurophysiology,* but the report

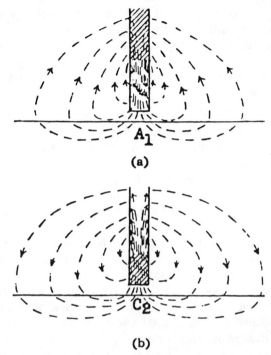

(a)

(b)

Figure 1 Diagrams of current flow at a schematic synapse with presynaptic impulse approaching synapse in (a), and at synapse in (b). Note reversal of current flow, the focal anodal A_1 effect being followed by the focal cathodal C_2 effect at the synaptic region of the postsynaptic membrane.

which reached you was certainly not wholly inaccurate. I shall, of course, look forward with additional interest to your paper in the *Journal of Physiology*, about the action of eserine on the ganglion.

I am told that John Fulton, in a recent number of *Science*, has begun to balance himself more carefully than before on the top of the hedge, so that eventually we may find you all on the same, safe side.

Here is an example of subtle comment with the wise and gentle irony of which Dale was a master!

Certainly for most of us the electrical-chemical controversy had been settled in the 1940s in favor of ACh for ganglionic and neuromuscular transmission. There are good accounts of the relative positions of the various disputants of a electrical versus chemical hypotheses in several symposia. In a 1939 symposium on the synapse that was published in the *Journal of Neurophysiology,* Lorente de Nó, Gasser, and Erlanger expressed themselves strongly in favor of electrical transmission. In the Paris symposium of 1949 there was good support for electrical transmission, but very largely for synapses in the central nervous system (CNS). Even as recently as the Brussels symposium of 1951 there was substantial support for electrical transmission, particularly by Fessard.

Concerning transmission at synapses in the CNS, the problem was still open through the 1940s (cf Eccles 1949). I continued to espouse the electrical theory both for excitation and inhibition. There were several reasons for this lingering faith!

1. The beautifully clear demonstrations of ACh action at peripheral synapses could not be carried out in the CNS. Certainly ACh was present in considerable amounts in different regions of the CNS, but, despite heroic attempts, there were no convincing experiments that ACh had an action mimicking synaptic action. It is now recognized that ACh action in most parts of the CNS is muscarinic and so very slow in onset and in decline that at the most it could form a background for synaptic action.
2. There was no bad mismatch between the electrical properties of the presynaptic and postsynaptic elements of a synapse, such as had been recognized at nerve-muscle synapses.
3. The ingenious ephaptic experiments of Arvanitaki, Katz, Lorente de Nó, and associates had given encouragement to the designing of models that would account for the two fundamental synaptic actions of excitation and inhibition, as described below.

From 1945 onwards I was deeply under the influence of Karl Popper, who stressed the necessity to formulate clear hypotheses and then test them by rigorous experiment. In 1945 and 1946 I had developed a model for electrical excitatory synaptic action, based on ephaptic studies, that is diagrammed in a simple form in Figure 1 (Eccles 1946). An impulse propagating into the nerve terminal at a synapse exerts an initial anodal (A_1) and later cathodal (C_2) current input across the subsynaptic membrane. It was assumed that C_2 sets up a local response postsynaptically, which acts as an amplifying mechanism. In 1948 Bullock published his elegant investigations on the single synapses in the stellate ganglion of the squid, and pointed out in conclusion "that the descriptions of properties given here corresponds exactly to expectations from the electrical theory elaborated by Eccles."

In 1947 I developed an electrical theory of synaptic inhibitory action which conformed with the available experimental evidence. Incidentally this theory came to me in a dream. On awakening I remembered the near tragic loss of Loewi's dream, so I kept myself awake for an hour or so going over every aspect of the dream, and found it fitted all experimental evidence. It was duly diagrammed and published in *Nature* (Brooks & Eccles 1947), and was known as the Golgi cell theory of inhibition. I still think it was an ingenious model because it used the current flow of an excited interneuron to generate anelectrotonic foci on neurons upon which their synapses were placed, as illustrated in Figure 2. Thus, there could be a close aggregation of many anelectrotonic foci on a neuron with the consequent inhibition of an excitatory synaptic action. For several years this electrical theory was tested experimentally and did very well with the not very challenging methods of extracellular recording.

The chance for more rigorous testing came when we had developed intracellular recording from motoneurons in mid-1951. After some preliminary experiments on synaptic excitation and antidromic responses the decisive tests were carried out in August 1951. A microelectrode had been inserted into a biceps-semitendinosus motoneuron and it was known that an afferent volley in the nerve to the quadriceps muscle had a powerful "direct" inhibitory action on biceps-semitendinosus neurons that was supposed to be monosynaptic (Lloyd 1946). Before the test was applied we had recognized that on the electrical model for inhibitory action the microelectrode would be in a brief positively going electrical field (x in Figure 2), whereas on the chemical hypothesis synaptic inhibition would be expected to be due to a brief increase in membrane potential, which means that it would record a brief negatively going potential. Thus it was a clear test. If the quadriceps volley caused the trace to go up it was electrical, if down it was chemical. It went down! The result was repeatable, it was graded with stimulus strength, and indubitable. In Figure 3 the initial records of that

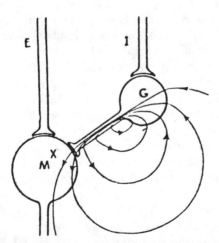

Figure 2   Diagram of current flow at a schematic synapse of Golgi cell G on a motoneuron M as postulated in the electrical theory of inhibition. E shows the excitatory line to M and I the inhibitory line that subliminally excites G and so generates the current flow producing an anelectrotonic focus on M.

At the Feldberg's, 1952, the evening after the British Physiological Society meeting. I have Feldberg and Lorente de Nó by the hand. (*Left to right, top to bottom*) Jung, Weiss, Feldberg, Fessard, Rose Eccles, M. Vogt, John Eccles, Mrs. Feldberg, Lorente de Nó, Murdoch Ritchie, Feldbergs' daughter, and Brenda Ritchie.

momentous experiment are published for the first time. We were momentarily stunned. It was in the early hours of the morning! But, on recovering from the shock, the decision was made. Inhibitory synaptic action was chemically mediated; and it was evident that the mirror image response (i.e. excitatory synaptic action) was also chemical (cf Figure 3). I went to England in January 1952 carrying the news of my belated conversion, and of course with all the enthusiasm of a neophyte. There was a paper to the Physiological Society and a fascinating Discussion Meeting of the Royal Society.

Although chemical transmission was now accepted, there were new problems that at first confused us! Dale and I both accepted the proposition that the same transmitter could act oppositely at different transmitter sites by virtue of specialization of the postsynaptic receptor membrane. It was accepted that the group Ia afferent fibers from muscle spindles monosynaptically excited the motoneurons in the same muscle, and monosynaptically inhibited the antagonist motoneurons (direct inhibition). At the Royal Society Discussion in February 1952, Dale quoted in support of the proposed double action of the transmitter at these synapses, the action of ACh on the urinary bladder in which some smooth muscle fibers are excited and others inhibited. That was the position until 1953.

The first crack in this monolithic structure appeared with the discovery by our group in Canberra (Eccles, Fatt & Koketsu 1953) that motor axon collaterals exerted a cholinergic excitatory action on Renshaw cells, which are small interneurones in the ventral horn that had been recognized by Renshaw about ten years earlier. Renshaw cells were shown to inhibit motoneurons and had apparently no excitatory action.

Naturally we informed Sir Henry of our discovery in advance of the publication and I quote from his very happy letter of reply:

H. H. D. to J. C. E.—1st October 1953
My dear Eccles,
   I am indeed grateful to you for giving me the very great pleasure of reading in advance the extremely interesting paper, which you are publishing with Fatt and Koketsu. I do congratulate you all, not only upon the beauty of the observations recorded, but on the very attractively clear and concise account of them in the paper. Your new-found enthusiasm is certainly not going to cause any of us embarrassment. It is extremely satisfactory to have the direct evidence of a cholinergic transmission from the ending of the collateral of a cholinergic axon, which, as you say, could have been predicted. I myself emphasized in 1934, in a Nothnagel Lecture which I gave in Vienna, the fact that the chemical function appeared to be a function, not merely of the nerve ending, but of the whole neurone, and speculated at the time concerning the possibility, that the identification of the peripheral transmitter of the so-called "antidromic vasodilatation" might give a clue to the transmitter, at the other, central synaptic ending, of what appeared to be normally an afferent nerve fibre.

EXPERIMENT 20-8-51

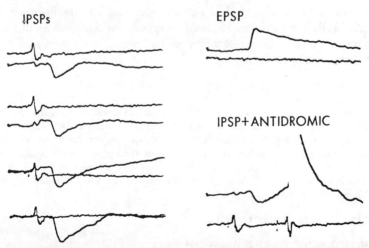

IPSPs

EPSP

IPSP+ANTIDROMIC

Figure 3   Intracellular recordings from a posterior-bicepts-semitendinosus neuron showing IPSPs generated by a group Ia quadriceps volley as downward (negative) deflections about 10 msec in duration. The other trace shows the action potential generated by this volley in the L_5 dorsal root. The EPSP is generated by a group Ia afferent volley in the semitendinosus nerve. The antidromic potential (truncated) is produced by stimulation of the L_7 ventral root.

Dale developed this idea in his Dundee Lecture (Dale 1952):

> The transmitter used by nerve-fibres of a particular kind, and concentrated at their
> endings in readiness for release, is also to be found along the whole length of the fibres;
> a transmitter is characteristic, then, not only of the endings, but of the whole neurone
> ... What shall we expect of such a substance? The synaptic endings in the central nervous
> system represent only one end of these dorsal-root fibres. Are we to expect that the
> transmitter at their central, synaptic endings would also be functional at their peripheral
> endings, and possibly be there concerned with transmitting the so-called antidromic
> vasodilator action?

On the basis of this and earlier statements, I proposed (Eccles 1957) that
Dale's Principle be defined as stating that at all the axonal branches of a
neuron the same transmitter substance or substances are liberated. Dale's
prescience has now been vindicated by the finding that substance P seems
to be the transmitter concerned both in the "antidromic vasodilatation" and
in the central synaptic transmission by dorsal root fibers in the spinal cord
(Hökfelt et al 1977, Nicoll et al 1980).

After the discovery of the cholinergic pathway through the inhibitory
Renshaw cells, a further discovery was made on the so-called "direct"
inhibitory pathway, under Paul Fatt's guidance. We had been bothered
about the longer latency of the inhibitory postsynaptic potential (IPSP) of
direct inhibition relative to monosynaptic excitation of the same motoneurons, the difference being about 0.8 msec. From 1951 to 1953 this was
explained as due to the longer conduction time in the presynaptic fibers
giving the IPSP of the antagonist motoneurons. However, this explanation
was called into question by Paul Fatt and so we set up experiments to test
it (Eccles, Fatt & Landgren 1953). We showed that there was an interneuron interpolated on the direct inhibitory pathway. I immediately informed
Sir Henry of the discovery.

> J. C. E. to H. H. D.—16 August 1954
> My dear Sir Henry,
>   There is a letter of yours of Oct. 1st of last year which I should have replied to earlier.
> You very kindly commented on our work and mentioned your very early reference to
> the chemical function of the whole neurone. You will now see that we have further
> evidence in support of that important generalization. We suggest that all central inhibi-
> tory action is due to the operation of small interneurones which are inserted as the final
> stage in all the inhibitory pathways that we have investigated. You will immediately see
> the general implication of these findings.

Sir Henry immediately replied:

> H. H. D. to J. C. E.—25–26 August 1954
>   ... It is, of course, an addition to knowledge of these matters of first rate interest, that
> you are now able to find evidence of a synaptic relay on the supposedly "direct" inhibi-
> tory path, and thus to account for the extra latent period of the inhibitory effect. I think

that I detect between the lines of your paper a feeling, as it were, of relief from the necessity of supposing, otherwise, that the same synaptic transmitter, acting on different motoneurones, might produce opposite effects. I confess that that did not worry me particularly, since we had become very familiar with the production at the periphery by the same transmitter, whether acetylcholine, or, in other cases, adrenaline, of opposite effects of augmentation or inhibition on the plain muscle of different organs, or even of different parts of the same organ. However, your evidence for the interneurones on the inhibitory pathway is very convincing, and I note your readiness to interpret it as possibly meaning that all excitatory effects in the CNS may be due to the release of a single transmitter and all inhibitory effects to the release of another. That, of course, would suit my particular prejudices or preconceptions extremely well; but I am sure that you will not mind my saying that I wonder whether, only a few years ago, you would not have seized rather eagerly upon the evidence for the inhibitory interneurones, as facilitating the explanation of the inhibition on electrical lines, and eliminating any excuse for bothering with such fantastic ideas as a chemical transmission!

(Here is another lovely example of his wise and gentle irony.)

I have two final comments to make. First, in the mammalian CNS it is now generally accepted that neurons are either excitatory or inhibitory and are never ambivalent. Most inhibitory synapses release glycine or GABA. Excitatory synapses are rarely cholinergic, but at many glutamate and aspartate appear to be the transmitters. There are also several other transmitters now recognized at special sites in the CNS: dopamine, noradrenalin, serotonin, and several peptides such as substance P. There is also evidence that several transmitters may be released at a synapse. For example, there may be a conventional synaptic transmitter and a slower acting neuron modulator. Second, the rejection of electrical transmission was correct for all synapses in the controversy, but electrical transmission has made a "come back," even in the vertebrate CNS, with the recognition of a special type of synapse, the gap junction. In 1959, Furshpan & Potter made the remarkable demonstration that there is one-way electrical transmission between crustacean giant fibers. There was very close apposition of the synaptic membranes, with, in addition, a polarization in the direction of current flow. Later, excitatory gap junctions were recognized in vertebrates for neurons that fire electric organs (Bennett et al 1963), but they are relatively rare in the mammalian brain and it has yet to be shown that they are functionally important in the brain. Electrical inhibition has been demonstrated in the goldfish Mauthner cell (Furukawa & Furshpan 1963, Korn & Faber 1975). It is interesting that the models for such electrical synapses resemble those of the original electrical synaptic story. This is particularly the case for the goldfish inhibition where interneurons generate a local anelectrotonus as in the original inhibitory model of my dream (Figure 2)! Korn & Faber have also demonstrated that an electrical inhibition briefly precedes the chemical inhibition exerted by basket cells on cerebellar Purkinje cells.

We are left with the impression that chemical transmission was an essential evolutionary advance which made possible the design of brains with all the infinite complexity required for higher nervous functions. Chemical transmission is much more effective and selective, and it gives a specificity of action from neuron to neuron, together with a large amplification factor in the transmission process. But, even more important, chemical transmission provides the opportunity for inhibition that plays such a key role in the elegant patterns of activity that we now are beginning to understand in the mammalian brain. Electrical inhibition requires complex neuronal design; it is much less effective, although quicker in action.

Literature Cited

Bennett, M. V. L., Aljure, E., Nakajima, Y., Pappas, G. D. 1963. Electrotonic junctions between teleost spinal neurones: Electrophysiology and ultrastructure. *Science* 141:262–64

Brooks, C. McC., Eccles, J. C. 1947. An electrical hypothesis of central inhibition. *Nature* 159:760–64

Brown, G. L. 1937. Action potentials of normal mammalian muscle. Effects of acetylcholine and eserine. *J. Physiol.* 87:220–37

Brown, G. L., Eccles, J. C. 1934. The action of a single vagal volley on the rhythm of the heart beat. *J. Physiol.* 82:211–41

Bullock, T. H. 1948. Properties of a single synapse in the stellate ganglion of the squid. *J. Neurophysiol.* 11:343–64

Dale, H. H. 1934. Chemical transmission of the effects of nerve impulses. *Br. Med. J.* 1:835

Dale, H. H. 1952. *Transmission of Effects from Nerve Endings.* London: Oxford Univ. Press

Dale, H. H., Feldberg, W. 1934. Chemical transmission at motor nerve endings in voluntary muscle. *J. Physiol.* 81:39

de Robertis, E. D. P. 1958. Submicroscopic morphology and function of the synapse. *Exp. Cell Res. Suppl.* 5:347–69

Eccles, J. C. 1935. Afterdischarge from the superior cervical ganglion. *J. Physiol.* 84:50P

Eccles, J. C. 1936. Synaptic and neuromuscular transmission. *Ergebn. Physiol.* 38:339–444

Eccles, J. C. 1946. An electrical hypothesis of synaptic and neuromuscular transmission. *Ann. NY Acad. Sci.* 47:429–55

Eccles, J. C. 1949. A review and restatement of the electrical hypothesis of synaptic excitatory and inhibitory action. *Arch. Sci. Physiol.* 3:567–84

Eccles, J. C. 1957. *Physiology of Nerve Cells.* Baltimore: Johns Hopkins Univ. Press

Eccles, J. C., Fatt, P., Koketsu, K. 1953. Cholinergic and inhibitory synapses in a central nervous pathway. *Aust. J. Sci.* 16:50–54

Eccles, J. C., Fatt, P., Landgren, S. 1953. The 'direct' inhibitory pathway in the spinal cord. *Aust. J. Sci.* 16:130–34

Eccles, J. C., Katz, B., Kuffler, S. W. 1942. Effect of eserine on neuromuscular transmission. *J. Neurophysiol.* 5:211–30

Eccles, J. C., Magladery, J. W. 1936. Pharmacological investigation on smooth muscle. *J. Physiol.* 87:81P

Feldberg, W., Gaddum, J. H. 1934. The chemical transmitter at synapses in a sympathetic ganglion. *J. Physiol.* 81:305–19

Feldberg, W., Vartiainen, A. 1934. Further observations on the physiology and pharmacology of a sympathetic ganglion. *J. Physiol.* 83:103–28

Furshpan, E. J., Potter, D. D. 1959. Transmission at the giant synapses of the crayfish. *J. Physiol.* 145:289–325

Furukawa, T., Furshpan, E. J. 1963. Two inhibitory mechanisms in the Mauthner neurons of goldfish. *J. Neurophysiol.* 26:140–76

Gray, E. G., Guillery, R. W. 1961. The basis for silver staining of synapses of the mammalian spinal cord. A light and electron microscope study. *J. Physiol.* 157:581–88

Henderson, V. E., Roepke, M. H. 1934. The role of acetylcholine in bladder contractile mechanisms and in parasympathetic ganglia. *J. Pharmacol.* 51:97

Hökfelt, T., Ljungdahl, A., Terenius, L., Elde, R., Nilsson, G. 1977. Immunohistochemical analysis of peptide pathways possibly related to pain and analgesia:

Enkephalin and substance P. *Proc. Natl. Acad. Sci. USA* 74:3081–85

Korn, H., Faber, D. S. 1975. An electrically mediated inhibition in goldfish medulla. *J. Neurophysiol.* 38:452–71

Lloyd, D. P. C. 1946. Facilitation and inhibition of spinal motoneurons. *J. Neurophysiol.* 9:421–38

Loewi, O. 1921. Über humorale Übertragbarkeit der Herznervenwirkung. *Pflügers Arch. Gesamte Physiol.* 189:239–42

MacIntosh, F. C., Paton, W. D. M. 1974. George Lindor Brown, 1903–1971. *Biographical Memoirs of Fellows of the Royal Society* 20:41–73 London: Royal Society

Monnier, A. M., Bacq, Z. M. 1935. Recherches sur la physiologie et la pharmacologie du système nerveux autonome. XVI. Dualité du mécanisme de la transmission neuro-musculaire de l' excitation chez le muscle lisse. *Arch. Int. Physiol.* 40:485–510

Nicoll, R. A., Schenker, C., Leeman, S. E. 1980. Substance P as a transmitter candidate. *Ann. Rev. Neurosci.* 3:227–68

Palay, S. L. 1958. The morphology of synapses in the central nervous system. *Exp. Cell Res.* 5:275–93

Ann. Rev. Neurosci. 1982. 5:341–62

DEVELOPMENTAL NEUROBIOLOGY AND THE NATURAL HISTORY OF NERVE GROWTH FACTOR*

Rita Levi-Montalcini

Institute of Cell Biology C.N.R., Rome, Italy

Introduction

> Those who refuse to go beyond fact rarely get as far as fact.
>
> T. H. Huxley (Taylor 1970)

I was invited to write about the discovery of Nerve Growth Factor and its relationship to developmental neurobiology. Rather than detail the life of NGF over the past 30 years, I wish to concentrate on what was the most attractive and also the most unusual feature in the investigation of NGF: that each finding has signaled a new turning point and opened up a new perspective. The story of NGF is therefore more like a detective story than a scientific enterprise, since science usually unfolds according to well-defined rules, along the route paved by previous findings. In retrospect, the previous experience with NGF holds out the promise that as many turning points still await us on the road ahead as we experienced on the road we have traveled so far. To me this is an encouraging rather than a depressing thought; the abrupt end of a scientific pursuit is more often synonymous with an intellectual dead end than with a goal successfully achieved. By way of introduction, I shall recount the early beginnings of this research, which are well known to old-timers but probably not to biochemically trained newcomers who were lured to take part in this game by the molecule itself and who are only vaguely acquainted with its natural history and with its problems.

*Dedicated to Viktor Hamburger.

341

0147-006X/82/0301-0341$02.00

1951: The Beginning of a Long Journey

In the Fall of 1951, Nerve Growth Factor, then still unnamed and chemically unidentified, was introduced to the public for the first time at a conference on "The Chick Embryo in Scientific Research" held at the New York Academy of Science (Miner 1952). One of the topics discussed at that meeting was the effects of chemical agents on tumors (from birds and mammals) that had been transplanted into the body wall or onto the chorioallantoic membrane of chick embryos. The tumor and the embryo were partners, each feeding on the limited resources of the egg and growing according to its own vastly different program. Of the two partners, the tumor was usually the main object of interest. Elmer Bueker, a former student of Viktor Hamburger (the foremost authority in the field of experimental neuroembryology), inverted this trend (Bueker 1948). He explored the changes that take place in the embryo that bear transplants of neoplastic tissues.

Bueker found that when dorsal root ganglia innervated a transplant of mouse sarcoma 180 that had been grafted into the body wall of three-day chick embryos, the ganglia became enlarged. This report did not stir the emotions of the neuroembryologists in attendance, accustomed as they had become to the extravagant performances of the developing amphibian and teleost nervous systems, that readily adapt even to the most bizarre demands that can be generated by the imagination of the investigator (Harrison 1935). Even the discoverer of this remarkable phenomenon did not consider it unusual. Rather, he thought it provided additional evidence for the important role played by peripheral structures in promoting the growth and differentiation of the nerve cells that innervate these tissues. The increase in volume of sensory ganglia that sent nerve fibers to the tumor seemed to support the prevailing concept that the size of a developing nerve center is under the control of its field of innervation and seemed to emphasize the parallel between the effects elicited by neoplastic tissue and those elicited by transplanting an extra limb rudiment. Bueker's observation also supported the hypothesis that nerve fibers channel the trophic agent synthesized by target tissues toward the cell bodies of the neurons.

Comparison of the transplanted supernumerary limbs and the transplanted tumors did bring to light some discrepancies, however, and suggested to me and to Viktor Hamburger that this phenomenon needed to be investigated in more detail. Using a silver staining technique specific for neurons, we traced the nerve fibers from their cell bodies of origin to their peripheral targets. By this means we uncovered new aspects of the effects elicited by the tumor that differed markedly from those produced by grafting additional limb rudiments. In two- to three-week old embryos bearing

Viktor Hamburger and Rita Levi-Montalcini in 1977.

transplants of mouse sarcomas 180 or 37, the sensory ganglia—and to an even greater extent the sympathetic ganglia—enlarged enormously compared to the normal contralateral ganglia innervating the host's limb. Many sensory and sympathetic nerve fibers emerging from both types of ganglia invaded the tumor and branched along the cells but did not establish synaptic connections with them (Levi-Montalcini & Hamburger 1951). Shortly afterwards, I discovered a new and startling effect elicited by the tumor. Fibers produced by sympathetic ganglia, too far removed from the neoplastic tissue to establish contact with it, invaded the viscera in large numbers. These fibers also invaded the blood vessels, and in some instances obliterated the lumen of these vessels. Similarly, chaotic and highly aberrant peripheral distribution of sympathetic nerve fibers was found in embryos in which the mouse sarcoma was transplanted onto the chorio-allantoic membrane in a way that prevented direct contact between the embryonic and the neoplastic tissues. Based upon these findings, I proposed the hypothesis (presented at the New York Academy of Sciences conference) that the tumor released a humoral factor which gained access to the ganglia through the circulatory system (Levi-Montalcini 1952). This initial hypothesis was soon supported by further experiments.

THE FIBRILLAR HALO   If the tumor releases a diffusible factor, tumor extract might be expected to work as well as the implanted tumor. We failed to replicate the effects of actively growing neoplastic cells by injecting tumor extract, however, and this suggested to me the alternative possibility of studying the effect in tissue culture. In 1952 tissue culture was still an amateur's tool. Rather than being used to study basic biological problems, it had been used primarily to satisfy the biologists' curiosity and esthetic pleasure by permitting one to watch cells proliferate and axons elongate and branch. I came to the idea of studying this effect in culture only after injection of extracts derived from the tumor had failed to produce results. We did not have tissue culture facilities at Washington University, but I knew that my friend, Hertha Mayer, had established an efficient culture facility at the Institute of Biophysics of the University of Rio de Janeiro, directed by Professor Carlos Chagas. In the Fall of 1952, following an exchange of letters and an invitation from Professor Chagas, I boarded a plane for Rio de Janeiro, carrying in my handbag two mice bearing transplants of mouse sarcoma 180 and 37.

The outcome of the experiments performed in that most attractive and hospitable milieu was so clear as to require only a few words of comment. As I wrote in a previous article, "The tumor factor had given a first hint of its existence in Saint Louis but it was in Rio de Janeiro that it revealed itself, and it did so in a theatrical and grand way, as if spurred by the bright atmosphere of that explosive and exuberant manifestation of life that is the Carnival in Rio" (Levi-Montalcini 1975).

When sensory and sympathetic ganglia from eight-day chick embryos were cultured in a semi-solid medium close to fragments of mouse sarcoma 180 or 37, they produced a dense halo of nerve fibers in 12–24 hours. This discovery marked still another turning point. It offered the possibility, unattainable in developing embryos, of performing biochemical studies on the tumor factor. This rapid in vitro bioassay also offered the possibility of screening other tissues and organic fluids for potential sources of growth-promoting agents. The discovery of two of these sources materialized shortly and fulfilled our hope far beyond expectation. The same technique later provided a valuable, and, in fact, the only way for discovering other specific growth factors, opening new and promising areas of research.

I reported the in vitro effects elicited by the two mouse sarcomas that same year at the Academy of Sciences in Rio de Janeiro and again one year later at a Cell Biology Symposium in Copenhagen (Levi-Montalcini 1953). I also presented a detailed description of the fibrillar halo in a subsequent article (Levi-Montalcini et al 1954). Stanley Cohen, a gifted, young biochemist, who joined our group in St. Louis in 1953, soon identified the humoral factor released by the two mouse sarcomas in a nucleoprotein particle. This factor was christened Nerve Growth Promoting Factor, a name we later shortened to Nerve Growth Factor (NGF) (Cohen et al 1954).

SNAKE VENOM AND THE MOUSE SUBMAXILLARY SALIVARY GLANDS   In 1956 and 1958 we discovered that the most potent sources for NGF were in snake venom and the submaxillary salivary glands of the mouse (Cohen & Levi-Montalcini 1956, Cohen 1958, Levi-Montalcini 1958). Cohen played a major role in these apparently casual but brilliant discoveries. He identified the NGF of snake venom and of the mouse salivary glands in two protein moieties, both of which were found to possess biological and biochemical properties remarkably similar to the NGF of mouse sarcomas. Recent studies have provided new evidence to support this point (Barklis & Perez-Polo 1981). The discovery that NGF is synthesized in the collecting tubules of the salivary glands (Levi-Montalcini & Angeletti 1961) and that the synthesis is controlled by testosterone (Levi-Montalcini & Angeletti 1964, Ishii & Shooter 1975) and by thyroxine (Aloe & Levi-Montalcini 1980) has contributed further to the studies of NGF.

Despite an active search for alternative sources rich in NGF (Harper et al 1979, Harper & Thoenen, 1980, Goldstein et al 1978), the mouse salivary gland remains the richest source. It has supplied the purified NGF needed for structural studies as well as for the exploration of its mechanism of action.

MAIN FEATURES OF THE GROWTH RESPONSE ELICITED BY THE SALIVARY NGF   We carried out morphological, ultrastructural, and biochemical experiments on the growth response elicited by the salivary NGF. The main features of these effects are summarized as follows:

NGF elicits the formation of the nerve fibrillar halo at a concentration of 1–10 ng/ml of culture medium within 10 to 24 hr. Higher doses result in a change from a radial to a circular pattern of nerve fiber outgrowth (Levi-Montalcini & Angeletti 1968). The recent experiments of Rutishauser & Edelman (1980) have offered an explanation of this effect. In the absence of serum and NGF, dissociated nerve cells in culture undergo deterioration and death within 24 hr. When 10–100 ng/ml of NGF are added daily to the medium, the cells survive indefinitely, and build a dense nerve fiber network (Levi-Montalcini & Angeletti 1963).

Daily injections of NGF into neonatal rodents for two weeks result in a six- to nine-fold increase in volume of sympathetic ganglia compared to untreated age-matched littermates (Levi-Montalcini & Booker 1960a). This overgrowth of the ganglia to prolonged exposure to exogenous NGF is due to enhanced differentiation, overproduction of axonal material, and prevention of natural death, which cause the loss of a large number of differentiating nerve cells in control rodent pups (Hendry, 1977). Electron microscopic studies in vitro and in vivo showed that the earliest effect elicited by NGF is a massive increase in microtubules, neurofilaments, and microfilaments (Levi-Montalcini et al 1968).

Metabolic studies on dissociated embryonic sensory cells exposed to NGF for 8 hr showed that all anabolic processes are markedly enhanced; there is increased incorporation of labeled precursors into proteins, lipids, and RNA, but not into DNA.

THE PRIMARY STRUCTURE OF NGF   Studies on the physico-chemical properties of NGF first performed by Cohen (1960) were continued in different laboratories. The biological and structural properties of two forms of NGF—a high molecular weight and a low molecular weight form (which became known as 2.5 S and 7S NGF on the basis of their sedimentation constants)—have been discussed in several articles (Angeletti et al 1967, Zanini et al 1968, Bocchini & Angeletti 1969, Varon et al 1967). The two populations of NGF result from different techniques used in purification: the high molecular form of NGF consists of other subunits besides the active NGF moiety; it does not differ from the 2.5 S NGF in its biological activity.

The procedure devised by Bocchini & P. Angeletti provided purified NGF that was used in 1971 by R. H. Angeletti & Bradshaw to elucidate the primary structure of the 2.5 S NGF. Their studies showed that it consists of two identical monomers held together by noncovalent bonds. Each monomer consists of 118 amino acid residues and has a molecular weight of 13,250 (R. H. Angeletti & Bradshaw 1971). The significance of this feat is apparent. It achieved a goal set two decades earlier when the growth effect elicited by the two mouse sarcomas was traced to an unidentified humoral factor released by the neoplastic cells. Purification of NGF signaled the beginning of investigations directed at answering an increasing number of questions that have not yet been totally resolved. Other features of the NGF-target cells interaction, not foreseen at the end of the second decade, were to emerge at the forefront of the field, opening new perspectives and initiating new studies.

DESTRUCTION OF THE PARA- AND PREVERTEBRAL SYMPATHETIC CHAIN GANGLIA WITH ANTISERUM TO NGF   In 1959 Cohen produced a specific antiserum to NGF (Cohen 1960). Adding the antiserum to the culture medium totally prevented the formation of the NGF fibrillar halo. The antiserum (AS-NGF) was next assayed by injecting it into neonatal and adult rodents and other mammals. In neonatal rodents AS-NGF produced massive destruction of the adrenergic neurons located in the para- and prevertebral chain ganglia (Levi-Montalcini & Booker 1960b). However, nerve cells of the sympathetic ganglionic complexes positioned close to their end organs (genital system, brown adipose bodies in rodents and heart tissue) are not receptive to the growth-promoting activity of NGF nor are they vulnerable to the NGF antiserum (Levi-Montalcini 1972). This treatment, known as immunosympathectomy, does not interfere with the vitality and the normal development of the injected pups (Levi-Montalcini & Angeletti 1966). The availability of the antiserum to NGF made it possible to study adult rodents deprived from birth of the function of the sympathetic para- and prevertebral ganglia (Steiner & Schönbaum, 1972).

Two candidate mechanisms were considered as causing the destructive effects of NGF antiserum on immature sympathetic nerve cells: (a) a cytotoxic complement-mediated cell lysis or (b) a deprivation of endogenous NGF through inactivation by the injected antiserum. The second mechanism has been strongly supported by the recent demonstration that administration of NGF (even when given 24–48 hours after antiserum) reverses the otherwise deleterious effects produced by NGF antibodies (Thoenen et al 1979, Harper & Thoenen 1980). Whereas immature sympathetic nerve cells are destroyed by antibodies to NGF, the same treatment produces only moderate and largely reversible damage to the sympathetic neurons of adult rodents (Levi-Montalcini & Cohen 1960).

1972–1981: The Road Ahead and the Everchanging Panorama

STRUCTURE-ACTIVITY RELATIONSHIP OF NGF   Studies of the structural properties of NGF have elucidated a number of basic features of the molecule, including the following:

1. The molecular size and subunit composition of the biologically active NGF.
2. The processing of active NGF from a large molecular weight precursor.
3. The effects of modifications of single amino acid residues on biological activity.
4. A comparison of the primary structure of NGF with insulin.
5. The non-enzymatic properties of the biologically active NGF.
6. The evolutionary history of the NGF molecule (as evaluated by structural and immunological studies).

The extensive literature on these topics is reviewed by investigators actively engaged in this area (Angeletti et al 1973, Varon 1975, Bradshaw 1978, Server & Shooter 1977, Mobley et al 1977, Thoenen et al 1979, Thoenen & Barde 1980, Greene & Shooter 1980, Harper & Thoenen 1980).

NGF-TARGET CELL INTERACTION: BINDING, INTERNALIZATION, AND MECHANISM OF ACTION   NGF target cells possess specific binding sites with an affinity in the range of the biological activity of NGF (1–10 ng/ml) (Banarjee et al 1973, Frazier et al 1973, Herrup & Shooter 1973). When NGF binds to receptors at the nerve endings of the target neurons, the NGF is internalized and carried by retrograde axonal transport to the cell soma where NGF is still biologically active (Hendry et al 1974, Paravicini et al 1975, Stoeckel et al 1976). There are also receptors to NGF on the perikaryon. NGF that binds to these cell body receptors also undergoes internalization (Yankner & Shooter 1979, Levi et al 1980, Calissano & Shelanski 1980, Biocca et al 1980). These findings raised the question: Is internalization instrumental in the activity of NGF? If so, what are the intracellular targets of NGF? Three main hypotheses are under investigation.

1. NGF acts on the plasma membrane and its effects are mediated within the cell through a second messenger. In this case, internalization might only serve to channel NGF toward lysosomes for degradation. The failure to find clearcut changes in the concentration of the classical second messengers (cAMP and cGMP) argues against this hypothesis.

2. NGF acts through multiple sites, some located on the plasma membrane and others present in intracellular organelles. Receptors located on the plasma membrane could alter membrane permeability or produce other surface events, a possibility favored by Varon and associates (Varon 1975, Skaper & Varon 1979, Varon & Skaper 1980). Intracellular sites either on the nuclear membrane or inside the nucleus (Yankner & Shooter 1979, Marchisio et al 1980) could mediate the effects of NGF on gene expression. In addition, NGF could act on receptors on the cytoskeleton (Calissano &

Cozzari 1974, Levi et al 1975, Calissano et al 1976, Calissano et al 1978, Calissano et al 1980) and produce postranscriptional molecular events that lead to axonal growth.

3. NGF directly controls gene expression. There is as yet no evidence to support this hypothesis, but Greene has recently proposed a model for the action of NGF consisting of transcription-dependent and transcription-independent events, the former occurring in pre-differentiated state, the latter occurring after differentiation of the target cell (L. A. Greene, D. Burstein, M. Black 1980).

The end of the third decade of studies on NGF marked a renewal of interest in problems that had been previously neglected. Previous efforts had been directed primarily toward elucidating the structural properties of NGF and its interaction with its target cells. Chance, rather than planned research, brought some of these problems back to the forefront opening new vistas and reviving the old, traditional adventerous spirit. The next section surveys the new areas that have a direct bearing on NGF and developmental neurobiology.

THE TROPHIC, TROPIC, AND TRANSFORMING EFFECTS OF NGF   The magnitude of the growth effects elicited by NGF on intact immature sympathetic nerve cells raised an important question. Would this molecule also exert beneficial effects in cells lethally injured with antiserum to NGF and to other pharmacological agents, such as 6-OHDA, which lead to their massive destruction? In all instances the protective effects of NGF were far above our expectation. The combined treatment of neonatal rodents with 6-OHDA (which destroys noradrenergic neurons) and with NGF not only protects the sympathetic neuron but actually resulted in a 30-fold increase in volume of sympathetic ganglia compared to the ten-fold increase produced by NGF alone (Levi-Montalcini et al 1975, Aloe et al 1975). Structural and histofluorescence studies provided an explanation for this paradoxical finding. NGF does not prevent accumulation of 6-OHDA in the adrenergic nerve endings and their destruction by this toxic compound (Thoenen & Tranzer 1968). (This effect is counteracted, however, in that NGF gains access to the cell perikarya, either directly through membrane receptors or through intact preterminal endings.) The exogenous NGF supply, far greater than the NGF that normally reaches the cell body through retrograde axonal transport in intact cells (Hendry et al 1974), enhances all metabolic processes and results in the exuberant production of collaterals from the proximal segment of the axon. This process also largely accounts for the enormous volume increase of the ganglia. A similar though lesser effect is elicited in surgically axotomized nerves of the superior cervical ganglion in neonatal rodents (Hendry & Campbell 1976). Surgical axotomy also prevents NGF entrance from peripheral tissues and leads to death of 90–95% of immature sympathetic nerve cells in superior cervical ganglia. This death is counteracted by exogenous NGF (Hendry 1975, Aloe & Levi-Montalcini 1979). NGF also counteracts the destructive effects of guanethidine in sympathetic ganglia by calling forth an overgrowth of these ganglia not exceeding that produced by administration of NGF alone (Johnson & Aloe 1974). A third agent, vinblastine, exerts its action by interfering with the assembly and function of microtubules. The destruction of sympathetic ganglia by vinblastine is likewise prevented by simultaneous administration of NGF (Calissano et al 1976, Menesini-Chen et al 1977, Johnson 1978).[1]

A unifying hypothesis that explains the ability of NGF to counteract antiserum to NGF, noxious agents, as well as the effect of axotomy, is to attribute cell death in all of these instances to deprivation of the cells of NGF. This deprivation is achieved by different mechanisms, however, such as inactivation of circulating NGF (AS-NGF), blockage of its access to adrenergic nerve endings (6-OHDA) and surgical axotomy, and prevention of NGF retrograde axonal transport (vinblastine). The only treatment in which the protective effect of NGF remains to be clarified is that with guanethidine (Johnson et al 1979).

THE NEUROTROPIC ACTIONS OF NGF   The initial discovery that nerve fibers from sensory and sympathetic embryonic ganglia placed near an NGF source (mouse sarcoma) in vitro direct their course toward it suggested that NGF could also exert a neurotropic action. This effect has been unequivocally proved in recent years in different in vitro and in vivo systems (Charlwood et al 1972, Chamley et al 1973, Levi-Montalcini 1976, Menesini-Chen et al 1978, Levi-Montalcini 1978, Campenot 1977, Letourneau 1979, Gundersen & Barrett 1979, Rutishauser & Edelman 1980). The main findings are outlined as follows:

Experiments in vivo   Daily intracerebral microinjections of NGF near the two loci coerulei in neonatal rodents repeated for ten days result in an increase in volume of sympathetic ganglia comparable to that elicited by systemic NGF injections. In addition—and at variance with the effects of systemic NGF—the intracerebral administration of NGF results in the penetration of large sympathetic fiber bundles into the spinal cord and their growth in the brain stem toward the point of injection. The existence of these adrenergic fiber systems was ascertained with his-

[1]The protective effects of NGF against AS-NGF were documented by Harper and Thoenen (1981).

tofluorescence techniques and their origin traced to markedly enlarged paravertebral ganglia. Stem axons or, more likely, collateral nerve fibers sprouting from the main axon, assemble in bundles soon after they emerge from the ganglia. These fibers gain access to the central nervous system at the level of the spinal cord by adhering to the sensory dorsal roots. Once inside the spinal cord, they assemble in the lateral and dorsal funiculi where they persist as long as NGF is supplied by intracerebral injections. This highly aberrant and parasitic fiber system establishes no connections within the spinal cord or the brain, and upon termination of treatment is reabsorbed. These adrenergic nerve fibers are thought to be channeled inside the spinal cord and brain along a rostro-caudal diffusion gradient of NGF. This idea is supported by studies of the spatial distribution of labeled NGF at different times after the intracerebral injections.

Experiments in vitro   In an elegant series of in vitro experiments, the growth of neurites from dissociated sympathetic nerve cells was shown to be controlled by NGF in the local environment. In a three-chamber culture system, nerve cells were cultured in the middle chamber, rich in NGF. The neurites produced by these cells crossed a fluid-impermeable barrier interposed between the central and the two lateral chambers. They branched only in the lateral compartment that also contained a medium rich in NGF. Removal of NGF from this compartment caused the distal portion of the neurite to stop growing and the fibers to undergo degeneration (Campenot 1977). In the experiments described by Gundersen & Barrett, axons from dissociated sensory neurons turned and grew toward the NGF solution released from a micropipette. Dislocation of the the NGF-filled micropipette resulted in a changed orientation of the growth cone, which directed its course toward the new position of the NGF source (Gundersen & Barrett 1979). The relevance of these findings to problems of neurogenesis is considered in the last section.

A TRANSFORMING EFFECT OF NGF   A new, most valuable model system for exploring the activity of NGF has resulted from the discovery that cells isolated from a rat pheochromocytoma tumor line (PC 12 cells) and treated with NGF undergo transformation into nerve cells that are indistinguishable from typical sympathetic neurons in their morphological, biochemical, and electrical properties (Tischler & Greene 1975, Greene & Tischler 1976, Dichter et al 1977). Studies of this PC 12 cell line are mentioned above, since this line is ideally suited for investigating the NGF-receptor binding and internalization processes (Yankner & Shooter 1979, Calissano & Shelanski 1980, Biocca et al 1980, Levi et al 1980). A particularly useful feature of PC 12 cells is that the actions of NGF appear to be reversible (Greene & Shooter 1980). Withdrawal of NGF causes neurites to disintegrate and cell division to recommence (Greene & Tischler 1976).

More recently, a transforming effect of NGF has been described on normal adrenal chromaffin cells obtained from 10 to 12-day-old rats and cultured in vitro (Unsicker & Chamley 1977, Unsicker et al 1978). This finding extended previous reports of the formation of nerve-like processes by adrenal medullary cells transplanted into the anterior chamber of the eye of a rat even in the absence of NGF (Olson & Malmfors 1970) and in adrenal medullary cells carried in routine culture media (Manuelidis & Manuelidis 1975).

Recent studies performed in our laboratory have shown that NGF injections in 16- to 17-day-old rat fetuses, administered through the intact tubes and resumed immediately after birth, result in the massive transformation of chromaffin cells into sympathetic nerve cells. This effect produces dramatic changes in individual cells as well as on the adrenal gland, the paraganglia, and the carotid bodies (Aloe & Levi-Montalcini 1979). The transformation is even more impressive when NGF injection starts in early embryonic stages (Levi-Montalcini & Aloe 1980).

Old Problems and New Vistas

The beginning of the fourth decade of NGF studies welcomes a revival of interest in old problems and the shaping of new ones. I shall consider two of the old ones that have been with us ever since the beginning. They concern (*a*) the role of NGF in the differentiation of sensory nerve cells and (*b*) the functional significance of NGF produced by the salivary glands for the development of sympathetic nerve cells.

THE ROLE OF NGF IN THE DIFFERENTIATION OF SENSORY NERVE CELLS

In the analysis of the multiple effects of nerve growth factor (NGF), the sympathetic ganglia have played the dominant role; the sensory ganglia which are the other targets of NGF have attracted much less attention (Hamburger et al 1981, p. 60).

In an initial attempt to study the role of NGF on sensory ganglia in vitro, Brunso-Bechtold & Hamburger (1979) confirmed and extended the studies of Thoenen & Barde (1980) on the selective retrograde axonal transport of ^{125}I NGF by nerve fibers of dorsal root ganglia in adult rats. Brunso-Bechtold & Hamburger showed that subcutaneous implantation of polyacrylamide gel pellets impregnated with ^{125}I NGF in the leg of ten-day chick embryos results in heavy labeling of sensory ganglia innervating the leg (Brunso-Bechtold & Hamburger 1979). In a subsequent study, Hamburger et al (1981) examined the effects of NGF on two neuronal cell populations found in dorsal root ganglia of chick embryos and distinguishable on a topographical basis: the dorso-medial (DM) and ventro-lateral (VL) nerve cells (Hamburger & Levi-Montalcini 1949). These populations are easily distinguishable from one another on the basis of their time course of differentiation, their morphology, and their response to experimental decrease or increase in their peripheral field of innervation. Early studies on implantation of mouse sarcomas 180 or 37 showed a marked differential response to the tumors: only the DM cells underwent an increase in number and size and showed enhanced differentiative processes in embryos bearing these transplants (Levi-Montalcini & Hamburger 1951). The concept that the DM but not the VL nerve cells are receptive to the tumoral NGF was now revised as a result of these new studies. Daily administration of NGF, supplied through injections in the egg yolk from the third to the twelfth days of incubation, counteracted neuronal death in both cell populations, (cell death in the two populations occurs at different times in development). The VL cells, however, unlike the DM cells, do not undergo striking morphological changes upon treatment with NGF in vivo or in vitro (Levi-Montalcini 1966), and consequently NGF exerts its anabolic effects by promoting changes that are more subtle than in the DM cells and not as easily detected visually. This finding is of great interest since it shows that all sensory neurons respond to NGF, and that NGF can promote effects other than enhancing nerve fiber growth. This hypothesis is supported by other recent studies. As described above, the precursors of chromaffin cell respond to NGF even before they have differentiated into glandular cells. Melanoma cells that do not undergo clear morphological changes upon NGF in vitro treatment have NGF receptors on their cell membranes and undergo some biochemical changes (Fabricant et al 1977).

THE MOUSE SALIVARY GLANDS AS AN ENDOGENOUS SOURCE OF NGF FOR SYMPATHETIC NEURONS   There is still no satisfactory answer to the much investigated question: Does the NGF synthesized and released from the salivary glands of adult mice play a role in the functioning of sympathetic nerve cells that respond in such a dramatic way to exogenous salivary NGF? Earlier results of several types of experiments (Levi-Montalcini & Angeletti 1968, Hendry & Iversen 1973, Thoenen & Barde 1980) do not reveal a functional role of salivary NGF with respect to sympathetic nerve cells. But recent data call for a revision of this hypothesis. NGF released from the salivary glands of adult male mice after injections of cyclocytidine, a pyrimidine nucleoside endowed with β-adrenergic stimulant properties, markedly raises the level of circulating NGF. Sympathetic ganglia of mice injected with cyclocytidine undergo an increase in volume due to neuronal hypertrophy. Protracted treatment with cyclocytidine also enhances the speed of regeneration of surgically transected postganglionic nerve fibers (Levi-Montalcini & Aloe 1981). Orally administered NGF also promotes overgrowth of sympathetic ganglia. This finding suggests that the NGF released into the saliva has a growth-regulating role not previously considered (Aloe et al 1982, submitted).

A NEUROENDOCRINE ROLE FOR NGF?   Although injections of NGF in rat fetuses lead to the transformation of immature chromaffin cells in sympathetic neurons (Aloe et al 1975), no somatic changes are apparent in newborn rats. However, injections of a specific antiserum to NGF in 15 to 17-day-old fetuses and continued postnatally resulted in dramatic somatic and behavioral changes indicative of specific alterations of sensory primary neurons (Aloe et al 1981). The effect of NGF antiserum in developing sensory neurons in turn produces a severe sensory deprivation syndrome consisting of unsteady gait and decreased sensibility to tactile and noxious stimuli (Aloe et al 1981). Electrophysiological studies in rat pups injected with AS-NGF during fetal life showed marked depression of excitability of neurons innervating the leg muscles. However, no apparent changes were found in those innervating cutaneous tissues (Arancio 1981, unpublished results). Destruction of sensory neurons in dorsal root ganglia in rats and guinea pigs, deprived of NGF during fetal life, was first described by Gorin & Johnson (1979) and Johnson et al (1980). By using a different technique —that of injecting a specific antiserum to NGF in rat fetuses—we found that administration of NGF antiserum during fetal life results in depletion of sensory nerve cells in the spinal ganglia and in widespread signs of damage in the residual neurons.

Of particular interest is the finding that the same pups exhibited pronounced behavioral changes in addition to these sensory deficiencies. Neonatal rats injected with AS-NGF did not differ at birth from controls. The following two weeks marked a postnatal period during which time the infants exhibited a progessive loss of weight and refused to suckle. This resulted in the death of about 60% of the treated infant rats. Those who survived this early postnatal period showed a sluggish apathetic behavior and also differed from age-matched controls in their smaller size and weight. Histological studies showed a marked decrease in volume of the thyroid gland. Studies in progress are aimed at identifying the neuronal or non-neuronal systems that were damaged by the administration of the specific antiserum to NGF during fetal life. I consider the significance of this neuroendocrine syndrome again in the following and final section.

NGF and Developmental Neurobiology at a Crossroad

In 1951 a phantasmal and elusive NGF took us on a byway that departed from the main road, running across a rather dull and flat land. The narrow lane of NGF was now all the more alluring as it started winding through picturesque and ever-changing panoramas. Now, three decades later, NGF has joined the main road once again, presented with a splendid landscape. Although still partly hidden from view, it has heightened our anticipation of the possibilities that lie ahead. In this landscape, NGF has at last "found its place in the ever-changing game of the neuroscience chessboard" (Levi-Montalcini 1975). It has provided a most profitable approach to basic questions that are the main considerations in today's fast-growing field of developmental neurobiology. I do not consider here the most apparent and generally acknowledged merit of NGF, that it has provided a unique system for the exploration of processes of nerve cell growth and differentiation at different levels. Rather, I relate other contributions of NGF-target cell interaction to our understanding of basic neurobiological problems.

THE HUMORAL AND AXONAL MEDIATED TRANSPORT OF TROPHIC FACTORS TO NERVE CELLS   Is the effect of NGF on growth radically different from those produced by implanting leg buds or other embryonic tissues? Although the action of NGF-producing tumors seemed at first to be quantitively and qualitatively different from those elicited by implantation of additional limb primordia, this impression was later revised in a 1964 article:

> In contrast to the tumor and the purified nerve growth factor, end organs have a very restricted field of action. End organs affect only the growth of nerve centers which provide their innervation. Such differences could be of a quantitative rather than qualitative order and could be correlated with differences in the production and release of growth factors in the two sets of experiments. Production and discharge may in fact be very limited in cases of implantation of limbs or additional organs. In suggesting that peripheral structures and the NGF might in the final analysis operate in a similar way, we do not imply that the release of growth factors should be the same; on the contrary, there is reason to believe that each nerve cell type might be receptive to only one specific factor (Levi-Montalcini 1964, p. 11).

Some years later NGF was found to gain access to its target cells through membrane receptors present on the perikarya of target cells as well as on nerve endings. In addition, it was found that NGF gains entrance to the cytoplasmic compartment of the cell through retrograde axonal transport following binding and internalization. These two discoveries established beyond doubt the dual access of this protein molecule to its receptive nerve cells. More recently, considerable interest has developed in the humoral nonaxonal-mediated effects as these have become one of the most intensively investigated problems in neurobiology. With the discovery that many nerve cells receive and dispatch messages over a distance in a hormonal fashion (besides communicating locally through specific connections), the boundary between "endocrine" and "synaptic" action has narrowed. It is debatable whether such a line indeed still exists. The view of the brain as an "endocrine organ" (Guillemin 1978) testifies to this evolution in one of the most firmly established dogmas of neurobiology. NGF made a most valuable contribution supporting this novel viewpoint in still another way, one that I consider in the final section.

THE NEUROTROPIC EFFECTS OF NGF AND THE ROLE OF ENVIRON-MENTAL FACTORS IN THE WIRING OF NEURONAL CIRCUITS   Is the formation of neuronal circuits in the central nervous system and the estab-lishment of specific connections between nerve fibers and peripheral end organs rigidly programmed and unmodifiable, or are nerve fibers endowed with sufficient plasticity to allow for deviation from predetermined routes, in response to chemical signals issued during neurogenesis and regeneration from neuronal and non-neuronal cells? The existence of neurotropic signals has not been convincingly proved because a valid experimental approach to this problem has been lacking. The isolation and characterization of bio-chemical agents endowed with such properties in the developing organism and in the mature nervous system represents a continuing challenge to biochemists. A novel and most valuable approach to this problem became available with the discovery of NGF. Thus, a century-old problem, the hypothesis first proposed by Cajal that exogenous humoral factors play a role in guiding nerve fibers toward their matching cells, is now receiving increasing attention and direct support from the demonstration that NGF exerts a clear neurotropic effect on adrenergic nerve fibers both in vivo and in vitro.

A WIDENING SPECTRUM OF ACTION FOR NGF AND THE NEUROEN-DOCRINE SYNDROME DUE TO ENDOGENOUS ANTIBODIES FOR NGF   A new and most revealing feature of the NGF properties has come to light with the discovery that neoplastic pheochromocytoma cells and immature chromaffin cells undergo transformation into sympathetic neurons upon administration of NGF in vitro and in vivo. This discovery has made available a unique model system to study nerve cell growth and differentia-tion processes under ideal environmental conditions. At the same time, the recognition that the action of NGF is broader than was at first envisaged prompted a search for other putative target cells.

Causes of stunted growth, hypothyroidism, and other morphological and behavioral alterations resulting from the treatment with NGF antiserum still cannot be traced to the cells or system(s) impaired by antibodies to NGF. In addition, it is unclear whether these multiple effects are due to inactivation of NGF or to inactivation of NGF-like cross-reacting protein molecules by the antiserum to NGF. The fact remains, however, that inject-ing antibodies to NGF produces this complex syndrome. That a large number of neuronal centers and circuits control or modulate the function of the neuro-endocrine system is a great incentive for pursuing these studies and to identify the neuronal or possibly non-neuronal cells responsible for the described effects.

These findings, in turn, raise another question of whether NGF may not display activity similar to that of the ever-increasing list of polypeptides endowed with important neurotransmitter or neuromodulator functions in the central nervous system and in the gastro-intestinal tract.

In concluding with this rather bold hypothesis (which at present has no support from experimental evidence), I find encouragement in the statement by Huxley quoted at the beginning of this article, "Those who refuse to go beyond fact rarely get as far as fact."

Concluding Remarks

In 1954, shortly after NGF had led us away from the main road, Viktor Hamburger, to whom I give the major credit for opening the field of experimental neuroembryology, for devising rigorously controlled experiments, and for clarifying and refining concepts of neuron-to-target cell interaction, abandoned NGF to its fate and continued his route on the main road, making in the next two decades new and extremely important contributions in the area of developmental neurobiology (Hamburger 1977). Two years ago, when NGF merged once again with the main road, Viktor, most welcomed, returned to NGF.

This biographic report of the prodigious deeds accomplished by NGF while we were chasing it along the side alley is dedicated with love and gratitude to Viktor.

Literature Cited

Aloe, L., Cozzari, C., Calissano, P., Levi-Montalcini, R. 1981. Somatic and behavioral postnatal effects of fetal injections of nerve growth factor antibodies in the rat. *Nature* 291:413–15

Aloe, L., Levi-Montalcini, R. 1979. Nerve growth factor induced overgrowth of axotomized superior cervical ganglia in neonatal rats: Similarities and differences with NGF effects in chemically axotomized sympathetic ganglia. *Arch. Ital. Biol.* 117:287–307

Aloe, L., Levi-Montalcini, R. 1980. Comparative studies on testosterone and L-thyroxine effects on the synthesis of nerve growth factor in the mouse submaxillary salivary glands. *Exp. Cell Res.* 125:15–22

Aloe, L., Levi-Montalcini, R., Calissano, P. 1982. Effects of oral administration of nerve growth factor and of its antiserum on sympathetic ganglia of neonatal mice. *Dev. Brain Res.* In press

Aloe, L., Mugnaini, E., Levi-Montalcini, R. 1975. Light and electron microscope studies on the excessive growth of sympathetic ganglia in rats injected with 6-OHDA and NGF. *Arch. Ital. Biol.* 113:326–53

Angeletti, P. U., Calissano, P., Chen, J. S., Levi-Montalcini, R. 1967. Multiple molecular forms of the nerve growth factor. *Biochim. Biophys. Acta* 147:180–82

Angeletti, P. U., Gandini-Attardi, D., Toschi, G., Salvi, M. L., Levi-Montalcini, R. 1965. Metabolic aspects of the effect of nerve growth factor on sympathetic and sensory ganglia: Protein and ribonucleic acid synthesis. *Biochim. Biophys. Acta* 95:111–20

Angeletti, P. U., Levi-Montalcini, R. 1970. Sympathetic nerve cell destruction in newborn mammals by 6-hydroxydopamine. *Proc. Natl. Acad. Sci. USA* 65:114–21

Angeletti, P. U., Levi-Montalcini, R., Calissano, P. 1968. The nerve growth factor: Chemical properties and metabolic effects. *Adv. Enzymol.* 31:51–75

Angeletti, P. U., Levi-Montalcini, R., Caramia, F. 1972. Structural and ulstructural changes in developing sympathetic ganglia induced by guanethidine. *Brain Res.* 43:515–25

Angeletti, R. H., Angeletti, P. U., Levi-Montalcini, R. 1973. The nerve growth factor. In *Humoral Control of Growth and Differentiation*, ed. J. Lobue, A. S. Gordon, pp. 229–47. New York: Academic

Angeletti, R. H., Bradshaw, R. A. 1971. Nerve growth factor from mouse salivary gland: Amino acid sequence. *Proc. Natl. Acad. Sci. USA* 68:2417–20

Arancio, O. 1981. Sul ruolo dell'NGF nello sviluppo dei neuroni sensitivi primari. PhD thesis. Pisa, Italy

Auerbach, R., Kubani, L., Knighton, D., Folkman, J. 1974. A simple procedure for long term cultivation of chick embryos. *Dev. Biol.* 41:391–94

Banerjee, S. P., Snyder, S. H. Cuatrecasas P., Greene, L. A. 1973. Binding of nerve growth factor in sympathetic ganglia. *Proc. Natl. Acad. Sci. USA* 70:2519–23

Barklis, E., Perez-Polo, J. R. 1981. S-180 cells secrete nerve growth factor protein similar to 7S nerve growth factor. *J. Neurosci. Res.* 6:21–36

Biocca, S., Levi, A., Calissano, P. 1980. Cell density-dependent regulation of NGF-receptor complexes in PC 12 cells. In *Control Mechanisms in Animal Cells*, ed. L. Jimenez, R. Levi-Montalcini, R. Shields, S. Iacobelli, pp. 43–52. New York: Raven

Bocchini, V., Angeletti, P. U. 1969. The nerve growth factor: Purification as a 30,000 molecular weight protein. *Proc. Natl. Acad. Sci. USA* 64:787–94

Bradshaw, R. A. 1978. Nerve growth factor. *Ann. Rev. Biochem.* 47:191–216

Brunso-Bechtold, J. K., Hamburger, V. 1979. Retrograde transport of nerve growth factor in chicken embryos. *Proc. Acad. Natl. Sci. USA* 76:1494–96

Bueker, E. D. 1948. Implantation of tumors in the hind limb field of the embryonic chick and developmental response of the lumbosacral nervous system. *Anat. Rec.* 102:369–90

Calissano, P., Castellani, L., Monaco, G., Mercanti, D., Levi, A. 1980. Studies on the mechanism of NGF-induced neurite growth. In *Nerve Cells, Transmitters and Behaviour*, ed. R. Levi-Montalcini, pp. 65–80. Amsterdam: North Holland/Eslevier

Calissano, P., Cozzari, C. 1974. Interaction of nerve growth factor with the mouse brain neurotubule protein(s). *Proc. Natl. Acad. Sci. USA* 71:2131–35

Calissano, P., Monaco, G., Castellani, L., Mercanti, D., Levi, A. 1978. Nerve growth factor potentiates actinomycin ATPase. *Proc. Natl. Acad. SCI. USA* 75:2210–15

Calissano, P., Monaco, G., Levi, A., Menesini-Chen, M. G., Chen, J. S., Levi-Montalcini, R. 1976. New developments in the study of NGF-tubulin interaction. In *Contractile Systems in Non-Muscle Tissues*, ed. S. V. Perry et al, pp. 201–11. Amsterdam: North Holland/Elsevier

Calissano, P., Shelanski, M. L. 1980. Interaction of nerve growth factor with pheochromocitoma cells—evidence for tight binding and sequestration. *J. Neurosci.* 5:1035–39

Campenot, R. B. 1977. Local control of neurite development by nerve growth factor. *Proc. Natl. Acad. Sci. USA* 74:4516–19

Chamley, J. H., Goller, I., Burnstock, G. 1973. Selective growth of sympathetic nerve fibers to explants of normally densely innervated autonomic effector organs in tissue culture. *Dev. Biol.* 31:362–79

Charlwood, K. A., Lamont, D. M., Banks, B. E. C. 1972. Apparent orientating effect produced by nerve growth factor. In *Nerve Growth Factor and its Antiserum*, ed. E. Zaimis, pp. 102–7. London: Athalon Press, Univ. London

Cohen, S. 1958. A nerve growth-promoting protein. In *The Chemical Basis of Development*, ed. W. D. McElroy, B. Glass, pp. 665–76. Baltimore: John Hopkins Press

Cohen, S. 1960. Purification of a nerve growth promoting protein from the mouse salivary gland and its neurocytotoxic antiserum. *Proc. Natl. Acad. Sci. USA* 46:302–11

Cohen, S., Levi-Montalcini, R. 1956. A nerve growth stimulating factor isolated from snake venom. *Proc. Natl. Acad. Sci. USA* 42:571–74

Cohen, S., Levi-Montalcini, R., Hamburger, V. 1954. A nerve growth stimulating factor isolated from sarcoma 37 and 180. *Proc. Natl. Acad. Sci. USA* 40:1014–18

Costrini, N. V., Bradshaw, R. A. 1979. Binding characteristics and apparent molecular size of detergent-solubilized nerve growth factor receptor of sympathetic ganglia. *Proc. Natl. Acad. Sci. USA* 76:3242–45

Dichter, M. A., Tischler, A. S., Greene, L. A. 1977. Nerve growth factor induced increase in electrical excitability and acetylcholine sensitivity of a rat pheochromocitoma cell line. *Nature* 268: 501–4

Eränkö, O., Eränkö, L. 1971. Histochemical evidence of chemical sympathectomy by guanethidine in newborn rats. *J. Histochem.* 3:451–56

Fabricant, R. N., De Larco, J. E., Todaro, G. 1977. Nerve growth factor receptors on human melanoma cells in culture. *Proc. Natl. Acad. Sci. USA* 74:565–69

Frazier, W. A., Boyd, L. F., Bradshaw, R. A. 1973. Interaction of the nerve growth factor with surface membranes. Biological competence of insolubilized nerve growth factor. *Proc. Natl. Acad. Sci. USA* 70:2931–35

Goldstein, L. A., Reynolds, C. P., Perez-Polo, J. R. 1978. Isolation of human nerve growth factor from placental tissue. *Neurochem. Res.* 3:175–83

Gorin, P. D., Johnson, E. M. 1979. Experimental autoimmune model of nerve growth factor deprivation: Effects on developing peripheral sympathetic and sensory neurons. *Proc. Intl. Acad. Sci. USA* 76:5382–86

Greene, L. A., Shooter, E. M. 1980. The nerve growth factor: Biomistry, synthesis, and mechanism of action. *Ann. Rev. Neurosci.* 3:353–402

Greene, L. A., Tischler, A. S. 1976. Establishment of a noradrenergic cell line of rat adrenal pheochromocitoma cells which respond to nerve growth factor. *Proc. Natl. Acad. Sci. USA* 73:2424–28

Guillemin, R. 1978. The F. O. Schmitt Lecture in Neuroscience 1977. The brain as an endocrine organ. *Neurosci. Res.* 16(S):1–25

Gundersen, R. W., Barrett, J. N. 1979. Neuronal chemotaxis: Chick dorsal-root axons turn toward high concentration of nerve growth factor. *Science* 206: 1079–80

Hamburger, V. 1977. The F. O. Schmitt lecture in Neuroscience 1976. The developmental history of the motor neuron. *Neurosci. Res.* 15(S):1–37

Hamburger, V., Brunso-Bechtold, J. K., Yip, J. W. 1981. *J. Neurosci.* 1:60–71

Hamburger, V., Levi-Montalcini, R. 1949. Proliferation, differentiation and degeneration in the spinal ganglia of the chick embryo. *J. Exp. Zool.* 111:457–501

Harper, G. P., Barde, Y. A., Burnstock, G., Carstairs, J. R., Dennison, M. E., Suda,

E., Vernon, A. 1979. Guinea pig prostate as a rich source of nerve growth factor. *Nature* 279:160–62

Harper, G. P., Thoenen, H. 1980. The distribution of nerve growth factor in the male sex organs of mammals. *J. Neurochem.* 77:391–402

Harper, G. P., Thoenen, H. 1981. Target cells: Biological effects and mechanism of action of nerve growth factor and its antibodies. *Ann. Rev. Pharmacol. Toxicol.* 21:205–29

Harrison, R. G. 1935. The Croonian lecture on the origin and development of the nervous system studies by the methods of experimental embryology. *Proc. R. Soc. London* 118:155–96

Hendry, I. A. 1975. The response of adrenergic neurons to axotomy and nerve growth factor. *Brain Res.* 94:87–97

Hendry, I. A. 1977. Cell division in the developing sympathetic nervous system. *J. Neurocytol.* 6:299–309

Hendry, I. A., Campbell, J. 1976. Morphometric analysis of rat superior cervical ganglion after axotomy and nerve growth factor treatment. *J. Neurocytol.* 5:361–70

Hendry, I. A., Iversen, L. L. 1973. Reduction in the concentration of nerve growth factor in mice after sialectomy and castration. *Nature* 243:500–4

Hendry, I. A., Stach, R., Herrup, K. 1974. Characteristics of the retrograde axonal transport for nerve growth factor in the sympathetic nervous system. *Brain Res.* 82:117–28

Herrup, K., Shooter, E. M. 1973. Properties of the beta nerve growth factor receptor of avian dorsal root ganglia. *Proc. Natl. Acad. Sci. USA* 70:3384–88

Ishii, D. N., Shooter, E. M. 1975. Regulation of nerve growth factor synthesis in mouse submaxillary glands by testosterone. *J. Neurochem.* 25:843–51

Johnson, E. M. 1978. Destruction of the sympathetic nervous system in neonatal rats and hamsters by vinblastine: Prevention by concomitant administration of nerve growth factor. *Brain Res.* 141:105–18

Johnson, E. M., Aloe, L. 1974. Suppression of the in vitro and in vivo effects of guanethidine in sympathetic neurons by nerve growth factor. *Brain Res.* 81: 519–32

Johnson, E. M., Gorin, P. D., Brandeis, L. D., Pearson, J. 1980. Dorsal root ganglion neurons are destroyed by exposure in utero to maternal antibodies to nerve growth factor. *Science* 210:916–18

Johnson, E. M., Macia, R. A., Andres, R. Y., Bradshaw, R. A. 1979. The effects of drugs which destroy the sympathetic nervous system on the retrograde transport of nerve growth factor. *Brain Res.* 171:461–72

Letourneau, P. C. 1979. Chemotactic response of nerve fiber elongation to nerve growth factor. *Dev. Biol.* 66:183–96

Levi, A., Cimino, M., Mercanti, D., Chen, J. S., Calissano, P. 1975. Interaction of nerve growth factor with tubulin. Studies on binding and induced polymerization. *Biochem. Biophys. Acta* 399:50–60

Levi, A., Shechter, Y., Neufeld, E. J., Schlessinger, J. 1980. Mobility, clustering, and transport of nerve growth factor in embryonal sensory cells and in a sympathetic neuronal cell line. *Proc. Natl. Acad. Sci. USA* 77:3469–73

Levi-Montalcini, R. 1952. Effects of mouse tumor transplantation on the nervous system. *Ann. NY Acad. Sci.* 55:330–43

Levi-Montalcini, R. 1953. In-vivo and in-vitro experiments on the effect of mouse sarcoma 180 and 37 on the sensory and sympathetic system of the chick embryo. *Proc. 14th Congr. Zool.,* Copenhagen, p. 309

Levi-Montalcini, R. 1958. Chemical stimulation of nerve growth. In *Chemical Basis of Development,* ed. W. D. McElroy, B. Glass, pp. 646–64. Baltimore: John Hopkins Press

Levi-Montalcini, R. 1964. Growth control of nerve cells by a protein factor and its antiserum. *Science* 143:105–10

Levi-Montalcini, R. 1966. The nerve growth factor: Its mode of action on sensory and sympathetic nerve cells. *Harvey Lect.* 60:217–59

Levi-Montalcini, R. 1972. The nerve growth factor. In *Immunosympathectomy,* ed. G. Steiner, E. Schönbaum, pp. 25–45. Amsterdam: Elsevier

Levi-Montalcini, R. 1975. NGF: An uncharted route. In *The Neurosciences Paths of Discovery,* ed. F. G. Worden, J. P. Swazey, G. Adelman, pp. 245–63. Cambridge: MIT Press

Levi-Montalcini, R. 1976. The nerve growth factor: Its role in growth, differentiation and function of the sympathetic adrenergic neuron. *Prog. Brain Res.* 45:235

Levi-Montalcini, R., Aloe, L. 1980. Trophic, tropic and transforming effects of nerve growth factor. In *Histochemistry and Cell Biology of Autonomic Neurons, SIF Cells and Paraneurons,* ed. O. Eränkö, S. Soinila, H. Päivärinta, pp. 3–15. New York: Raven

Levi-Montalcini, R., Aloe, L. 1981. Synthesis and release of the nerve growth factor from the mouse submaxillary salivary glands: Hormones and cell regulation, ed. J. E. Dumont, J. Nunez, 5:53–72. Amsterdam: North-Holland Elsevier

Levi-Montalcini, R., Aloe, L., Mugnaini, E., Oesch, F., Thoenen, H. 1975. Nerve growth factor induces volume increase and enhances thyrosine hydroxylase in sympathetic ganglia of newborn rats. *Proc. Natl. Acad. Sci. USA* 72:595–99

Levi-Montalcini, R., Angeletti, P. U. 1961. Biological properties of a nerve growth promoting protein and its antiserum. In *Regional Neurochemistry. Proc. 4th Int. Neurochem. Symp.,* ed. S. S. Kety, J. Elkes, pp. 362–76. New York: Pergamon

Levi-Montalcini, R., Angeletti, P. U. 1963. Essential role of the nerve growth factor in the survival and maintenance of dissociated sensory and sympathetic nerve cells in vitro. *Dev. Biol.* 7:655–59

Levi-Montalcini, R., Angeletti, P. U. 1964. Hormonal control of the NGF in the submaxillary salivary glands of mice. In *Salivary Glands and Their Secretion,* ed. L. M. Sreenby, J. Mayer, pp. 129–41. New York: Pergamon

Levi-Montalcini, R., Angeletti, P. U. 1966. Immunosympathectomy. *Pharmacol. Rev.* 18:619–28

Levi-Montalcini, R., Angeletti, P. U. 1968. Nerve growth factor. *Physiol. Rev.* 48:534–65

Levi-Montalcini, R., Angeletti, P. U. 1965. Biological aspects of the nerve growth factor. *Ciba Found. Symp.* pp. 126–47

Levi-Montalcini, R., Booker, B. 1960a. Excessive growth of the sympathetic ganglia evoked by a protein isolated from mouse salivary glands. *Proc. Natl. Acad. Sci. USA* 46:373–84

Levi-Montalcini, R., Booker, B. 1960b. Destruction of the sympathetic ganglia by an antiserum to the nerve growth protein. *Proc. Natl. Acad. Sci. USA* 46:384–91

Levi-Montalcini, R., Caramia, F., Luse, S. A., Angeletti, P. U. 1968. In-vitro effects of the nerve growth factor on the fine structure of the sensory nerve cells. *Brain Res.* 8:347–62

Levi-Montalcini, R., Cohen, S. 1960. Effects of the extra of the mouse salivary glands on the sympathetic system of mammals. *Ann. NY Acad. Sci.* 85:324–41

Levi-Montalcini, R., Hamburger, V. 1951. Selective growth stimulating effects of mouse sarcoma on the sensory and sympathetic nervous system of the chick embryo. *J. Exp. Zool.* 116:321–62

Levi-Montalcini, R., Meyer, H., Hamburger, V. 1954. In-vitro experiments on the effects of mouse sarcoma on the sensory and sympathetic ganglia of the chick embryo. *Cancer Res.* 14:49–57

Malmfors, T., Theonen, H., eds. 1971. *6-Hydroxydopamine and Catecholmine Neurons,* p. 368. Amsterdam: North Holland/Elsevier

Manuelidis, I., Manuelidis, E. F. 1975. Synaptic boutons and neuron-like cells in isolated adrenal gland cultures. *Brain Res.* 96:181–86

Marchisio, P. C., Nalidini, L., Calissano, P. 1980. Intracellular distribution of nerve growth fractor in rat pheochromocitoma PC 12 cells: Evidence for a perinuclear and intranuclear location. *Proc. Natl. Acad. Sci. USA* 77:1656–60

Menesini-Chen, M. G., Chen, J. S., Calissano, P., Levi-Montalcini, R. 1977. Nerve growth factor prevents vinblastine destructive effects on sympathetic ganglia in newborn mice. *Proc. Natl. Acad. Sci. USA* 74:5559–63

Menesini-Chen, M. G., Chen, J. S., Levi-Montalcini, R. 1978. Sympathetic nerve fibers ingrowth in the central nervous system of neonatal rodents upon intracerebral NGF injection. *Arch. Ital. Biol.* 116:53–84

Miner, R. W., ed. 1952. The chick embryo in biological research. *Ann. NY Acad. Sci.* 55:37–344

Mobley, W. C., Server, A. C., Ishii, D. N., Riopelle, R. J., Shooter, E. M. 1977. Nerve growth factor. *N. Engl. J. Med.* 297:1096–1104, 1149–58, 1211–18

Olson, L., Malmfors, T. 1970. Growth characteristics of adrenergic nerves in the adult rat. *Acta Physiol. Scand. Suppl.* 348:1–112

Paravicini, U., Stoeckel, K., Thoenen, H. 1975. Biological importance of retrograde axonal transport of nerve growth factor in adrenergic neurons. *Brain Res.* 84:279–91

Rutishauser, U., Edelman, G. M. 1980. Effects of fasciculation in the outgrowth of neurites from spinal ganglia in culture. *J. Cell Biol.* 87:370–78

Server, A. C., Shooter, E. M. 1977. Nerve growth factor. *Adv. Protein Chem.* 31:339–409

Skaper, S. D., Varon, S. 1979. Sodium dependence of nerve growth factor regulated hexose transport in chick embryo sensory neurons. *Biochem. Biophys. Res. Commun.* 88:563–68

Steiner, G. , Schönbaum, E. ed. 1972. *Immunosympathectomy,* p. 253. Amsterdam: Elsevier

Stoeckel, K., Guroff, G., Schwab, M., Thoenen, H. 1976. The significance of retrograde axonal transport for the accumulation of systematically administered nerve growth factor (NGF) in the rat superior cervical ganglion. *Brain Res.* 190:271–84

Taylor, A. M. 1970. *Imagination and the Growth of Science,* p. 110. New York: Shocken

Thoenen, H., Barde, Y. A. 1980. Physiology of nerve growth factor. *Physiol. Rev.* 60:1284–335

Thoenen, H., Otten, U., Schwab, M. 1979. Ortograde and retrograde signals for the regulation of neuronal gene expression: The peripheral sympathetic nervous system as a model. In *The Neurosciences. Fourth Study Program,* ed. F. O. Schmitt, F. G. Worden, pp. 911–28. Cambridge: MIT Press

Thoenen, H., Tranzer, J. P. 1968. Chemical sympathectomy by selective destruction of adrenergic nerve endings with 6-hydroxydopamine. *Naunyn-Schmiedebergs Arch. Exp. Pathol. Pharmakol.* 261:271–88

Tischler, A. S., Greene, L. A. 1975. Nerve growth factor-induced process formation by cultured rat pheochromocitoma cell. *Nature* 258:341–2

Tranzer, J. P., Richards, J. G. 1971. Fine structural aspects of the effect of 6-hydroxydopamine on peripheral adrenergic neurons. In *6-Hydroxydopamine and Catecholamine Neurons,* ed. R. Malmfors, H. Thoenen, pp. 15–32. Amsterdam: Elsevier/North-Holland

Unsicker, K., Chamley, J. H. 1977. Growth characteristics of postnatal rat adrenal medulla in culture. *Cell Tissue Res.* 177:247–68

Unsicker, K., Krisch, B., Otten, U., Thoenen, H. 1978. Nerve growth factor-induced fiber outgrowth from isolated adrenal chromafin cells: Impairement by glucocorticoids. *Proc. Natl. Acad. Sci. USA* 75:3498–502

Varon, S. 1975. Nerve growth factor and its mode of action. *Exp. Neurol.* 43(2):75–92

Varon, S., Bunge, R. P. 1978. Trophic mechanisms in the peripheral nervous system. *Ann. Rev. Neurosci.* 1:327–61

Varon, S., Nomura, J., Shooter, E. M. 1967. The isolation of the mouse nerve growth factor protein in a high molecular weight form. *Biochemistry* 5:2203–9

Varon, S., Skaper, S. D. 1980. Short-latency effects of nerve growth factor: An ionic view. In *Tissue Culture in Neurobiology,* pp. 333–47. La Jolla: Univ. Calif., San Diego, Sch. Med., Dept. Biol.

Yankner, B. A., Shooter, E. M. 1979. Nerve growth factor in the nucleus: Interaction with receptors on the nuclear membrane. *Proc. Natl. Acad. Sci. USA* 76:1269–73

Zanini, A., Angeletti, P. U., Levi-Montalcini, R. 1968. Immunochemical properties of the nerve growth factor. *Proc. Natl. Acad. Sci. USA.* 61:835–42

Reference added in proof:

Green, L. S., Burnstein, D., Black, M. 1980. Can nerve growth factor play a transcription-dependent "instructive" role during developement? Evidence from studies with clonal pheochromocitoma cells. *Nerve Cells, Transmitters and Behavior,* ed. R. Levi-Montalcini, pp. 45–57. Amsterdam: Elsevier North-Holland

Ann. Rev. Neurosci. 1982. 5:363–70

CORTICAL NEUROBIOLOGY:
A Slanted Historical Perspective

David H. Hubel

Department of Neurobiology, Harvard Medical School, Boston,
Massachusetts 02115

I have been asked to compose an essay on the development of ideas in cortical neurophysiology over the past two decades. That, it seems to me, amounts to writing a paper that is all introduction—no methods, no results —perhaps some discussion. Not a task one undertakes lightly, and I make no guarantees as to the outcome.

When asked to reminisce about the good old days, two thoughts occur to me. One: the "good old days" are right now. Competitive and over-crowded and at times almost vicious as the field has become, we have nevertheless to keep pinching ourselves to wake up and appreciate how lucky we are to be in a field of science that is moving and breathlessly exciting. And unlike the situation in those "good old days," it is now at least possible to earn a living at it. We are now where molecular biology was 10 to 15 years ago and where physics was when I was in college. To be in a field of science at just the right time takes, among other things, a lot of luck.

My second thought is a completely different one. A few years ago Torsten Wiesel and I decided to put together a set of reprints with forethoughts and afterthoughts, a book of readings like so many others, except that with characteristic modesty these readings would be restricted to our own pa-pers. (An average of one paper a year seems meager to Medical School Promotions Committees, but if you live long enough it adds up.) Well, we partitioned the work in some fair manner such as me writing two-thirds and Torsten one-third, but when we got to the 1963–1965 visual-deprivation papers, through some misunderstanding we each independently wrote some pages of introduction, describing how and why we started closing cats' eyes.

0147-006X/82/0301-0363$02.00

When I saw Torsten's version I was flabbergasted, and even a bit angry. His version was that we had wanted to examine postnatal development of the visual cortex, and so closed one eye to hold up development in that pathway. According to me, we simply wanted to see whether we could produce an amblyopic eye and then look physiologically for the point or points in the pathway at which the failure occurred. These motivations, to be sure, aren't mutually exclusive, but I would have sworn that at the time the work was done our motives, whatever they were, were the same. In any case, if I describe how I think things were in cortical neurophysiology twenty years ago, you have to realize that my memory of things past may be a far cry from the way things really were. As in Akutagawa's story, *Rashomon,* even at a given time no two people have the same assessment of current events.

So much for my introduction to my introduction. I now turn to how things really were in the good old days, and subsequent developments. A glance at Howell's *Textbook of Physiology,* vintage about 1950, will convince anyone that we have come a very long way. Even then, a half-century after Cajal's *Histologie du Système Nerveux,* despite the work of people like Adrian, Woolsey, Jasper, and Penfield, the question of cortical localization was still hotly debated. One of the first scientific papers I ever heard, in 1953, was by Lashley, at the Montreal International Physiological Congress. I have no recollection at all of what Lashley said at that plenary session (one almost as crowded as a Presidential Symposium of the Society for Neuroscience) but he then was one of the strongest antagonists of the concept of precise cortical localization, and I must confess that when I try to detect the difference between Brodmann's areas 18 and 19 in a Nissl section of the cerebral cortex I'm not unsympathetic to Lashley. But in those dark ages it was worse than that. Elsewhere in this volume Eccles has reviewed the medieval problems of electrical vs chemical synapses at nerve-muscle junction and spinal cord; then everyone was at least agreed that the business was done at the synapses. When it came to the cortex, the concepts of nebulous electrical field effects were still taken seriously. It is hard to believe, today, that as recently as 1955 no less a person than Roger Sperry saw fit to dice up the cortex with vertically placed sheets of mica or tantalum wires in order to stop the supposed current flow or short circuit it, and thus quash those concepts. And along with ephaptic mechanisms one had glia and all kinds of ideas as to what *they* might be doing; one had reverberating circuits and suppressor strips. Any sensible person who was well read in the literature, and had physics or biochemistry as options, would surely avoid a field in such a sad state.

There were of course healthy streams, and things had begun to change. Part of the awakening was related to technical progress. Until the late 1950s most work in physiology of the cortex depended on the EEG and evoked

David H. Hubel and Torsten N. Wiesel.

potentials. Not surprisingly the main thrust then was towards under-standing the part that the cortex plays in sleep and waking and in attention. (Learning was also mentioned *sotto voce;* Hebb was decades ahead of his time; the idea of memory-and-macro-molecules that later polluted the field was still not yet hatched.) No one could see the changes in the EEG of a person falling asleep and not be fascinated—it is still just as fascinating. Unfortunately as far as our understanding goes, the EEG, and for that matter the problems of sleep and arousal, are now not so much better understood than they were then. For the EEG one can see little progress. For sleep, we indeed now do have REM sleep, the locus ceruleus and the raphé nuclei, and a host of transmitters to think about today, but that only says that we have in 1981 a better grasp of the magnitude of the problem, not that we are one iota nearer to understanding what happens to the cortex in sleep or why it happens. The problem then of course was that the tool available (and I say the tool, because the EEG was about the only one at hand) was completely empirical. Some of the discoveries made with it, such as the recognition of slow-wave sleep stages and REM sleep, were momen-tous, but the boot-strap operation, of coming to understand the EEG by using the EEG, was not possible. So the subject languished.

The closely related evoked potential method produced more fundamental results because the problem it addressed, a definition of the sensory areas and a working out of the topography within some of them, were simpler. The stream of work that began with Adrian, Talbot and Marshall, Bard, Woolsey and others has continued right to the present, though microelectrodes have largely replaced the saline wicks and coarse wires that registered the evoked potentials. It's a closely guarded secret that of the two techniques, evoked potentials and single-cell recording, the evoked potential technique is far more difficult to carry out and to interpret: that is what makes Talbot and Marshall's 1941 paper still such a wonder to contemplate.

It was the microelectrode that made the difference, between 1955 and 1965, but it took time and a lot of technical development before it came to be used effectively. Early in that era a number of groups succeeded in inserting micropipettes into cortex, and even in recording spikes. A crucially important technical innovation was the Davies chamber (and its variants), which allowed one to keep the cortex from bouncing around with the heart beat and respiration, long enough to allow one to find what kinds of stimuli, if any, would influence the occurrence of those spikes. Metalic electrodes—indium, gold, platinum, tungsten, which didn't break and didn't plug and had remarkably low resistance at the frequencies that counted for extracellular work—were a great help. Curiously, why they work as well as they do, especially the big dishmop-like platinum-black ones, is still not understood at all. I think that the most important advance was the strategy of making long microelectrode penetrations through the cortex, recording from cell after cell, comparing responses to natural stimuli, with the object not only of finding the optimal stimulus for particular cells, but also of learning what the cells had in common and how they were grouped. This method came, I believe, mainly from the Johns Hopkins Medical School and was the result of combining the talents of physiologists such as Mountcastle with those of neuroanatomists, especially Jerzy Rose. It led to a clear proof of localization of function in the brain stem, thalamus, and cortex, and led directly to Mountcastle's discovery in 1957 of cortical columns. This fusion of the methods and ideas of physiology and anatomy was a new thing. Previously, physiologists had rarely looked to see where their electrodes were. I'm not sure why, unless perhaps they were convinced that histology was a next-to-impossible art. Conversely, anatomists have been slow to use physiology to monitor placement of their instruments of injection or destruction, doubtless also because of terror over the idea of using an amplifier or oscilloscope. It took physiologists some time to realize that you can't study function without studying form. Physiologists also had to get over their love affairs with electronic gadgetry, and learn that if you want to study pain a good start is to pinch the animal's tail and see if cells respond.

Until the late 1950s neurobiology suffered seriously from the separation of its main component parts, anatomy, physiology, and chemistry. To earn a living neurophysiologists often had to be competent in renal and cardiac physiology, and neuroanatomists in gross anatomy, with little chance or time to learn the other person's field. I count myself blessed to have been trained first in Montreal at the Montreal Neurological Institute and then at Walter Reed, where Fuortes, Galambos, and Nauta all worked on the same corridor. It took universities and medical schools years to learn that these three components of neurobiology are closely intertwined, and that they all suffer when separated.

The sedulous anatomical-physiological business of studying exhaustively not just one cell but many, in marathon penetrations, had its epitome in Vernon Mountcastle, already a famous person when I first got into the field. I went once to visit him in Baltimore from Walter Reed (where I was lucky enough to be drafted in 1955) to discuss plans for possible postdoctoral training. I arrived around 3 P.M. Vernon was doing an experiment, by himself, and looking reasonably fresh, or perhaps just slightly weary. I said, "Is this the first penetration?" He said, "Yes, I started yesterday morning."

Some months later, around the time that Torsten and I started working together at Hopkins, Vernon gave a paper reporting results from, I think it was, 900 cells, for those days an astronomic number. We knew we could never catch up with that, so we did the next best thing and called our first cell No. 3000 and numbered them from there. When Vernon visited our lab we made sure that we mentioned several times the i.d. number of any cell we described to him. Vernon was most encouraging and suitably excited about our first results. I remember his comment: "What a wonderful system! It will give you enough work to last a good five years."

But to return to the quantitative propensities of neurophysiologists, I remember a momentous occasion when at Walter Reed, in 1957, or so, I felt pleased enough at finding directionally selective cells in area 17 of a purring cat to drag Bob Galambos down to my lab for a demonstration. Bob was suitably impressed—he always gave wonderful positive feedbac* and was very tolerant of a stubborn and moody postdoc—but said, "David, this is fine, but what is the latency of this cell?" Well, I did feel a bit sheepish, to be preoccupied with postgraduate stuff like movement and not to have even established the latency, so I found a strobe light, and quickly found that this cell's latency was 83 msec. That was odd, considering a myelinated pathway a few inches long, and allowing for a delay of about 1 msec per synapse. I took the lesson to heart: that was the first latency I ever measured, and also the last.

The leading neurophysiologists—those who worked in the periphery and spinal cord—had rather firm ideas about what neurophysiology was. They tried to be quantitative; they expressed results in milliseconds and used graphs. Once around 1962 Jack Eccles came to visit Stephen Kuffler, and Torsten and I got to show him some oriented receptive fields. Again we were quizzed about latency, and the virtues of intracellular methods and electrical stimulation were pointed out to us. In leaving Jack said, "You know, sooner or later you have to start doing neurophysiology." Jack's attitude was of course correct; we all know the success story that came out of Eccles's work on the cerebellum a few years later. But for better or worse the course of events has been quite different in cerebral and cerebellar cortex. In cerebellum we now know the physiology of the circuit better than anywhere except possibly the retina (or the aplysia abdominal ganglion, to do justice to our host at the Presidential Symposium!), but without the faintest idea of overall information processing. In contrast, for the striate cortex we do know, at least in rough outline, the difference between the meaning of a spike discharge in the input as opposed to the output: the information processing is in some sense understood. But the circuit, in terms of excitation and inhibition and transmitters, is still like midnight in central Africa. The emphasis has been in studying responses of cells in terms of the optimal natural sensory stimuli. In vision this trend began with Stephen Kuffler; Torsten and I simply extended his approach further centrally.

That one still has not got around to doing neurophysiology in the cortex, in Eccles's sense of the word neurophysiology, is of course partly because the anatomy of the cerebral cortex is still far from worked out. The cerebellum, though not exactly childishly simple, does have (and has had since Cajal) a pellucid quality to its anatomy—its cortex has five or six kinds of cells, whose connections have been rather well known for a long time. In the cerebral cortex, Cajal and Lorente de Nó's work had made it clear that the major intrinsic connections run vertically. But until the work of Jennifer Lund, no one had ever taken one cortical area in one animal species and looked long and hard at the Golgi anatomy, an amazing thing given that the Golgi method has been around for more than a century. Lund's recent Golgi work makes it clear how little was really known. Moreover, the half dozen or so major advances in neuroanatomical methods that have been developed in the past ten years or so, especially methods based on axonal transport, and the techniques just appearing for identifying enzymes immunohistochemically, are all accelerating the pace of progress in cortical anatomy, but again making it clear how much there still is to do. It does seem probable that the cerebral cortex will turn out to have stereotyped sets of connections, intricate yet repetitive like a crystal, but an order of magnitude (I'm never sure if that expression just means "ten times") more complex than the cerebellum.

If in the 1950s there were doubts about the existence of specificity and topography in the cortex, there seemed also to be a profound lack of ideas about what the structure could possibly be doing. About the only thing one could say was that it "analyzed," but no one had any clear thoughts as to what kind of things the analysis might entail. I suppose that perhaps the main contribution of work over the past decades on the visual cortex has been the demonstration that something does indeed happen between the input and output of area 17. This kind of information, as opposed to mapping, has been harder to obtain in other cortical areas, probably for a variety of reasons. In the auditory system it is harder to imagine what the appropriate biological stimulus should be in a cat that hears so well up to 50 kilocycles; the pathway leading to the cortex is more complex; the response properties of medial geniculate cells are less thoroughly understood than those of lateral geniculate cells. Lately considerable progress is being made in mapping the auditory cortex, and this knowledge may be necessary before one can get at the problem of comparing input and output.

In the 1950s when at Walter Reed I decided once to take advantage of Galambos' laboratory full of audio stimulators and soundproof rooms, and set about to record from the auditory cortex of cats. I remember vividly how one could turn every dial of the stimulator, with sine waves, white noise and clicks, with no effects at all on cortical cells. Then on going into the chamber to check on the cat, I would find that the cell would fire like a machine gun when I rattled the doorknob. Shaking keys was also often very effective. But I never got anywhere in trying to pin down what was so good about these stimuli. I suspect that among other things they were simply interesting to the cat.

An important advance that has come out of work on visual cortex is the realization, first from the physiological work of Toyama, and then anatomical studies by a number of groups, that different layers send their outputs to different places. Meanwhile receptive field mapping has made it possible to compare the properties and the information transmitted by the cells in these layers. While cells in the different layers do things in common, they exhibit marked differences that are presumably tailored to the functions of the structures to which they project. For example the large field motion-sensitive cells of layer V presumably provide just what the superior colliculus needs. The common properties, like responding to oriented lines or favoring one eye, define the several column systems. It is this intricate Chinese puzzle aspect that one finds so challenging and aesthetically pleasing. One can only guess (and there are already many hints) at what kinds of beauty lie in the fifty or more still unexplored cortical areas.

In case anyone thinks I'm at all complacent about area 17 being well in hand, let me reassure you; there are still problems enough even to keep all 5000 (or whatever the number is) area-17 physiologists busy. One has only to think of apical dendrites and that crowning mystery, layer 1. Or if that leaves you cold, there are those reciprocal connections—18 to 17, 17 to the lateral geniculate—which, along with every other reciprocal connection in the nervous system, except for the gamma efferents, are utterly unununderstood.

It seems to me that another quite different development that I have alluded to earlier is the snowballing or mushrooming of a number of areas in a number of different systems—the Talbot-Marshall-Woolsey-Kaas-Allman school of anatomy-physiology. Grossly it looks as if, far from a Lashleyian subdivision into just a few major functional areas, the cortex is made up of areas whose total number may yet put Brodmann to shame. We really don't know quite what to think about the multiplicity of these areas. In recent years neurobiologists have had their hands full just mapping them out; there has not yet been time to address the physiology except in a preliminary way. Certainly a possibility is that each area will deal with a submodality, one for movement, one for stereopsis, one for color. The idea is not without some experimental support and may turn out to be correct, but it nevertheless to me has a kind of naivete, like the notion that stripped of our cortex—a kind of double scalping—we became alligators, or the notion that our right hemisphere is for art and music and other nice things and our left is rational and analytic and propositional, in short a bore. In the studies that have been done outside of area 17 in monkeys, it must be freely admitted that the avalanching increase in optimal-stimulus complexity which one might have predicted from an extrapolation of the increase that occurs in going from geniculate to cortex has simply not been evident. The first disappointment came when we found nothing very interesting by way of higher order complexity in the Clare-Bishop area in the cat, or the superior temporal sulcus in the macaque. It is too early to guess whether this is just a temporary setback that will pass when one sets about to study these prestriate areas with adequate momentum and determination and patience. At the moment—and I would never have guessed this twenty years ago—we still have our hands full working in area 17.

In an essay that is supposed to have been dealing with the development of ideas about the cortex in general, you may have detected how little I've said about the huge no-person's land beyond the occipital lobe: the speech areas, motor areas, parietal, temporal, and frontal lobes. For most of these I suspect there is first a huge job of blocking out to do, analogous to the job the evoked potentials and long electrode tracks did and are still doing for the main sensory areas. We're full of hope—I think justifiably—that methods like deoxyglucose and the PET scan may begin a revolution here; that will presumably have to take place before any cell-level approach will be possible. One has the feeling of being only in the foothills of some gigantic mountain range. I can only say that the foothills aren't especially boring!

AUTHOR INDEX

371

SUBJECT INDEX

A

Acetic anhydride
  acetylcholinesterase
    aggregation and, 63
Acetylcholine
  acetylcholinesterase synthesis
    and, 93–94
  muscarinic and nicotinic
    actions of, 326
  urinary bladder and, 334
  vasal inhibition of heart and,
    326
Acetylcholinesterase
  aggregations and soluble
    forms of, 66–69
  biosynthesis and secretion of,
    79–80
  catalytic properties of, 58
  cholinesterase and, 82–84
  collagen-tailed, 59–66
  developing skeletal muscle
    and, 89–90
  distribution and localization
    of, 73–78
  evolution in nerve cells, 85
  forms in different species, 70
  hydrophobic globular forms,
    65–66, 69–71
  inactivation of, 328
  insertion in skeletal muscle,
    75–78
  molecular forms in fish
    electric organs, 59–66
  molecular forms in solution,
    71–73
  in motor nerves, 87–89
  muscular activity and, 94–95
  muscular synthesis of, 92–93
  in nervous tissue, 78
  nonhydrophobic globular
    forms of, 69–71
  in peripheral ganglia, 85–87
  regulation and localization
    of, 92–95
  in skeletal muscle, 75–78,
    90–92
  in vertebrate tissue, 74–75
Acetyl-β-methylcholine
  acetylcholinesterase synthesis
    and, 93–94
3-Acetylpyridine
  olivocerebellar neurons and,
    283
ACTH
  multiple sclerosis therapy
    and, 234
2-Acylamido-p-nitrophenol
  Niemann-Pick disease
    diagnosis and, 49

Adrenal chromaffin cells
  nerve growth factor
    transformation of, 352
Agnathi
  electroreception and, 124
Alcohol abuse
  Korsakoff's syndrome and,
    242
Alcoholism
  information-processing
    deficits and, 256
trans-4-Aminomethylcylohex-
  ane-1-carboxylic acid
  experimental allergic
    encephalomyelitis and,
    224
4-Aminopyridine
  multiple sclerosis therapy
    and, 235
Amnesia, 242–67
  anatomy of, 262–67
  anterograde, 243–52
  bitemporal, 265–67
  declarative knowledge and,
    259–61
  diencephalic, 263–65
  global, 242–43
  intellectual ability and,
    242–43
  nature of memory deficit in,
    247–52
  perceptual-motor skills and,
    258–59
  preservation of learning and
    memory in, 258–62
  procedural knowledge and,
    256–61
  reference vs working memory
    in, 261–62
  retrograde, 252–58
  temporally limited, 253–55
  semantic vs episodic memory
    in, 261–62
Ampullae of Lorenzini,
  123–24
Ampullary receptors, 126–37
  anatomy of, 128–31
  physiology of, 131–37
Amyotrophic lateral sclerosis
  demyelinating factors and,
    227
Anguilla
  electroreception and, 125
Anoxia
  amnesia and, 243
Anterograde amnesia, 243–52
Antiadrenalin drug 933F, 328
Anticholinesterase agents
  curarized muscle endplate
    potentials and, 330

Antigens
  multiple sclerosis therapy
    and, 234
Antilymphocyte sera
  multiple sclerosis therapy
    and, 234
Antimyelin, 227
Antiviral agents
  multiple sclerosis therapy
    and, 235
Articular mechanoreceptors,
  172–74
Aspartate
  excitatory synapses and, 337
Aspartylglycosaminuria, 34, 39
Astroscopus
  electroreception and, 125
Atropine
  acetylcholine mediation on
    heart activity and, 326
Axonal transport
  acetylcholinesterase and, 87–89
Azathioprine
  multiple sclerosis therapy
    and, 234

B

Batten's disease, 41, 48–49
B cell antigens
  multiple sclerosis and,
    220–21
Biceps tendon vibration
  sensation of extension and,
    199
Bitemporal amnesia, 265–67
Brain lesions
  Korsakoff's syndrome and,
    242
Butyrylcholine
  hydrolysis of, 57

C

Calcium conductance
  in vertebrate neurons, 114–16
Calcium currents, 111–14
Catecholamines
  antibodies labeling, 2
Cellular responses
  multiple sclerosis and,
    228–29
Central nervous system
  electroreceptor input and,
    134–37
  joint direction information
    and, 184
  joint position information
    and, 183–84
  joint speed information and,
    185

CUMULATIVE INDEXES

CONTRIBUTING AUTHORS, VOLUMES 1–5

CHAPTER TITLES, VOLUMES 1–5

ORDER FORM   ANNUAL REVIEWS INC.

Please list the volumes you wish to order. If you wish a standing order (the latest volume sent to you automatically each year), indicate volume number to begin order. Volumes not yet published will be shipped in month and year indicated. Prices subject to change without notice.

ANNUAL REVIEW SERIES Prices Postpaid, per volume	Regular Order Please send:	Standing Order Begin with:
Annual Review of ANTHROPOLOGY Vols. 1−8 (1972−79): $17.00 USA; $17.50 elsewhere Vols. 9−10 (1980−81): $20.00 USA; $21.00 elsewhere Vol. 11 (avail. Oct. 1982): $22.00 USA; $25.00 elsewhere	Vol(s). _____	Vol. _____
Annual Review of ASTRONOMY AND ASTROPHYSICS Vols. 1−17 (1963−79): $17.00 USA; $17.50 elsewhere Vols. 18−19 (1980−81): $20.00 USA; $21.00 elsewhere Vol. 20 (avail. Sept. 1982): $22.00 USA; $25.00 elsewhere	Vol(s). _____	Vol. _____
Annual Review of BIOCHEMISTRY Vols. 28−48 (1959−79): $18.00 USA; $18.50 elsewhere Vols. 49−50 (1980−81): $21.00 USA; $22.00 elsewhere Vol. 51 (avail. July 1982): $23.00 USA; $26.00 elsewhere	Vol(s). _____	Vol. _____
Annual Review of BIOPHYSICS AND BIOENGINEERING Vols. 1−9 (1972−80): $17.00 USA; $17.50 elsewhere Vol. 10 (1981): $20.00 USA; $21.00 elsewhere Vol. 11 (avail. June 1982): $22.00 USA; $25.00 elsewhere	Vol(s). _____	Vol. _____
Annual Review of EARTH AND PLANETARY SCIENCES Vols. 1−8 (1973−80): $17.00 USA; $17.50 elsewhere Vol. 9 (1981): $20.00 USA; $21.00 elsewhere Vol. 10 (avail. May 1982): $22.00 USA; $25.00 elsewhere	Vol(s). _____	Vol. _____
Annual Review of ECOLOGY AND SYSTEMATICS Vols. 1−10 (1970−79): $17.00 USA; $17.50 elsewhere Vols. 11−12 (1980−81): $20.00 USA; $21.00 elsewhere Vol. 13 (avail. Nov. 1982): $22.00 USA; $25.00 elsewhere	Vol(s). _____	Vol. _____
Annual Review of ENERGY Vols. 1−4 (1976−79): $17.00 USA; $17.50 elsewhere Vols. 5−6 (1980−81): $20.00 USA; $21.00 elsewhere Vol. 7 (avail. Oct. 1982): $22.00 USA; $25.00 elsewhere	Vol(s). _____	Vol. _____
Annual Review of ENTOMOLOGY Vols. 7−25 (1962−80): $17.00 USA; $17.50 elsewhere Vol. 26 (1981): $20.00 USA; $21.00 elsewhere Vol. 27 (avail. Jan. 1982): $22.00 USA; $25.00 elsewhere	Vol(s). _____	Vol. _____
Annual Review of FLUID MECHANICS Vols. 1−12 (1969−80): $17.00 USA; $17.50 elsewhere Vol. 13 (1981): $20.00 USA; $21.00 elsewhere Vol. 14 (avail. Jan 1982): $22.00 USA; $25.00 elsewhere	Vol(s). _____	Vol. _____
Annual Review of GENETICS Vols. 1−13 (1967−79): $17.00 USA; $17.50 elsewhere Vols. 14−15 (1980−81): $20.00 USA; $21.00 elsewhere Vol. 16 (avail. Dec. 1982): $22.00 USA; $25.00 elsewhere	Vol(s). _____	Vol. _____
Annual Review of MATERIALS SCIENCE Vols. 1−9 (1971−79): $17.00 USA; $17.50 elsewhere Vols. 10−11 (1980−81): $20.00 USA; $21.00 elsewhere Vol. 12 (avail. Aug. 1982): $22.00 USA; $25.00 elsewhere	Vol(s). _____	Vol. _____
Annual Review of MEDICINE: Selected Topics in the Clinical Sciences Vols. 1−3, 5−15, 17−31 (1950−52, 1954−64, 1966−80): $17.00 USA; $17.50 elsewhere Vol. 32 (1981): $20.00 USA; $21.00 elsewhere Vol. 33 (avail. Apr. 1982): $22.00 USA; $25.00 elsewhere	Vol(s). _____	Vol. _____
Annual Review of MICROBIOLOGY Vols. 15−33 (1961−79): $17.00 USA; $17.50 elsewhere Vols. 34−35 (1980−81): $20.00 USA; $21.00 elsewhere Vol. 36 (avail. Oct. 1982): $22.00 USA; $25.00 elsewhere	Vol(s). _____	Vol. _____
Annual Review of NEUROSCIENCE Vols. 1−3 (1978−80): $17.00 USA; $17.50 elsewhere Vol. 4 (1981): $20.00 USA; $21.00 elsewhere Vol. 5 (avail. Mar. 1982): $22.00 USA; $25.00 elsewhere	Vol(s). _____	Vol. _____
Annual Review of NUCLEAR AND PARTICLE SCIENCE Vols. 9−29 (1959−79): $19.50 USA; $20.00 elsewhere Vols. 30−31 (1980−81): $22.50 USA; $23.50 elsewhere Vol. 32 (avail. Dec. 1982): $25.00 USA; $28.00 elsewhere	Vol(s). _____	Vol. _____
Annual Review of NUTRITION Vol. 1 (1981): $20.00 USA; $21.00 elsewhere Vol. 2 (avail. July 1982): $22.00 USA; $25.00 elsewhere	Vol(s). _____	Vol. _____

(continued on reverse)

Annual Review of PHARMACOLOGY AND TOXICOLOGY
  Vols. 1–3, 5–20 (1961–63, 1965–80): $17.00 USA; $17.50 elsewhere
  Vol. 21 (1981): $20.00 USA; $21.00 elsewhere
  Vol. 22 (avail. Apr. 1982): $22.00 USA; $25.00 elsewhere   Vol(s). _____ Vol. _____

Annual Review of PHYSICAL CHEMISTRY
  Vols. 10–21, 23–30 (1959–70, 1972–79): $17.00 USA; $17.50 elsewhere
  Vols. 31–32 (1980–81): $20.00 USA; $21.00 elsewhere
  Vol. 33 (avail. Nov. 1982): $22.00 USA; $25.00 elsewhere   Vol(s). _____ Vol. _____

Annual Review of PHYSIOLOGY
  Vols. 18–42 (1956–80): $17.00 USA; $17.50 elsewhere
  Vol. 43 (1981): $20.00 USA; $21.00 elsewhere
  Vol. 44 (avail. Mar. 1982): $22.00 USA; $25.00 elsewhere   Vol(s). _____ Vol. _____

Annual Review of PHYTOPATHOLOGY
  Vols. 1–17 (1963–79): $17.00 USA; $17.50 elsewhere
  Vols. 18–19 (1980–81): $20.00 USA; $21.00 elsewhere
  Vol. 20 (avail. Sept. 1982): $22.00 USA; $25.00 elsewhere   Vol(s). _____ Vol. _____

Annual Review of PLANT PHYSIOLOGY
  Vols. 10–31 (1959–80): $17.00 USA; $17.50 elsewhere
  Vol. 32 (1981): $20.00 USA; $21.00 elsewhere
  Vol. 33 (avail. June 1982): $22.00 USA; $25.00 elsewhere   Vol(s). _____ Vol. _____

Annual Review of PSYCHOLOGY
  Vols. 4, 5, 8, 10–31 (1953, 1957, 1959–80): $17.00 USA; $17.50 elsewhere
  Vol. 32 (1981): $20.00 USA; $21.00 elsewhere
  Vol. 33 (avail. Feb. 1982): $22.00 USA; $25.00 elsewhere   Vol(s). _____ Vol. _____

Annual Review of PUBLIC HEALTH
  Vol. 1 (1980): $17.00 USA; $17.50 elsewhere
  Vol. 2 (1981): $20.00 USA; $21.00 elsewhere
  Vol. 3 (avail. May 1982): $22.00 USA; $25.00 elsewhere   Vol(s). _____ Vol. _____

Annual Review of SOCIOLOGY
  Vols. 1–5 (1975–79): $17.00 USA; $17.50 elsewhere
  Vols. 6–7 (1980–81): $20.00 USA; $21.00 elsewhere
  Vol. 8 (avail. Aug. 1982): $22.00 USA; $25.00 elsewhere   Vol(s). _____ Vol. _____

SPECIAL PUBLICATIONS	Prices Postpaid, per volume	Regular Order Please send:

Annual Reviews Reprints: Cell Membranes, 1975–1977
  (published 1978)   Soft cover: $12.00 USA; $12.50 elsewhere   _____ copy(ies)

Annual Reviews Reprints: Cell Membranes, 1978–1980
  (published 1981)   Hardcover $28.00 USA; $29.00 elsewhere   _____ copy(ies)

Annual Reviews Reprints: Immunology, 1977–1979
  (published 1980)   Softcover $12.00 USA; $12.50 elsewhere   _____ copy(ies)

History of Entomology
  (published 1973)   Clothbound $10.00 USA; $10.50 elsewhere   _____ copy(ies)

Intelligence & Affectivity: Their Relationship During Child Development, by Jean Piaget
  (published 1981)   Hardcover $8.00 USA; $9.00 elsewhere   _____ copy(ies)

Telescopes for the 1980s
  (avail. Aug. 1981)   Hardcover $27.00 USA; $28.00 elsewhere   _____ copy(ies)

The Excitement & Fascination of Science, Volume 1
  (published 1965)   Clothbound $6.50 USA; $7.00 elsewhere   _____ copy(ies)

The Excitement & Fascination of Science, Volume 2
  (published 1978)   Hardcover $12.00 USA; $12.50 elsewhere   _____ copy(ies)
                       Soft cover $10.00 USA; $10.50 elsewhere   _____ copy(ies)

To: ANNUAL REVIEWS INC, 4139 El Camino Way, Palo Alto, CA 94306 USA   (Tel. 415-493-4400)

Please enter my order for the publications checked above.

Amount of remittance enclosed $ _____   California residents, please add applicable sales tax.
Please bill me ☐   Prices subject to change without notice.
Institutional purchase order #_____

Name _____

Address _____

_____ Zip Code _____

Signed _____ Date _____

☐ Please send free copy of the current *Prospectus* each year.

☐ Send free brochure listing contents of recent back volumes for Annual Review(s) of _____